ryk AP
(Pep)

MODERN ENVIRONMENTALISM

Modern Environmentalism presents a comprehensive introduction to environmentalism, the history of Western attitudes to nature and environment, and how these ideas relate to modern environmental ideologies.

Examining key environmental ideas within their social and historical context, the book outlines radical environmentalist approaches to valuing nature, to economics, Third World development, technology, ecofeminism and social change.

The author surveys pre-modern ideas about nature and humankind's relationship to it, the developments in science during the Enlightenment and the roots of radical environmentalism in nineteenth- and twentieth-century movements. The main influences include Malthus, Darwin and Haeckel, utopian socialism, romanticism, and organic and holistic systems thinkers. Science is placed at the heart of the society–nature debate, as the major constituent of our cultural filter. The book explains how postmodern ideas of subjectivity and the breakdown of scientific authority have developed, and scientific 'truths' about nature have become divorced from their social and ideological context.

Modern Environmentalism offers a comprehensive understanding of environmentalism and the environmental debate, and of different approaches to establishing the desired ecological society. This entirely new account interprets and synthesises the explosion of writing on society–environment relations since the appearance of Pepper's earlier work, *Roots of Modern Environmentalism*.

David Pepper is Professor of Geography at Oxford Brookes University.

To Alan, Derek, George, John,
Judy, Martin and Peter,
with thanks for their comradeship

MODERN ENVIRONMENTALISM

An introduction

David Pepper

London and New York

First published 1996
by Routledge
11 New Fetter Lane, London EC4P 4EE

Simultaneously published in the USA and Canada
by Routledge
29 West 35th Street, New York, NY 10001

Reprinted in 1997 and 1999

© 1996 David Pepper

Typeset in Garamond by
J&L Composition Ltd, Filey, North Yorkshire
Printed and bound in Great Britain by
Clays Ltd, St Ives PLC

British Library Cataloguing in Publication Data
A catalogue record for this book is available from the British Library

Library of Congress Cataloguing in Publication Data
A catalogue record for this book has been requested

ISBN 0–415–05744–2
0–415–05745–0 (pbk)

CONTENTS

FIGURES

TABLES

INTRODUCTION: WHY HISTORY AND IDEOLOGY MATTER

GOING BELOW THE SURFACE

I must confess that I know nothing whatever about true underlying reality, having never met any. For my own part I am pleased enough with surfaces – in fact they alone seem to me to be of much importance. Such things, for example, as the grasp of a child's hand in your own, the flavor of an apple, the embrace of a friend or lover, the silk of a girl's thigh, the sunlight on rock and leaves, the bark of a tree, the abrasion of granite and sand, the plunge of clear water into a pool, the face of the wind – what else is there? What else do we need?

These lines were written by that leading light of modern green romanticism, Edward Abbey, in his novel *Desert Solitaire*. They were quoted with obvious approval in the mainstream green magazine *Real World* (1994, issue 8), the 'voice of ecopolitics'. They represent what tends to be an abiding green sentiment: an impatience with arguments that political, economic and social processes that operate below the surface of society must be understood and confronted if the green call for fundamental social change is to have any chance of success. After all, isn't it obvious that there is an environmental crisis, that greed and arrogance are causing humans to try as never before to exceed physical limits to growth, and that the consequent destruction of the natural world cannot go on? Scientific evidence surely shows that this is objectively so, and provided that enough people can see the evidence then they must and will act differently. While many greens now acknowledge that it is *not* as simple as this, because environmental degradation goes on even though people know the consequences, the fact is that the more simplistic, impatient note still strikes a chord with the movement.

One consequence of the surface-only syndrome is that for most greens,

> even the committed activist, the Green movement has no history. Worries about environmental destruction seem very modern . . . Green activists, their opponents and the ever watchful news media proclaim the novelty of an ecological outlook.
>
> (Wall 1994, 1–2)

But it is an illusory novelty, for

> although the level of public interest in global environmental degrada-
> tion is something new, the history of anxiety over the environment is
> not; on the contrary, the origins and early history of contemporary
> Western environmental concern and attempts at conservation lie far
> back in time.
>
> (Grove 1990, 11)

But, failing to set their anxieties into historical perspective, modern environmentalists often do not 'save themselves the need laboriously to
reinvent intellectual wheels'. This is the view of Derek Wall (1994, 3), an
intensely practical Green activist who argues that 'Greens have a history . . .
[and] they would do well to learn from it.' This is because 'by tracing the
origins and social context of ideas, it becomes easier to understand their
practical implications and significance'.

A study of the history of 'green' ideas about the relationship between
society and nature also reveals that these ideas are, and always have been,
part of deeper *ideological* debates. Ideologies are sets of ideas that form the
basis of a personal or group 'world view': a particular perspective on how
the world is, and ought to be. Ideologies usually contain hidden assumptions which may go unchallenged – they seem obvious 'common sense' and
not worth debating. But really these assumptions *are* challengeable, often
being little more than a way of rationalising and justifying the material
position in society of the person or group who uses them – they are
weapons in a political battle.

John Short (1991) has described some of the ideologies, based on myths,
associated with nature, countryside and wilderness. One is that the countryside is, or was, a place of harmony, peace and serenity, and, as such, is
where national identity particularly resides. Romantic images of countryside
with these associations have been used as powerful tools in persuading
ordinary people to go to war – essentially defending something (e.g. an
idealised rural English landscape) that they actually had no stake in.
Notions of countryside were here clearly used for political ideological
purposes. Conversely, countryside has been portrayed as a place of hard
work, harshness and 'idiocy': a symbol of conservatism and backwardness.
Marx and Engels made this association in their ideological battle against the
rich classes who had their power base in the countryside. As Short (p. 67)
puts it:

> It is ironic that the typical English countryside, the supercharged image
> of English environmental ideology, which can still conjure up notions
> of a community, unchanging values and national sentiments, is in reality
> the imprint of a profit-based exercise which destroyed the English

peasantry and replaced a moral economy of traditional rights and obligations with the cash nexus of commercial capitalism.

Marx and Engels' representations of countryside sought to bring this fact out, the representations of the authorities and popular media during wartime sought to conceal it.

Similarly, how people represent 'wilderness' can have political ideological dimensions. In the Quabbin area of central Massachusetts, for instance, proponents of deer hunting portray wild nature as dynamic, violent, disruptive and fiercely competitive. By contrast the anti-hunting lobby defines wilderness as a place of balance, harmony and order (Dizard 1993).

And the city has been used ideologically: alternatively to symbolise civilised life, individualism, wisdom, sophistication and hope for the future – or to epitomise crime, disease and fecklessness: an 'unnatural' destroyer of innocence and community. Greens today often promote the latter image of urbanism – as a 'cancer'.

A historical perspective on ideas about nature, which also bears in mind what was happening materially (especially economically) in society when they were current, helps us to realise that when we hear them reiterated today we would do well not to accept them immediately at their face value but to evaluate them against the ideological position of those who advance them. Bertrand Russell (1946, 58) made a similar point:

> When an intelligent man expresses a view which seems to us obviously absurd, we should not attempt to prove that it is somehow not true but we should try to understand how it ever came to *seem* true. This exercise of historical and psychological imagination at once enlarges the scope of our thinking, and helps us to realise how foolish many of our own cherished prejudices will seem in an age which has a different temper of mind.

Above all, a historical and ideological perspective teaches us that there is no one, objective, monolithic truth about society–nature/environment re-lationships, as some might have us believe. There are different truths for different groups of people in different social positions and with different ideologies. One 'truth', for instance, has nature as capable of quick recovery from human interference, especially if aided by wise management policies. Schwarz and Thompson (1990) call this the 'myth' of 'nature benign', which believers in free market economics often favour. Contrastingly, radical environmentalists, who often express egalitarian and communal values, embrace the 'truth' of nature as very vulnerable and potentially damaged by any human activity. Hence it is wise to be cautious about development. This is the 'myth' of 'nature ephemeral'. A third 'truth' – the 'myth' of 'nature perverse/tolerant' – holds that development is acceptable as long as it observes laws and limits of nature. 'Hierarchists' believe in this: those who put their faith in the authority of scientific experts as those most suited to tell us about the laws and limits. We need to appreciate how the material – social and economic – positions of individuals and groups might condition which of these 'obvious truths' about the society–nature relationship they lean towards, for this will colour their arguments about specific environmental issues. This can be seen, for instance, in the discourse in 1989–90 between would-be developers and conservationists over proposals to build a 1600 acre commercial and entertainment complex (Universal City) on the site of the Inner Thames grazing marshes at Rainham, east London (Harrison and Burgess 1994). The developers argued that the environmental integrity of the site (1200 acres of which had been officially protected as a 'site of special scientific interest') could be maintained by careful management and stewarding, applying the expertise of scientific ecologists. This position was underpinned and legitimated principally by the 'nature perverse/tolerant' and 'nature benign' concepts. Conservationists, however, emphasised the vulnerability of all of the site, asserting that the development was beyond the limits of tolerance of its ecosystems. Their position drew on the 'nature ephemeral' concept. Harrison and Burgess stress (p. 298) that these concepts are more than just competing ideas passing before open minds:

> Each myth functions as a cultural filter, so that adherents are predisposed to learn different things about the environment and to construct different knowledges about it. In this way beliefs about nature and society's relation with it are linked with particular rationalities, that support the modes of action appropriate for sustaining the myths.

Such studies, of the 'social construction' of nature and society's relationship to it, all underline, then, the need to get below the surface in order to be thinking and acting effectively: to see ideas about nature in social and historical context.

Environmentalism as a rejection of modernism

Recently, 71 per cent of respondents in a UK poll thought that government should give a higher priority to environmental policy even if it meant higher prices for some goods (ICM, *Guardian,* 17.9.93). Only 16 per cent disagreed. There clearly is popular concern about the hazards of environmental degradation. But, again, this movement of concern should be set against its cultural and historical context. You cannot evaluate ideas and messages about our relationship to nature as if they were free-standing – just the result of sitting down and thinking about them. Wilfred Beckerman, member of the Royal Commission on Environmental Pollution, says (BBC 1994):

there is far too much attention . . . devoted to glamorous, melodramatic, posh issues about the world living on the edge of an abyss . . . the green message is easy to sell for the same reason that apocalyptic prophesies have been the staple of evangelical crusades throughout the ages.

Rosalind Coward (1989), too, sets environmentalism (particularly the alternative health movement) against the backdrop of a public penchant for millennialist, doom-laden thinking. Her purpose is not to deny the seriousness of environmental concerns, but to point out that they may be part of bigger late twentieth-century insecurities. In the BBC programme to which Beckerman contributed, the author Brian Appleyard expressed these insecurities as people suffering a 'future shock' to do with *modernisation:* the globalisation of the world into one big commercial, electronic and cultural system. They experience insecurity at lacking any sense of control over this system. And Bryan Wynne, a researcher on risk perception, commented that people's worries were less about precise, physical environmental risks and more about modernity itself.

Indeed, a 'postmodern' mistrust of the high science and technology which Enlightenment (eighteenth-century) thinkers championed is central to green ideology. The Enlightenment promise to control and manipulate nature to improve everyone's lot seems now to have produced mass war, violence and repression, nuclear and environmental threats, and technologies that ordinary people feel they cannot explain or control. Greens often express, too, a fashionable mistrust of the grand political theories of the 'modern' period, liberalism and socialism. These talked of changing underlying social, political and economic realities to benefit society, but many now think they have done more harm than good. Edward Abbey's view that the only real things are what can be experienced at surface level is very much a postmodernist – and green – sentiment.

Denis Cosgrove (1990) has described how this reaction against modernism has caused many of us to hark back to pre-modern ideas about nature, our relationship to it and our place in the cosmos. Holism, Gaianism and nature worship, for instance, which have resurfaced in deep ecology (Chapter 1.2) have reiterated medieval and older traditions. This is not to say that ideas as old as this constitute exactly the *roots* of modern environmentalism, since there are no directly traceable lines from them to Friends of the Earth. But these ancient perspectives never died out completely during the modern period: they persisted as minor, countercultural strands into the nineteenth and early twentieth centuries, and formed the basis of movements and ideas that *can* be directly traced into modern environmentalism.

Science as a cultural filter

One such minority strand is a mistrust of what this book calls 'classical' science (that developed from the sixteenth century onwards), and of its conception of nature as a manipulatable machine from which human society is separate and distinct. Greens, like Romantics in the nineteenth century, reject this way of thinking about the natural world.

Hence science should be at the centre of the debate about society–nature relationships. From the sixteenth century science has developed into a, perhaps the, principal source informing our perceptions of nature. It is as if most of us in the West have been wearing similar pairs of glasses to look at nature – indeed at everything outside of us. As Jeans (1974) expressed it (see Figure 1.1):

> The real environment . . . is seen through a cultural filter, made up of attitudes, limits set by observation techniques, and past experience. By studying the filter and reconstructing the perceived environment the observer is able to explain particular options and actions on the part of the group being studied.

'Real' and perceived environments differ. The latter is the important influence on decision making. Environmental perception is different in different cultures – humans perceive nature through their cultural filter.

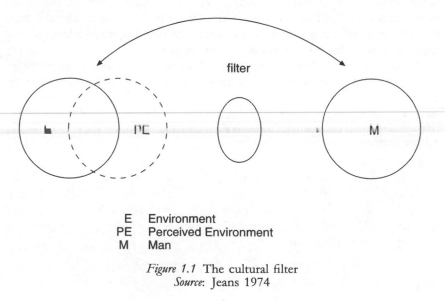

E   Environment
PE   Perceived Environment
M   Man

Figure 1.1 The cultural filter
Source: Jeans 1974

Before the modern period the lenses in the cultural filter – in the 'glasses' with which we looked at nature – were composed of religious myths and teachings. During that period, however, they were gradually replaced by lenses consisting of the scientific world view. Despite today's scepticism, science is still imagined to be a leading source of authority; of 'truth' about what the world is like. If you want to win an argument over someone, you can short-circuit the need to grapple with what they actually say by calling their view 'unscientific': such an insult instantly denies them credibility.

So if you want to understand the environmental debate you must understand the assumptions and perceptions concerning nature which are contained in the scientific world view. This, again, necessitates investigating the history of ideas, and what people were doing which encouraged them to believe in particular ideas as against others.

The scope and purpose of this book

This book is basically an anatomy and history of the ideas about nature and environment that appear in modern environmentalism, both reformist ('technocentric') and radical ('ecocentric'). In Chapter 1 it broadly outlines these ideas and how they relate to different elements in the environmental movement. In Chapter 2 certain key environmentalist themes are further examined, to illustrate how environmentalism revives many issues and problems that are part of longer established political, economic, social and cultural debates: debates which arose particularly during the nineteenth century and were essentially about the problems of modernism. The book then turns specifically to changing perceptions of society–nature relationships. Chapter 3 first outlines some elements of pre-modern thought about the place of humans in nature. It then describes the development of the 'classical' scientific world view, which largely displaced pre-modern cosmologies and still informs technocentric thinking about the environment today. The roots of radical, ecocentric environmentalism seem to lie mainly in the late eighteenth, nineteenth and early twentieth centuries and they are described in Chapters 4 and 5. Chapter 5 also considers how the very notion of science and scientific experts as neutral, objective authorities about nature has come under critical scrutiny, a trend which greens have sometimes eagerly seized on even though they themselves regard the science of ecology as embodying universal principles that all should follow. Chapter 5 is intended to drive home the points made above concerning science as a principal part of our cultural filter, and how the ideas about nature which it conveys should not be divorced from their social, ideological context.

Lastly, no account of environmentalism would be complete without reference to questions of social change and how to establish the desired ecological society. Chapter 6 explores these questions from different viewpoints within radical environmentalism.

The book is first and foremost intended as a multidisciplinary textbook for university students, but also, I hope, will appeal to the general environmentally interested reader. I try to achieve in a more thorough way what Chapters 1–5 of the first edition of my *Roots of Modern Environmentalism* attempted. Chapters 6 and 7 become a separate book, published in 1993 as *Eco-socialism*, and it is in that volume that most of the polemic of the original has been placed. This volume is also effectively a new book, and little of the original remains.

Since that original appeared in 1984 there has been an explosion of written material from the social sciences on the environment and environmentalism, reflecting what radical environmentalists had already recognised: that solving environmental problems is more than a matter of technical fixes and reformist management of the same social and economic system. My aim here is to offer readers a taste of that literature. I write less as a disciplinary expert; more as a synthesiser and interpreter of the work of others, laying out the territory, therefore attempting something very broad in scope. As an interdisciplinary 'fool' rushing in where more scholarly angels fear to tread I again run the risk of misrepresenting others, especially those whose work I have lent on heavily. I apologise in advance for any such errors, but I hope by compensation to have interested readers in reading the originals, and to help this process I have referenced the text as fully as I can and have provided an annotated bibliography. I also hope that my text is as straightforward and intelligible to the non-specialist as it could be. That, above all, has been the aim and, to help, a glossary of terms is appended, in case explanations in the text are inadequate or have been forgotten.

ACKNOWLEDGEMENTS

I am particularly grateful for the help, advice and constant encouragement of Tristan Palmer and Sarah Lloyd, Routledge editors; John Perkins and Martyn Youngs, whose own work I have again drawn on and used in Chapters 3.1 and 3.2, and 4.4 respectively; Pyrs Gruffudd, John O'Neill, Paul Ekins, Peter Keene, Martin Haigh and Nickie Hallam.

The chapter title illustrations, which appeared in Keith Thomas's *Man and the Natural World* (1983, Allen Lane: London) are originally from Robert Hooke's *Micrographia* (1665); Moses Rusden's *A Further Discovery of Bees* (1679); and Conrad Gesner's *Historiae Animalium* (1551–8). They are reproduced by permission of the BBC Hulton Picture Library, the British Library and the Bodleian Library and with the agreement of Mr Thomas and Allen Lane – Penguin Ltd.

Thanks to the following for permission to reproduce their material: Table 1.2, *Clean Slate*, the journal of Alternative Technology Association, Table 2.8, the Centre for our Common Future; Figures 2.1 and 2.2, Earthscan Publications Ltd; Figure 3.2, Open University Press; Figures 6.1, 6.2 and 6.3, Letslink UK.

Chapter 1

DEFINING ENVIRONMENTALISM

DEFINING ENVIRONMENTALISM

1.1 WHAT GREENS ARE FOR AND AGAINST

'Diffuse, incoherent, a hotch potch': these are but some of the epithets often flung at people who call themselves 'green'. Especially when they talk about what *society* is and should be like. These beliefs often seem to come from 'all over the place'. They are a *mélange* of ideas associated traditionally with the political right, left and centre, mingled with principles drawn from the science of ecology. Greens themselves may deny incoherence, claiming a distinctiveness based on 'biocentrism' (see p. 23). Or they may argue that a diversity of ideas is anyway a political strength rather than a weakness.

However, although there are many positions associated with environ-mentalism, it is also true that there is a core of beliefs common to most greens – especially *radical* rather than reformist greens (those who believe that fundamental social change is necessary to create a proper, sustainable, environmentally sound society). Other books (Dobson 1990, Goodin 1992) have described these core beliefs in detail, so it is necessary here only to summarise them.

Inevitably, much of the green world view is about society, since it is concerned about the relationship between (Western) society and nature. It often says that we experience environmental problems because, at root, we have undesirable values about nature. These link with the undesirable way that, individually and in groups, we value and behave towards each other. Hence there is a specifically green *critique of existing society* and conventional values – what greens are against – together with beliefs about what *future society should be like* if it is to be sustainable and environmentally benign – what greens are for.

What greens are against

Greens often say that at the heart of the world's problems of poll resource depletion and environmental deterioration are domin exploitative attitudes to nature. Western culture is though

10

Table 1.1 Green values compared with conventional values

Conventional values	Green values
About nature	
1 Humans are separate from nature.	Humans are part of nature.
2 Nature can and should be exploited and dominated for human benefit.	We must respect and protect nature for itself, regardless of its value to us, and live in harmony with it.
3 We can and should use the laws of nature (scientific laws) to exploit and use it.	We must obey the laws of nature (e.g. the law of carrying capacity, which means that there's a limit to the number of people that the earth can support).
About humans	
1 Humans are naturally aggressive and competitive.	Humans are naturally cooperative.
2 Human societies naturally organise themselves hierarchically, and must do.	Social hierarchies are unnatural, undesirable, and avoidable.
3 You can measure our social standing by our material possessions. Society mainly progresses by making more goods for people to own, and inventing more complex technology.	Spiritual quality of life and loving relationships are more important than material possessions. We reject the latter, and live simply.
4 Logical, rational thought is more valid and reliable than what our emotions and intuitions tell us. You can only trust facts and scientific evidence.	Emotions and intuitions are at least as important and valid as any other form of knowledge. There's no such thing as objective 'facts' anyway.
About science and technology	
1 Science and technology can solve environmental problems, so we must go on perfecting them.	Science and technology can't be relied on: we must find other ways to solve environmental problems.
2 It's progress in technology that largely determines social and economic changes, and there's not much we can do to control it.	We can change society and economics as we like: technology should be servant not master. We don't *have* to have technology that harms us.
3 Large-scale 'high' technology (e.g. nuclear power) is a mark of progress.	Intermediate, appropriate and democratically-owned technology (e.g. renewables – solar, wind etc.) is a mark of progress.
4 You solve problems by analysis – splitting them up into component parts.	You solve problems by synthesis – seeing all the parts as a whole and related to each other.
5 You understand nature by knowing about the basic building blocks of matter and the forces that control them.	You must take a holistic view. There is more to nature (and society) than the sum of its parts.

Table 1.1 continued

About production and economics

1 The main object of producing goods and services is to make capital to invest in more goods and services, benefiting everyone eventually.

We should produce goods and services that society needs, regardless of whether they are profitable or not.

2 The lower the cost of producing goods and services compared with their selling price, the more economically 'efficient' is the production process.

Economic 'efficiency' should be measured by how many (fulfilling and environmentally benign) jobs are produced, and how much the material needs of people (for food, clothing, transport, communication and leisure without excessive consumption) are satisfied with minimal resource depletion. Social and environmental damage is economically inefficient.

3 Economic growth of any kind is good, and it can go on forever. It need not harm the environment.

Indiscriminate economic growth is bad. It can't continue because it uses up finite resources and creates pollution.

4 To maximise growth, you must limit how much you recycle materials and control pollution – otherwise industry will be uncompetitive.

All production must use minimal materials, and recycle them. This is more efficient in the long run. If we had local economies we could worry less about competition.

5 Economic planning can't usually be for more than 5–10 years ahead, because investors have to see a reasonable return by then.

The time-scale for economic planning should be several hundred years.

6 Nations and regions develop and progress by building trade between them.

Trade relations between nations and regions should be reduced: the goal should be self-sustaining regions and communities.

7 It's better and more efficient to make products on a large scale with central control and production-line techniques.

It's better and more efficient (see 1 above) to make products on a small scale, with local control, and in craft production.

8 It's better and more efficient to mechanise and automate production: doing away with boring jobs.

It's better and more efficient to put labour back into jobs, and make them less boring: we all need work in order to be fulfilled.

9 Full employment is the ideal.

Everyone should have work, but this doesn't necessarily mean a conventional job.

Table 1.1 continued

About politics

1	The nation state is the most important political unit.	The local community is most important but as part of an international community (think globally: act locally).
2	We can solve environmental problems without changing our social-economic-political system: though we'll have to regulate that system and intervene in the free market.	The only way to solve environmental problems is by wholesale social, economic and economic-political change – we must get rid of the industrial way of life.
3	The greens want to take us back to pre-industrial stone age, or a romantic rural vision.	To create a 'non-industrial' society, centred on small-scale production for local economies and social need, and in greater touch with nature, is going *forwards.*
4	In the end we must leave environmental decisions to experts best fitted to take them: politicians advised by scientists.	We must *all* take decisions, as much as we can – 'experts' should advise us, but should not command extra authority or power.
5	The way forward is through representative (parliamentary) democracy.	The way forward is through direct democracy, that is, by consensus decisions and delegates.
6	A strong central state will continue to be needed to make national and global economies and social systems work – and ensure law and order in a democracy.	The state should have as little influence as possible: mainly functioning to help local communities to do what they want to do. In a green society people should organise themselves as they wish, but there should be strongly enforced laws protecting the environment.

particularly pernicious global influence, because Westerners see nature as an instrument to be used for endless material gain (see Table 1.1). We take this perspective partly because we imagine we are *separate* from nature; a view inherent in our science and technology as it developed from the seventeenth century. This followed Francis Bacon's creed that by observing nature analytically (splitting it into parts) and reducing everything to its basic components (e.g. biology is a matter of chemistry, which is a matter of physics, which is a matter of mathematics), we can know and manipulate nature's laws for our own ends (Table 1.1). This gives us tremendous technological power, now used, say greens, for ignoble ends based on an ignoble view of *human* nature – that it is aggressive, selfish and competitive (Table 1.1). Wasteful consumerism is now the false god against which we measure both individual and social 'progress'. Our spiritual, emotional, artistic, loving and cooperative sides are neglected for this cold materialism,

13

which overplays the role of rationality, 'hard facts' and calculating economic utilitarianism in deciding what is good or bad. We lack any deeper moral standards.

Our over-linear thinking leads to the false conclusion that if something is good, more of it is necessarily better. Hence more complex technology and economic growth are, unwisely, advocated as the way to cure the social and environmental ills which have been side-effects of technological and economic advancement. Indeed the very idea of progress is now equated with the latter rather than with moral or spiritual advancement. By extension, we regard what is technically most complicated, like nuclear power or weapons, as most *progressive*, notwithstanding how much it destroys or pollutes. So we think we should not reject high technology, even though it often destroys the environment and does not seem to relate to ordinary people.

Greens think that 'industrial' society is founded on the too-narrow objective of profit maximisation, encouraging overconsumption. Blindly pursing profit, industries 'externalise' their waste by-products to society at large rather than paying to make themselves clean. Given today's large-scale industrialisation, pollution becomes unacceptably great, while materials recycling and pollution control are limited in the interests of cost cutting and competition (Table 1.1). Resources are treated as limitless, though clearly, greens maintain, they are finite – a fact never appreciated in the short time perspective of conventional economics. Giantism, profit maximisation, division of labour, the production line, mechanisation and de-skilling combine to produce uncreative, unfulfilling and alienating work, and drab, uniform living environments. Cities and suburbs are huge and impersonal, while the countryside is dominated by ecologically monotonous agribusiness-produced landscapes that give us poisoned and low-value food and water.

The search to expand markets and command resources and cheap labour has extended the industrial-consumer society across the globe, destroying rainforests and changing climate. The 'overpopulated' Third World is polluted and materially and culturally impoverished by this international trade system, which most people still see as essential to 'development'. It produces a political system dominated by both narrow nationalism and uncontrollable multinational corporations. Each country needs a centralising state to make its economic and political arrangements work. But this state interferes with individual and community rights, inhibiting freedoms, self-determination and self-responsibility and producing undemocratic politics (Table 1.1).

This critique is not new. It is more than just a protest against the immediate effects of a polluted, 'overpopulated' world where natural resources are thought to be running out. It is a discontent at the alienation of urban-industrial capitalism and some of its central institutions such as the nuclear family, or hierarchical power relationships. The critique has affinities with most of the dissenting voices that accompanied the rise of modern capitalism, with its political philosophy of *laissez-faire* liberalism, over the past three hundred years; ranging from Romanticism, traditional conservatism and anarchism to the many varieties of socialism. Its most immediate ancestor is probably the 'countercultural' movement of the 1960s, which was intellectually sustained by, among others, 'neo-Marxists' concerned with social and spiritual alienation in our society (as distinct from traditional orthodox Marxism's preoccupation with economic alienation).

What greens stand for

The core green values are *ecocentric*, that is, they start from concern about non-human nature and the whole ecosystem, *rather than* from humanist concerns. They invoke, in 'deep' ecology, the idea of *bioethics*. Bioethics say that nature has intrinsic worth, in its *own* right, regardless of its use value to humans. Humans are therefore morally obliged to respect plants, animals and all nature, which has a right to existence and humane treatment.

This ethic may be linked to the notion of Gaia: the whole earth behaving like a living, self-regulating organism. Though humans depend on the rest of nature, are an intimate part of it, and are not separate from or above it, it follows from both bioethics and Gaia that if humans were removed from the earth, the rest of nature should and could continue to thrive. But our presence has caused an 'environmental crisis', threatening large parts of nature including the human population. This crisis requires us to be humble, not arrogant, towards the Earth. We must abide by the laws governing all nature.

Such laws as carrying capacity lay down limits to growth; economic, population and technological. And they tell us that both social and ecological systems derive strength from diversity; sameness (whether in agricultural monocultures, or in uniform Western industrialism spreading to the detriment of local cultures) leads to lack of robustness and a destructive instability.

An anti-urban bias may follow from this, since cities often contravene ecological 'laws'. This bias may manifest as love, respect and even reverence for countryside and wilderness as the imagined repository of simple and honest values, and reverse feelings towards cities and suburbia as the home of corrupting 'civilisation'.

And there is a call for holistic thinking, recognising the full implication of our place in the global ecosystem, which is that whatever we do to one part of that system will affect all other parts, eventually reverberating on ourselves. Greenhouse and ozone layer effects are prime examples of this principle.

Social implications

If we are to live 'in harmony with nature' then social behaviour and personal morality should observe ecological laws. Indeed, the way that the rest of nature is organised should be a model for human social organisation. To make fewer demands on the planet's resources we must reject materialism and consumerism and accept population control and low-impact technology based on renewable energy. Economics must henceforth incorporate environmental criteria in measuring value, efficiency and the costs and benefits of development. And all development must be *sustainable*, that is, it must not reduce environmental and economic options open to future generations. Geographical reorganisation is essential, into small economic, political and social units, comprising self-reliant regions and local communities (Table 1.1), since these are socially and ecologically most rewarding and stable.

Environmental degradation and social injustice are inextricably linked, so present trade, 'aid' and debt relationships between North and South, which encourage, for instance, tropical deforestation, must be replaced by more independent development.

As part compensation for lower material standards for the rich, and to improve the lot of the majority, *quality of life* must be improved. Environmental quality, personal well-being, rewarding relationships, creativity, art and sheer enjoyment – all these are part of quality of life. Economics must make quality of life a large part of the measure of wealth and value. Wealth indicators, like gross national product, should include only socially useful activity (whereas at present even pollution-generating activities are counted as part of 'productivity' and so are desirable).

A social wage, paid to everyone, would mean that useful work, e.g. housework, caring for the sick and needy and tending the allotment or garden, would not be economically second-class activities. Manufacture and services should be geared to social needs rather than wants expressed solely through the market. Many wants are anyway 'artificial', induced through advertising for the sake of profit. And, since meaningful work is a basic human need, work should be designed wherever possible not to be degrading, boring or alienating, emphasising instead craft and creativity.

Quality of life can also be enhanced through people having control over their own lives in a genuine participatory democracy. Individuals and local communities, rather than the state or large private corporations, should own resources. Individuals must also feel that their views are heard and respected, whether they are 'expert' or not, and that they can tangibly influence decision making.

Obviously most of this is unachievable without deep-rooted changes in Western values and social organisation. 'Feminine' values – contractive, responsive, cooperative, intuitive, synthesising – should be encouraged, while currently dominant 'masculine' values – demanding, aggressive, competitive, rational, analytic – should be de-emphasised. A green society will be less hierarchical and more participatory than present society, and very definitely more communal.

However, individual self-fulfilment is also vital – for respect and love for others, and for nature, must be founded in self-respect. A non-aggressive individualism is a cornerstone of radical green political ideology. Many greens emphatically reject traditional political ideologies and approaches as part of the problem rather than the cure. Instead they adopt the 'personal is political' maxim, which says that by changing our personal lifestyles, attitudes and values we make a powerful contribution to general political change. Here, inner-directed philosophies and practices and enlightened education and socialisation are crucial.

Although in the mid-1990s the political popularity of green parties has waned (see Bramwell 1994), none the less green ideas have made progress in the quarter century since the rise of popular environmentalism. From his perspective as someone involved in the alternative technology movement for most of that time, Peter Harper (1990) composed a personal and impressionistic schema of degrees of social acceptability of green ideas (Table 1.2). Readers can judge whether, some years later, any of these ideas have nudged their way further towards the 'mainstream' end of the spectrum.

1.2 DEEP ECOLOGY, SOCIAL ECOLOGY AND THE NEW AGE TENDENCY

Basic beliefs of deep ecology

The distinction between 'deep' and 'shallow' ecology was first made by Arne Naess (1973). Since then, several leading writers have promoted deep ecology as the philosophical basis of truly green practices and lifestyles (Devall and Sessions 1985, Tokar 1987, Naess 1989, see also the description of deep ecology in Bramwell 1994). Naess says that 'deep' ecologists are so because they do not discuss technicalities without asking *basic* questions first. Before asking how to secure a supply of material goods, deep ecologists would question whether we really need so many goods in the first place, with a view to lowering our demands on the planet's resources.

Deep ecology fundamentally rejects the *dualistic* view of humans and nature as separate and different. It holds that humans are intimately *a part* of the natural environment: they and nature are *one*. The view of what a green society should be like stems from a firm belief in bioethics and nature's intrinsic value. Following Commoner's (1972) 'third law of ecology', that 'nature knows best', and his principle that any human-induced

Table 1.2 The fate of early green ideas

A	*The fate of early green ideas in public opinion*	Mainstream: have now achieved orthodox status:	Respectable: still not quite orthodox, but widely discussed or practised, and with an aura of incipient success:

A *The fate of early green ideas in public opinion*

Mainstream: have now achieved orthodox status:

- Green issues are worth paying attention to
- The greenhouse effect is real and needs to have something done about it
- Ditto acid rain
- Lead as an environmental poison
- Whales should not be hunted
- Environmental education
- Equal roles for women
- Catalytic converters
- Combined Heat and Power (CHP) systems
- Nuclear power, no thanks

Respectable: still not quite orthodox, but widely discussed or practised, and with an aura of incipient success:

- Renewable energy
- Energy conservation and efficiency
- Green products
- Organic farming and gardening
- Osteopathy and homeopathy
- Recycling
- Intermediate technology for developing countries
- City farms
- Natural childbirth

No longer cranky: once-wild notions that can now be discussed at dinner parties or with strangers in the pub, or which can be practised without causing too much of a stir in public places:

- Health and diet; vegetarianism; wholefoods
- Additives in food worth worrying about
- Cycling to work
- Fewer cars, more public transport
- Acupuncture
- Ecological investment makes economic sense
- Zero-growth economics
- The Gaia hypothesis
- Holistic worldview
- Healthy building
- More trains
- New work patterns, jobsharing, etc.
- Workers' coops
- Opposition to factory farming
- Cruelty-free cosmetics
- Family life without marriage
- Radical feminism
- Use of hydrogen as a fuel
- Therapeutic massage

Still too way out: incomprehensible, controversial or embarrassing outside the circles of the faithful:

- Severely frugal lifestyles
- Veganism
- Macrobiotics
- Communal living
- Unilateralism
- Domestic sewage recycling; alternative toilets generally
- Pacifism
- Decentralisation
- Local currencies
- Wage parity
- Colourful New Age beliefs
- Utopian socialism
- DIY schools
- Foreign aid is counterproductive and should be scrapped

18

Table 1.2 continued

B *The fate of early green ideas within the green movement*	Abandoned: things which just haven't stood the test of time, and seem to be on the way out, although they may continue to command support in some quarters:	Adopted: ideas or practices we wouldn't have been seen dead with ten years ago, but which somehow we seem to be taking on board (or is it just that we are getting old?):
	• Autonomous houses • Self-sufficiency • Radical arcadianism • De-industrialisation • Beards and sandals • No high-tech • Small is necessary and always the best ecological solution • Recreational drugs • Mineral resources are scarce and about to run out • Don't-need-experts-it's-all-mystification-anyone-can-do-anything • Don't bother about organisational structures, things will work themselves out spontaneously • State socialism • The apocalypse is imminent • China leads the way • Capitalists invariably have top hats and cigars and their sole purpose is to exploit workers	• Acceptance of the national economy and its industrial base • The national grid • The necessity of (some) large-scale energy systems • International trade • Banks and other financial institutions • Profits • Interest • Markets; marketing • The importance of good management • Presentation and appearance • Plastics • Electronics; computers, etc. • Representative democracy • The bourgeois virtues of thrift, punctuality, order, cleanliness, due process and efficiency • Collective wishes overriding individual ones

Source: Harper 1990

change to a natural system is likely to be detrimental to that system, deep ecologists propose humble acquiescence to nature's ways: trying to 'live with' and not against natural rhythms. They oppose *anthropocentrism*, defined as (a) seeing human values as the source of all value, and (b) wanting to manipulate, exploit and destroy nature to satisfy human material desires.

This view of nature revives ideas of philosophers such as Baruch Spinoza (1632–77) and Martin Heidegger (1889–1976). They proposed that every being has the right to express its own nature, while the ultimate goal of humans is to contemplate nature. This implies what Watson (1983) describes as an 'inactivism' towards nature. Consequently there is much interest in Eastern philosophies like Taoism, Buddhism and Hinduism (interpreted by such writers as Alan Watts (1968, and Watts and Huang

1975) and Gary Snyder (1969, 1977)), which contemplate nature passively in the face of the forces of the universe – going 'with the flow' rather than struggling against it. (For a review of the links between Islam, Buddhism, Confucianism and deep ecology, see Engel and Engel 1990 and Ferkiss 1993; for discussions of the links between ecocentrism and Buddhism, Hinduism, Taoism, Confucianism, Jainism, Islam, Baha'ism, and Christianity, see Tucker and Grim 1993.)

Naess (1988) considers that deep ecology combines its concerns for nature with a desire to transform society. But it shuns social change by confrontation, where one side in an argument considers the other 'wrong'. There is value in all viewpoints, and consensus is possible when all sides shift their positions a little. Hence a plurality of world views is tolerated – atheism, paganism, Buddhism, Christianity, or whatever. There are no right or wrong religions – rather there are some basic principles which all religions share. These include the most important principles of deep ecology. They argue, among other points (see Table 1.3) that the richness of life on the planet is greatest when the *diversity of life forms* is greatest.

While deep ecologists may thus be relativist about social values, they do advocate one absolute set of 'correct' attitudes, defined by the laws of ecology, which all cultures should share. 'Although humans have unique characteristics as a species, they are still subject to the same ecological laws and restraints as other organisms' (Merchant 1992, 89), implying depen-

Table 1.3 The eight basic principles of deep ecology

1	The well-being and flourishing of human and non-human life on Earth have value in themselves (synonyms: intrinsic value, inherent value). These values are independent of the usefulness of the non-human world for human purposes.
2	Richness and diversity of life forms contribute to the realisation of these values and are also values in themselves.
3	Humans have no right to reduce this richness and diversity except to satisfy *vital* needs.
4	The flourishing of human life and cultures is compatible with a substantial decrease of the human population. The flourishing of non-human life requires such a decrease.
5	Present human interference with the non-human world is excessive and the situation is rapidly worsening.
6	Policies must therefore be changed. These policies affect basic economic, technological, and ideological structures. The resulting state of affairs will be deeply different from the present.
7	The ideological change is mainly that of appreciating life *quality* (dwelling in situations of inherent value) rather than adhering to an increasingly higher standard of living. There will be a profound awareness of the difference between big and great.
8	Those who subscribe to the foregoing points have an obligation directly or indirectly to try to implement the necessary changes.

Source: Devall and Sessions 1985

dency on 'finite' natural resources and a need to press less heavily on earth's carrying capacity by population control. This implies radical social change – society–nature relationships cannot be fundamentally transformed within the existing social structures. Those who say they can are dubbed mere 'shallow' ecologists; technocrats and managers who allege that there is not yet conclusive proof that our present ways are destroying the earth.

Deep ecology's approach to social change focuses on transformation at the level of *individual consciousness*. The prime need is for each individual to change attitudes, values and lifestyles to emphasise respect for and peaceful cooperation with nature. When enough people have done this, then all society will change. The feminist slogan, 'The personal is political', expresses this approach (see Chapter 6.1).

Deep ecologists do not totally rely on ecological science (involving data, logic and proof) for their conclusions. They also value emotional and intuitive knowledge. They say that we can never know enough for certain, but intuition should tell us that we should not do anything that *might* do long-term environmental damage.

When this 'eco-wisdom' is applied to everyday lifestyle, we have the holistic philosophy which Naess calls *ecosophy*. Ecosophical principles are what guide deep ecologists, not purely information gained through the methods and philosophies of classical science.

Gaia

Donald Worster (1985) describes how deep ecologists have *chosen their values first*. If they look to science, it is mainly to get its 'seal of approval' or stamp of value. That seal is available from two main sources. There are the findings of twentieth-century physics (Chapter 5.3), and there is the Gaia hypothesis. In the latter James Lovelock (1989) argues that the earth can be regarded as *if* it were a *single living organism* – a 'superorganism', as geologist James Hutton termed it in 1785, composed of interrelated parts. All parts help to regulate and balance the planet via feedback mechanisms, thus sustaining life as we know it. In this sense, only, earth is 'alive' because it is 'autopoietic', that is, self-renewing: it can repair its own 'body' and grow by processing materials. This happens neither by pure chance nor by outside design, but only by virtue of earth's own make-up and laws (Sahtouris 1989). It is the living organisms on earth which process chemicals so as to produce a non-equilibrium atmosphere. This has 'too much' oxygen and nitrogen and 'not enough' carbon dioxide compared with stable planets like Mars. But it supports life, which Mars does not. Living things themselves produce the conditions most conducive to their thriving. So they are not passive: they manipulate and radically change environments in ways most favourable to their further development. This makes the whole planet a self-sustaining system: a discrete entity able to maintain its own integrity by responding appropriately to changes via feedback mechanisms. Life and non-life are complementary and collaborating.

This is Lovelock's *analogy*, of earth *as if* it were alive, which is not the same as the medieval metaphor of living earth. That said that the earth literally *was* a living organism (Mills 1982). Lovelock himself (1990) stops short of this, but he does not discourage Gaianists who wish to go further and explore the Gaia concept's mystical, theological and feminist potential. They may infer that the whole planet is literally alive, even proposing a planetary intelligence beyond that of human intelligence, which controls the system as a whole.

Such interpretations of Gaia theory constitute effectively a deification of the holistic systems approach. They resuscitate pre-modern ideas and spiritual notions, for instance the *Great Chain of Being* (Chapter 3.1), that sees all earthly things linked together as in a chain. There is spirit and life in all components of the chain, coming from the one Supreme Being at the top (God, in the Christian version). This being has so much 'goodness' and life force that it overflows into everything (this is 'pantheism': God is everything and everything God – *OED*).

One leading Gaia theorist, Michael Allaby, stresses that Lovelock's concept does *not* regard Gaia as intelligent, nor as a god, even though the name is taken from the Greek earth goddess. Pseudo-religions which revive pre-Christian beliefs are harmful, Allaby (1989, 109) declares, because they falsely perceive reality. Not founded in proper knowledge of the natural world, they revive a belief in magic, that is, a world manipulated by unseen and intelligent forces with which we communicate by ritual. This 'excludes serious thought, replacing it by a sentimental cosiness'. The earth system is entirely automatic. It does not evolve in response to external design, or even via the altruism of its creatures. They collaborate merely because they know that collaboration is the best way to survive. They are entirely selfish.

There is a parallel here with Adam Smith's economic doctrine of the 'invisible hand', which underpins liberal capitalist ideology. This says that maximum social good unconsciously follows from allowing each individual consciously to maximise their selfish interests. And when Allaby stresses that Gaia is a mere machine, with no consciousness, he also affirms the mainstream Western view of nature (see Chapter 3). In view of this, and the fact that 'Gaia has no need of humans . . . we are no more than supernumaries . . . filling the stage while real actors get on with the play' (1989, 147) it is perhaps surprising that the Gaia hypothesis attracts so many environmentalists.

Holism, monism and the 'total field view'

Whereas 'shallow' ecologists consider that humans and nature are separate, and humans are most important, deep ecologists deny any separation. They claim a *'total field view'*, where every living being is part of Gaia, and has intrinsic value. As Naess puts it, all organisms are 'knots in the biospherical net or field of intrinsic relations', and the very notion of a world composed of discrete separate things is denied.

What this means is not easy to grasp, but it does mesh with Eastern mysticism and some interpretations of contemporary physics, both of which influence deep ecologists. It implies that the universe is made of one basic spiritual or material entity or 'stuff', and that different organisms or parts of nature are but different forms of this. Such a belief, called *monism*, follows Spinoza in replacing René Descartes' (1596–1650) idea of a split, or dualism, between mind and body by notions of a single substance or spirit, known as 'God' or 'nature'. It also follows the sixth-century BC Chinese philosopher Lao-Tzu, who, in the *Tao-Te-Ching*, presented a view of the universe as having an ultimate wholeness. This, the Tao ('the Way'), was the reality underlying surface appearances. It was ceaselessly moving, expanding and contracting. Thus what appeared to be opposites (male and female, heaven and earth, cold and hot) were really different facets of the same thing, as in *yin* and *yang*, needing to be in balance.

As Fox (1984) describes it: 'The central intuition of deep ecology is that there is no firm ontological divide in the field of existence.' 'Ontological' means concerned with the nature of being, so this means that there is no difference between what humans *fundamentally are* and what the rest of nature is. Hence there is no division into independent human and non-human realities. Devall and Sessions (1985) say 'deep ecology begins with the unity rather than the dualism which has been the dominant theme of Western philosophy' (see Chapter 3.2).

Bioethic, nature rights and the spiritual dimension

If humans are part of a total reality composed, basically, of the same 'stuff', then, deep ecology says, they are just *one constituency among others* in the biotic community: one strand in the web of life. 'The person is not above or outside of nature, but is part of creation on-going', say Devall and Sessions (1985). Therefore the value of other creatures and things in nature is not to be fixed by humans.

This is *biological egalitarianism*, where it follows that if all creatures belong to the same unified whole then they deserve equal consideration. The 'shallow ecology' idea of humans as the source of all values is therefore arrogant. Even humanistic perspectives which might otherwise be ecologically sound are to be rejected *because* of their human-centredness (Eckersley 1992).

Thus, the evolutionary cosmic philosophy of Catholic priest Teilhard de Chardin (1965a) has much about it which attracts ecocentrics (Chapter 6.1), including the concept of a 'biosphere'. But this is not considered as strictly 'Gaian' or deep ecological, 'because of its strong anthropocentrism'

(Grinevald 1988). Even the World Conservation Strategy, launched in 1980 by UN and conservation bodies, is denounced 'because of its ultimate concern for humans' rather than valuing nature in itself (Naess 1990).

The roots of biocentrism are many and they include Taoism, Buddhism, Hinduism, some Christian nature mystics (e.g. St Francis), witchcraft and paganism, American Indian spirituality, ecofeminism, bioregionalism (Chapter 6.3), ecological science and populist American politics (Taylor 1991). The militant Earth First! organisation in America displays all these influences (less so its European offshoots). It wants to resacralise nature, so that, for instance, trees are seen as gods and it is a sin to cut them. Its primal spirituality therefore is pantheistic and pagan (nature and nature goddess worship). It views pre-colonial America as sacred and present American culture as profane, advocating a 'future primitivism' (Merchant's phrase) where people live 'naturally' in a reinstated wilderness.

Native American land wisdom, incorporating beliefs that nature is alive and sacred, is a controversial plank of American deep ecology:

> In the native peoples of California, who lived here for so very long before the whites appeared, we can see the *true ecological man* – people who were truly a part of the land and the water and the mountains and valleys in which they lived.
>
> (Heizer and Elsasser 1980, cited in Short 1991, 102; emphases added)

A popular ecocentric expression of this imagined land wisdom was in a speech ostensibly made in 1854 to the US Government by Chief Seattle of the Susquamish Indians, on being threatened by land annexation. Published in 1976 by British Friends of the Earth, it contained such deep ecological sentiments as 'How can you buy or sell the sky or the warmth of the land?' However, this 'testament' was a forgery: actually a film script written in 1970 for Southern Baptists who wanted to sugar-coat their fundamentalist message with ecologically attractive sentiments (Church 1988).

None the less the Romantic and anarchist idea that 'traditional', or 'primitive' cultures were and are more ecologically benign than (corrupt) Western industrialised nations – they were 'noble savages', as Rousseau had it – persists among ecocentrics. Young (1990) considers that Australasian aboriginal ecological soundness is a sort of by-product of their social beliefs. Goldsmith (1988) is unreflectively enthusiastic about South American Indians, as is von Hildebrand (1988), who considers that the rituals and ceremonies of Amazonian tribes reinforce a general lifestyle of harmony and balance with the natural environment. Bunyard (1988) is euphoric about 'the world's indigenous peoples', declaring them 'vastly superior' to the West environmentally. Similarly, Callicott (1982) declares the 'native Americans' more noble than 'civilised' Europeans. They saw the rest of nature as alive, believing that the 'Great Spirit' was in all natural things. Dreams were contiguous with the everyday world, and were a way of communicating with the rest of nature. In this idea, as with their notion of an extended community of animals and humans, their beliefs were the non-exploitative antithesis of Western attitudes.

But there are those who argue the reverse, emphasising how Amerindians overculled mammals at the end of the Ice Age (Martin 1973), then later slaughtered buffalo and traded their skins, or how they may have been motivated by fear of nature rather than appreciation of any 'intrinsic value' (Regan 1982) or how their stewardship of land was a purely technical rather than moral issue in their culture (C. Martin 1978). Wall (1994, 21) puts the true position:

> An attitude that all contemporary hunter-gatherers are 'primitive' and awaiting progress must surely be rejected. The idea that such groups, together with our Neolithic, Mesolithic and Paleolithic forbears, were purely and simply Green is equally naive and ill-informed.

But this tendency to look to cultures that lived, or live, closer to nature than does the West, to try to establish the correctness and greater legitimacy (greater 'naturalness') of a particular set of values or a way of life is a long established strategy in the West. It has in the past been used to argue for values and lifestyles that deep ecologists would denounce, as much as for those they support.

Coward (1989) suggests that it is part of a larger syndrome: one that bemoans the destruction of some imagined original wholeness, and is therefore really an extension of the Christian notion of original sin. This tendency is popularised in the 'alternative health' movement, which is suspicious of technology, considers that nature, unlike society, is safe, gentle and kind (the reverse of a Darwinian interpretation), and regards health as a normal state which can be regained by unblocking vital forces and rebalancing energies that are imagined to flow through the body from the cosmos outside. This body had a natural state of 'innocence', now constantly corrupted by the ways of Western industrialised society.

Earth Firsters and other deep ecologists, like the 'Council of All Beings', perceive a kinship of humans with the rest of nature in a community of moral concern, following Aldo Leopold (1949). This kinship is spiritual and mystical as well as practical, symbolised by animal totems and dramatised in rituals where people adopt the roles of animals or inorganic nature, lamenting their abuse by 'industrial society', or symbolise the cycling of ancient elements – earth, air, fire and water – through the body and the biosphere. For less 'extreme' deep ecologists, spiritual experience of nature is still vital, often achieved through meditation.

Earth First! proclaims an 'ecovangelist' desire to convert non-pagans, and to resist the military–industrial state by acts of 'ecotage', such as blowing up dams in the south west US desert, as proposed in fiction by Edward Abbey's (1975) 'Monkey Wrench Gang'. Such violent tactics have encouraged violence from the state in jailings or worse. This, together with the extreme irrationalist tendency, or 'woo woo' ('eco-la-la' as Bookchin calls it) has caused splits in Earth First! – splits which mirror the wider deep-versus-social ecology debate (see p. 29).

Geography and deep ecology

Deep ecology says much about the geography of an ecological society. It features decentralised, small-scale, autonomous, self-reliant regions and communities. It emphasises 'reinhabitation', that is, re-learning a sense of *place* – feeling part of that place and its community, caring for it, and appreciating and enhancing its unique sense of identity. It revives Heidegger's concept of 'dwelling' in a place, which implies protecting landscape and ecosystems from disturbance: acting as 'guardians of Being' (Sikorski 1993). All this is epitomised in 'bioregionalism': this advocates that we should abandon the nation state as our basic economic-political unit and adopt *bioregions* instead. These are allegedly 'natural' regions (e.g. hydrological catchments) with common characteristics of soil, flora and fauna, physiography, etc. Each bioregion has a specific human carrying capacity, which should not be exceeded. This is the number of people whose basic needs can be satisfied by available resources without undue disruption of the environment (Chapter 6.3).

The New Age tendency

Capra (1982) takes an evolutionary view of 'world' history, suggesting that it is now entering a period of shift (a turning point) to a holistic, ecological world view. His *New Age* perspective regards the world as an interdependent network of all sentient and non-sentient elements, where actions in any one part affect the whole. Besides ecology, New Ageism includes feminist–anarchist perspectives on social organisation and social change combined with a Gandhian approach to peace and social relations. It emphasises the *spirituality* in social and environmental relationships.

New Ageism is in many respects an extension of deep ecology. Though some deep ecologists specifically reject New Ageism, e.g. Devall and Sessions or Earth First!, they appear to be fixing on a narrow aspect of it that welcomes new technology (and could therefore be accused of anthropocentric, 'technofix' prescriptions). But in most other respects New Ageism is highly compatible with deep ecology. In fact much radical environmentalism displays New Age tendencies to an extent, though not accepting it in full. New Ageism is

a spiritual hotch-potch of abstruse theories and esoteric snippets . . . a vast umbrella movement embracing countless groups, gurus and indi-

26

viduals bound together by a belief that the world is undergoing a transformation or shift in consciousness which will usher in a new mode of being.

(Storm 1991)

This 'millennialist' element invades most ecocentrism. In groups like Earth First! it can become an 'apocalyptic eschatology' (doctrine of death, judgement, heaven and hell – *OED*) (Taylor 1991). It maintains that the Age of Aquarius, which includes ecological consciousness, is about to dawn. Astrologers say that we enter a new age about every 2000 years, establishing whole new civilisations and cultures. At present the earth is emerging from the Age of Pisces (the fish) – an age initiated by Jesus and his teachings, which, however, has put few of them into practice. Pisces has been dominated by polarisation and conflict – *between* cultures, civilisations, religions, races, and *within* Western consciousness (the mind–body split, the masculine–feminine and society–nature polarities). This notion of duality is embodied in the Piscean symbol, of two fish swimming in opposite directions.

By contrast Aquarius symbolises harmony, holism, balance and high moral and spiritual awareness – it will be an age when human consciousness is not split from nature. The idea of *global* consciousness features strongly in New Ageism (see Chapter 6.1). It starts with each individual in the movement seeking to experience oneness with the planet, to bring an inner tranquillity and strength where 'the mountains and the sea and the stars are part of one's body and one's soul is in touch with the souls of all creatures' (Edward Carpenter, nineteenth-century social scientist and poet). Such experiences are then shared with others. Eventually the aggregate consciousness of New Age ideas stretches around the globe, becoming the majority world view.

This is the New Age theory of social change (Chapter 6.1), which identifies a 'leaderless but powerful network working to bring radical change in the US' – a 'conspiracy', literally a 'breathing together' (Ferguson 1981). Such conspiracies involve scattered disparate groups working on different causes (ecology, feminism, community politics, spiritual healing, consciousness raising, etc.) which none the less have common elements. Unconscious of each other at first, their political strength, it is argued, grows as they 'recognise' their common interests, eventually coming together consciously. New Agers saw Gorbachev as their prophet, stimulating a burst of New Age thinking in Eastern Europe.

Henderson's (1981) term for the change is the coming of the 'Solar Age', whereby the world adopts massive solar power as present fuel sources become exhausted. Underlying this will be substantial social changes, as industrial economies based on capital intensive production are forced to decentralise under the pressure of an 'inexorable energy crunch'.

The Findhorn light centre

All these New Age perspectives – astrological, spiritual and political-economic – share a monistic, deep-ecological viewpoint. This includes the doctrine, going back to the Greeks, of *pantheism*. And it embraces the pagan belief in *animism* ('attribution of living soul to plants, inanimate object and natural phenomena' – *OED*). Pantheism and animism figure strongly in parts of the deep ecology movement, like Earth First! in the USA or the Wrekin Trust, the Soil Association, and the Findhorn Foundation in Britain. This last is based on an 'alternative community' started in 1962 by Peter and Eileen Caddy at Findhorn, near Forres, in Scotland. Today the Findhorn Foundation has 130 resident adults (Coates *et al*. 1993). The Caddys were penniless and jobless when they first drove their caravan onto the sand dunes at Findhorn. They started to grow their own food on the infertile soil, helped by a third person, Dorothy McLean. Both she and Eileen Caddy claimed to be in touch with the spirit of God, or Christ-consciousness, and the 'divas' or plant-spirits. Soon, the name of Findhorn spread, with stories of hugely successful gardens, with gigantic plants, growing on the sand.

Gardening at Findhorn today is still done on this basis of communication with plant spirits (the basis also of 'bio-dynamic' agriculture developed by Rudolph Steiner and the Soil Association in the 1930s and onwards). Findhorn sees itself as a 'light centre' through which the vital energy of the New Age can enter the planet and diffuse all over the globe. The community concentrates more on people than plants – fostering personal awareness and development as a means of gaining New Age consciousness. It is an education centre based on cooperation, consensus decision making and spiritual awareness.

The common elements in Findhornians' New Age views are indicated in Table 1.4. They include the idea that 'god' does not lie separate and outside of ourselves and nature. She/he/it is everywhere; in us and the rest of nature. If God is in us, not 'out there', in a way we are *all* God. And we are all *good* – there is no such thing as 'original sin'. Seen from this perspective, the public 'revelation' in 1991 by British Green Party spokesman David Icke, to the effect that the Christ Spirit was speaking through him, was entirely thought out – not perhaps quite the gratuitously 'silly belief' which many commentators (e.g. Goodin 1992, 03) took it to be. However, it was heretical as far as Christianity was concerned, for in Christianity there is only one God, and 'He' is somewhere 'up there' in the spiritual realm. He is not of this earth, not of earthly materialism, and not therefore to be found *in* nature, although nature may be seen as His design.

Critiques of deep ecology

Despite its influence throughout the green movement, deep ecology has taken a veritable pasting from critics in and outside that movement. There is argument about whether nature has intrinsic value, which carries over

Table 1.4 Principal tenets of the Findhorn consensus

---

1 The New (Aquarian) Age is on its way – a technological and cooperative age.
2 Findhorn is a 'light centre' where people transform themselves and tap into the new type of energy of the Aquarian Age (light centres work towards the good of the planet; 'light' is life-giving force, growth).
3 We are more than just material: but part of a greater spiritual reality, which is loving and intelligent.
4 Everything has purpose, love and unity (three principles behind all great religions).
5 Everything is part of a whole – not separate ('planetary consciousness').
6 All people are the same – a oneness behind all divisions.
7 Everything, including all daily tasks, should be done with 'love'.
8 'Love' includes spiritual commitment and a sense of purpose. It involves being a perfectionist in eveything because everything is good.
9 'Love' involves working cooperatively in groups, by consensus. *Findhorn's* 'sacred culture' features groups attuning to a common goal.
10 The process by which you do things is as important as the end product. The process should bring life, and spirit into everything and involve always being open and honest.
11 The individual is where all social change starts.
12 Individuals must love themselves, discover themselves and take time to see where they are, emotionally. These processes, together with intense emotional crises, lead to spiritual growth and transformation.
13 As transformed individuals we call all be light centres.
14 Individuals are fullfilled by working together and being subsumed in the community.

---

into the animal liberation debate (Chapter 2.1). And there are many other points of disagreement between deep ecologists and their critics.

Foremost is the charge that deep ecology is politically naive at best, and at worst politically reactionary. The naivety comes from over-obsession with the idea that changes in the values, attitudes and lifestyles of *individuals* constitute the motor of social change. ('Idealistic' approaches to social change are outlined in Chapters 2.7 and 6.1.) This then leads to a corresponding failure to confront the problem posed by the huge power to block change possessed by state institutions and business corporations. A realistic politics of how to get to the decentralised bioregional society, is therefore lacking, along with any analysis of why capitalism specifically (as opposed to 'industrialism' in general) befouls the environment.

And, despite a professed species egalitarianism which includes humans, problems of poverty, inequality, the inner city (which is 'the environment' for so many) and racism are never addressed in a sustained way. Yet such issues are inextricably linked to ecological problems, so failure to address them makes 'deep' ecology in reality shallow (Bradford 1989).

The potentially reactionary nature of deep ecology's message can be seen in its very language. 'Wholeness, balance, harmony' and 'traditional ways', Coward (1989) reminds us, are concepts that, when applied to society, smack of traditional conservatism and its suspicion of change. Indeed, says Bookchin (1990), only a minor ideological tilt is needed to send deep ecology's obsession with 'community' and 'oneness with nature' into the camp of Nazism, with its nationalistic 'blood and soil' nature philosophy (Chapter 4.6).

While this is a typical Bookchin overstatement, it none the less makes us reflect that singing the praises of 'traditional' or 'primitive' folk cultures is a characteristic right-wing trait. And the vision of 'natural living' associated with such cultures is

> hopelessly romantic and deluded; it is a fantasy of a wholesome 'peasant' past where working in the fields, eating only fresh vegetables produced the healthy man . . . Monty Python's *Jabberwocky* and the *Holy Grail* are probably . . . closer to the constant struggle against hunger, lack of sanitation and disease which accompany such a lifestyle.
>
> (Coward 1989, 29)

Underlying such criticisms is unease about the potential anti-human stance of deep ecology. This stance follows from an unwillingness to see the built, fabricated environment of the city and the developed countryside as 'natural'. Deep ecology tells us that such development destroys 'nature' rather than being part of a natural evolution, where humans and non-human nature constantly transform each other. So human activity becomes an aberration: we become 'a kind of malignancy, multiplying too rapidly, destroying too much, a threat to the commonwealth of life' (Allaby 1989). Some Gaianists think we may be a 'pathogenic parasite on the whole planetary organism [which] had done quite well without us for a long time' (Ravetz 1988, 135). Such a pessimistic evaluation of humanity is not new, but, Ravetz thinks, it now has a basis in science: 'The possibility that we are on balance a bad thing for our planet can now be stated in a precise, even testable form . . . we are perhaps an unnatural part of nature.' This despite the apparent paradox that the Gaia hypothesis itself argues that organisms constantly and radically change their environment. Humans alone, it seems, do this to the detriment of Gaia.

Merchant considers that such general anti-humanism is compounded by sexism: Arne Naess consistently uses the term 'man', while failing to see any connection between domination of nature and domination of women in Western society. Warwick Fox (1989), however, strenuously denies that deep ecology is anti-human. It is merely *anti-human-centredness*.

Certainly, an anti-human image militates against deep ecology having mass appeal. This is compounded by its tendency to be inaccessible to ordinary people: the 'woo woo'. Arne Naess lapses 'into almost impenetrable philosophical jargon . . . quantitative mysticism,' says Morris (1993), often obfuscating what would otherwise be fairly obvious to most people. Bookchin (1990, 138–40) denounces deep ecology's 'potpourri' of Eastern mystical traditions overlaying a systems framework. It is a mix of Western 'mind' in its worst aspects with Eastern 'heart' in its most 'vaporous and squamous form', hybridised, incoherent and eclectic.

Morris' further concern is that deep ecology actually uses the language of capitalism and market economics, that is, of the mindset which it ostensibly opposes. Other writers make similar points. Sylvan (1985a) argues that in its concern to allow nature to maximise its 'self-realisation', deep ecology really subscribes to the old, not new, value system, where utilitarianism maintains that that which is best is that which brings greatest benefit/happiness to most people. This is simplistic, as is deep ecology's monism, which reduces all reality into a single, homogeneous force, agency, substance or energy source: 'A night in which all cows are black' in Hegel's words. This, argues Bookchin (1990), betrays thinking that is as reductionist as the world view it is supposed to replace.

Social and socialist ecology

Social ecology and eco-socialism react against the perceived shortcomings of deep ecology outlined above. Like deep ecology, social ecology contains elements of anarchism. The latter is particularly influenced by Peter Kropotkin's (1842–1921) anarcho-communism, being developed most conspicuously today by Murray Bookchin. Eco-socialism employs Marxist perspectives, especially those of William Morris (1834–96) and twentieth-century 'humanist Marxists'. Social and socialist ecology's fundamental message is that our ecological problems stem from social problems.

The social problems which social ecology dwells on are those of hierarchy and domination, as expressed in state-dominated and patriarchal society. Social ecology's solution is to eliminate hierarchy and patriarchy, recreating a 'natural' society, that is, an anarchist-communist one. This would fuse the spontaneous, non-hierarchical social relations of pre-literate peoples with modern scientific society, so as to make the latter truly democratic, communal, unobsessed with consumerism but adequately provided for, and ecologically benign. Its prevailing attitudes and values would fully acknowledge how nature shapes human economic, social and cultural activity. But, unlike deep ecology, it would not overemphasise this. Nor, unlike capitalism, would it overemphasise human society's power to transform non-human nature. Social ecology claims to be neither biocentric nor anthropocentric. Instead it wants humanity to 'situate its good within the context of the planetary good' (Clark 1990). This can come about only via small-scale, decentralised geographical organisation based on local and regional autonomy and, as far as possible, self-sufficiency. Cultural diversity and living within nature's constraints can come through organising into communes for living, cooperatives for working (probably in a near-barter and perhaps moneyless economy), local neighbourhoods for political

31

Table 1.5 Some socialist–anarchist differences

Socialism	Anarchism
Social injustice, environmental degradation caused by class exploitation	Social injustice, environmental degradation caused by hierarchical power relations
Class is defined by economic criteria	Class is also defined by non-economic criteria (race, sex)
Explanations and analyses are historical	Explanations and analyses tend to be ahistorical
Ambiguity about the state – favours at least localised forms of it	Total opposition to the state
Abolish capitalism first and the centralised state will wither away, because capitalism creates the state	Abolish the state first, as an independent act from abolishing capitalism, because the state creates capitalism
The state is the representative and defender of the bourgeoisie	The state represents its own interests, independently of other economic classes
Participation in conventional politics is permissible in the path to revolution	No participation in conventional politics is permissible
Revolution by subverting and confronting capitalism – experimental communities etc. are naive and utopian	Revolution through by-passing capitalism and creating 'prefigurations' of the desired society, such as alternative communities and economies
Emphasise strength of collective political action	Tend to emphasise personal-is-political maxim and individual lifestyle reform
Revolution particularly via our collective power as producers, i.e. unions withdrawing labour, especially in a general strike	Syndicalists advocate union organisation and action – other anarchists stress civil disobedience by community and other non-economically defined groups
Working class will be leading actors in social change	New social movements and community groups will be leading actors in social change
Tendency to vanguardism (Marxism-Leninism)	No revolutionary vanguards
Tendency to dictatorship of the proletariat (transitional stage)	Any 'dictatorship' or government is anathema
Materialist philosophy and approach to social analysis	Tendency to idealism
Modernist politics	Tendency to 'postmodernist' politics
Need for a planned economy	Communes should self-organise, within limits, because spontaneity is important
Limited support for decentralisation	Decentralisation is vital
Individual freedom may be circumscribed by the collective	Individual autonomy is vital
Necessary international exchange, based on reciprocity is an important aspect of international socialism	Opposed to most international trade. Applauds local self-sufficiency

Table 1.5 continued

Ambiguity abut a money economy. Only a few oppose	Most oppose a money economy
Urban-centred	Has a prominent anti-urban element, as well as urban anarchism
Conceives of nature as socially constructed	Tends to see nature as external to society but the latter should conform to nature's laws and regard nature as a template
Anthropocentric (but not in the same way as capitalism-technocentrism)	Advocates (in social ecology) neither anthropocentrism nor biocentrism
Advocates socialist development	Advocates various development models, inc. socialism, environmental determinism and independent development (bioregionalism)
Deep structures (especially economic) condition surface structures, such as spatial organisation	Spatial organisation is a determinant of economics, society, politics

organisation and bioregions for larger needs that cannot be met locally (Chapter 6.3).

Most radical greens, including deep ecologists, would subscribe to this kind of organisation (e.g. Schumacher 1973, Kemp and Wall 1990, 179). But social ecology and eco-socialism put greater emphasis on eliminating social injustice through carefully controlled development rather than any 'future primitivism'. And their strategy for social change involves collectively over-coming political and economic obstacles to the ecological society, including the deeply entrenched capitalist system, rather than concentrating on edu-cating 'wrong' ideas, attitudes, values and lifestyles out of individuals.

I have dealt with eco-anarchism (social ecology) and eco-socialism at length elsewhere (Pepper 1993). While it is true that both desire similar sorts of society, there are also many differences between them, as suggested by Table 1.5. There are, too, differences within anarchism, and therefore eco-anarchism, ranging from more liberal-individualist to more socialist-collectivist perspectives. Liberal anarchism takes an anti-state, consumer movement perspective, which most modern 'new social movements' (e.g. civil rights, greens, feminists) share (Scott 1990). Anarchist-communism, which approximates to Bookchin's social ecology, has much in common with the decentralist socialism which William Morris especially described, pre-echoing what most greens would recognise as an ecological society (Coleman and O'Sullivan 1990).

But eco-socialism's position is unashamedly humanist, rather than eco-centric, looking particularly to the structural features of capitalism to explain why there are ecological problems today. Correspondingly it requires the overthrow of capitalism and establishing true socialism (or communism: the two terms are interchangeable in this context) as the basis of an ecological society.

This does not, however, mean recreating the 'socialism' of the former Iron Curtain countries and China, which eco-socialists see as nothing more than state-managed capitalism. Many ecocentric commentators (e.g. Ferkiss 1993) persist in equating the socialism of eco-socialists with Leninism; consequently to them socialism represents nothing more than crude attempts to dominate nature via giantism and the state. In fact eco-socialists emphatically reject the Orwellian state, though they do argue for considerable planning for ecologically wise production, and some advocate state coordination and facilitation of what happens within regions, localities and neighbourhoods (Frankel 1987, Ryle 1988, and see Chapter 6.2).

In general, while greens tend towards 'postmodernism' – rejecting the culture, assumptions and aims of the modern period as the cause of rather than the cure for 'ecological crisis' – eco-socialists still believe in the Enlightenment promise of universal material progress, providing sustainable development and adequate living standards for all. Indeed, they argue, a satisfactory relationship with nature together with spiritually, intellectually and emotionally satisfying lifestyles – what all greens want – cannot happen *until* this basic level of development and social justice is attained.

1.3 CLASSIFICATIONS

Typologies

Despite claims that radical ecology is 'above' ideology (e.g. from Ash 1987), in fact, says Andrew Vincent (1993) it *is* an ideology, and one with a 'subtle and elusive internal complexity'. This makes classifying different elements of it extremely difficult. Any attempt has to stress the difficulty of drawing boundaries, how categories shade into each other, and how all individual and group ideologies are in any case normally an eclectic mix of different ideas.

There have been classifications of environmentalism which seem to leave nature out in the cold: Gandy (1992), for example, divides fundamentally between market-based and non market-based approaches. This is rare, however, and most classifications are centrally about differences in conceiving of nature and environment.

Vincent's typology of typologies of environmentalism underlines this. There are typologies which fasten on *attitudes* to nature, those which distinguish between different *approaches to knowledge* about it (e.g. scientific, Romantic), *political* ideology typologies (which often involve notions of the 'natural society'), different *philosophical* approaches to nature (e.g. dualistic versus monistic thought), and those which mix these approaches.

The needs of this book can best be met bearing in mind the fundamental distinction between environmental perspectives about where value, worth and good come from. Sylvan (1985a) describes a 'shallow' as opposed to 'deep' position on this, and one which is intermediate between the two.

Anthropocentrism and biocentrism

'Shallow' ecology is anthropocentric, since it makes earth instrumental to human ends. Humans are acknowledged to be the sole reference point of value. They are what confers 'value', 'rights', obligations and moral duty, and they decide what is and is not to be valued. Human concerns are to be met by *using* nature.

Deep ecology, however, as Goodin (1992, 8) clearly describes, gives nature an *independent* role in the creation of value. The green theory of value (Chapter 2.1), Goodin believes, is the 'single moral vision' at the core of radical environmentalism. It gives Gaia *intrinsic value*, making humans inextricably part of 'her'. While shallow ecology implies substantial 'intervention' by humans in the rest of nature, deep ecology implies minimal intervention. Shallow ecology's metaphor for nature is that of a machine, while deep ecology's metaphor is an organism (see Table 1.6).

Most positions in environmentalism, as both Vincent and Sylvan note, are intermediate between these two. Sylvan says that they place serious human concerns first, but attribute to higher animals value in their own right. Rather than the '*sole* value assumption' of shallow ecology this shows a '*greater* value assumption' saying that the rest of nature may have intrinsic value, but the value of humanity is higher.

Vincent describes an intermediate position which he calls 'weak anthropocentric'. Weak anthropocentrics are prepared to extend what is clearly recognised as a *human* set of moral attitudes (not intrinsic in nature) towards the rest of nature. They include 'ecological humanists' (Brennan 1988), who, recognising that benign attitudes towards nature are human-derived and -centred values, none the less argue in favour of such attitudes. This is because, they say, most *humans* prefer this – partly for materially pragmatic reasons (if we pollute ecosystems then this lowers their capacity to provide resources for us), and partly for non-materially pragmatic reasons (*humans* get spiritual and emotional satisfaction and enjoyment out of nature in a relatively 'undegraded' state: many *humans* are uncomfortable with the idea of factory farming animals, and it is *our* discomfort on account of the animals that is a compelling reason not to do it). This is the position of eco-socialists, among others.

Weak anthropocentrism also embraces 'evolutionary eco-naturalism'. This regards nature as constantly evolving, through its capacity for self-organisation, into more complex forms. 'Nature', in this context, includes humans, so natural and social history grade into and are part of each other.

Table 1.6 Mechanistic and organic ethics contrasted

Mechanism	Organicism (Holism)
1 Matter and society are made up of individual parts (e.g. atoms, people)	1 Everything is connected to everything else and each part is defined by its relationship to the whole, and by everything else in the whole.
2 The whole (of society, of any material entity) equals the sum of the parts.	2 The whole is more than the sum of the parts (ecological systems experience synergy, and society has more to it than just the contributions of the individuals in it).
3 Objects are independent of their context: rules of society and laws of nature are and should be universally valid and obeyed by all, regardless of individual differences resulting from different contexts (e.g. different cultures, different experimental conditions).	3 Knowledge and being depend on context. Therefore universal, 'objective' scientific laws cannot be derived unless some working assumptions (which are essentially unreal) are made, while human behaviours and ethics can only be understood in relation to their cultural contexts.
4 Change is essentially the rearrangement of parts in the whole. Energy is neither created nor destroyed, just redistributed and changing form. Individuals change society by associating and dissociating in different corporate bodies.	4 Processes have primacy over parts. That is, biological and social systems are open, constantly exchanging energy and matter with their surroundings. Therefore 'things', 'objects' and 'parts' of the whole, are really temporary structures within the continual flow of energy – the flux of the universe (see Chapter 5).
5 Thinking is dualistic, seeing mind and matter, society and nature, etc., subject and object as separate. Objective thought is possible. Nature and society can be described, controlled and repaired, as could the parts of a machine, by a separate human mind acting objectively according to rational laws.	5 Monistic thinking: humans and non-human nature are a unity – different facets of the same thing. All opposites are different facets of the same thing, or basic 'stuff' (like energy), or spirit (as in God or a universal being).

Source: After Merchant 1992, 68–9, 76–7

This perspective places *mind* at the highest point of evolution (non-living nature gave rise to simple organisms, from which came complex organisms, from which came organisms with minds). Since it is humans which have the highest development of mind, they are the high point towards which this process has worked: the entities by which earth has become self-conscious (Chapter 6.1). Murray Bookchin's (1990) social ecology notably embraces this position.

Ecocentric and technocentric, egocentric and homocentric

While it is clearly useful to think in the above terms (remembering always not to make them hard, fast and mutually exclusive categories), there are also shortcomings in the idea of opposing 'shallow' to 'deep' ecology. For instance, such terms lack precise meaning: deep in relation to what? Marxists for example, accuse 'deep' ecology of being in fact shallow because it does not place at the centre of its analysis the deep economic structures of society without which the workings of cultures and belief systems cannot be fully understood. Second, the terminology belittles those who do not subscribe to the self-styled 'deep' ecologists' beliefs. Deep ecologists such as Eckersley certainly do dismiss other approaches, not because they are necessarily ecologically damaging – she concedes that eco-

Table 1.7 European perspectives on environmental politics and resource management: contemporary trends in environmentalism

Ecocentrism		*Technocentrism*	
Gaianism	*Communalism*	*Accommodation*	*Intervention*
Faith in the rights of nature and of the essential need for co-evolution of human and natural ethics.	Faith in the cooperative capabilities of societies to establish self-reliant communities based on renewable resource use and appropriate technologies.	Faith in the adaptability of institutions and approaches to assessment and evaluation to accommodate environmental demands.	Faith in the application of science, market forces, and managerial ingenuity.
'Green' supporters; radical philosophers.	Radical socialists; committed youth; radical-liberal politicians; intellectual environmentalists.	Middle-ranking executives; environmental scientists; white-collar trade unions; liberal-socialist politicians.	Business and finance managers; skilled workers; self-employed; right-wing politicians; career-focused youth.
0.1–3 per cent of various opinion surveys.	5–10 per cent of various opinion surveys.	55–70 per cent of various opinion surveys.	10–35 per cent of various opinion surveys.
Demand for redistribution of power towards a decentralised, federated economy with more emphasis on informal economic and social transactions and the pursuit of participatory justice.		Belief in the retention of the status quo in the existing structure of political power, but a demand for more responsiveness and accountability in political, regulatory, planning and educational institutions.	

Source: O'Riordan 1989

socialism, for instance, may not be so – but because they do not share the core belief of supposed 'biocentrism'. In fact, as Sylvan notes, 'shallow' approaches are not necessarily bad for the environment: pragmatism and human-centredness are very good motivations for caring for nature. Barry (1994) rightly says that anthropocentrism can form a legitimate, strong and flexible 'actual, practical, moral basis of green theory'. Indeed Sylvan regards the whole 'shallow–deep' debate as about a false dichotomy: not recognising those important intermediate positions described above.

The kind of terminology proposed and developed by O'Riordan (1989) may be more useful (see Table 1.7). *Ecocentrism* views humankind as part of a global ecosystem, subject to ecological laws. These, and the demands of an ecologically based morality, are seen to constrain human action, particularly through imposing limits to economic and population growth. There is also a strong sense of respect for nature in its own right, as well as for pragmatic reasons.

Ecocentrics lack faith in modern large-scale technology and technical and bureaucratic élites, and they abhor centralisation and materialism. If politically to the right they may emphasise the idea of limits, wanting to restrain growth in population and resource consumption, and access to nature's 'commons'. If to the left, they emphasise decentralised, democratic, small-scale communities.

The ecocentric position on technology is complex. It is not anti-technology, though it is 'Luddite', given that the Luddites did not protest against technology of itself but against its ownership and control by an élite. Ecocentrism advocates 'alternative', that is, 'soft', 'intermediate' and 'appropriate', technologies partly because they are considered environmentally benign, but also because they are potentially 'democratic'. That is, unlike high technology they can be owned, understood, maintained and used by individuals and groups with little economic or political power.

Technocentrism recognises environmental problems but believes either unreservedly that our current form of society will always solve them and achieve unlimited growth (the interventionist 'cornucopian' view) or, more cautiously, that by careful economic and environmental management they can be negotiated (the 'accommodators'). In either case considerable faith is placed in the usefulness of classical science, technology, conventional economic reasoning (e.g. cost–benefit analysis), and the ability of their practitioners. There is little desire for genuine public participation in decision making, especially to the right of this ideology, or for debates about values. The technocentric's veil of optimism can be stripped away to reveal 'uncertainty, prevarication and tendency to error' (O'Riordan 1981). Technocentrics envisage no radical alteration of social, economic or political structures, although those on the left are gradualist reformers.

Typical technocentric perspectives emerge from the pages of *Atom*, the

UK Atomic Energy Authority's magazine. A critique of 'green science' (Grimston 1990) contrasts it with 'real science', where

> a potential problem is identified and analysed, an imaginative suggestion is made, a resulting course of research and possible action is taken, and underlying it all is a sense of technical optimism.

Grimston is optimistic about the 'imaginative suggestion' of injecting iron into the seas to stimulate plankton growth, thus increasing carbon dioxide absorbtion:

> If it works we can save the world while continuing to help the less developed world to industrialise . . . and yet still enjoy our fabulous lifestyles . . . In principle only energy cannot be recycled: I cannot see any way round the second law. But everything else is only atoms . . . there are no technical obstacles to sustaining ten billion people in comfortable conditions without destroying the environment . . . Greens want de-industrialism, but this would be disastrous. We have tasted of the tree of the video machine and what we now have – long lives, good health, leisure time, we would certainly miss . . . Mankind has actually largely defeated nature's callousness and we have done it through industry.

And he declares war on green abuse of his 'objective' science:

> I am a scientist. I love my subject, and it pains me when it is abused simply to lend credibility to a preconceived moral or political viewpoint . . . It is time for the technical community to abandon its attempts to accommodate irreconcilable opponents [greens] and instead aim to re-establish the idea of scientific authority . . . [we] plead guilty with heads held high to the charge of having been trained to solve problems without imposing colossal social change.

Clearly, mainstream technocentrism is politically reformist compared with ecocentrism, which requires radical redistribution in political power.

O'Riordan says that technocentrism is 'manipulative': it sees humanity's destiny as manipulating and transforming nature into a 'designed garden' to improve both nature and society. But there are differences between those who would freely intervene in nature ('interventionists', who, paradoxically, are non-interventionist in the market economy, for example Simon and Kahn (1984)) and those who recognise a need to accommodate natural constraints. Accommodation involves environmental management based on cost–benefit and risk analysis, together with manipulating the economy via environmental taxes and penalties, standard setting and the like (Pearce et al. 1989).

Ecocentrism also contains differences in emphasis within its general perspective of nurturing nature rather than intervening destructively in it.

There is what O'Riordan (1981, 89–90) calls 'communalism' (see Table 1.7), where 'economic relationships are intimately connected with social relationships and feelings of belonging, sharing, caring and surviving'. Communalism stems from nineteenth-century anarchism and seeks to live socialistically in cooperative networks of community organisations. It is largely equivalent to 'social ecology', while O'Riordan's 'Gaianism' equates with deep ecology.

O'Riordan's classification is widely used, but its usefulness is limited by not addressing positions between deep and shallow ecology. Merchant's (1992) classification does map this middle ground, at least in terms of its ethics and ideological antecedents (see Table 1.8). She contrasts 'egocentrism', which is equivalent to the ideologies of *laissez-faire* capitalism and a mechanical view of nature (as in cornucopian technocentrism) with ecocentrism, including deep and spiritual ecology, cultural feminism, bioregionalism, organic farming and indigenous people's movements – all of which see nature as an organism rather than a machine. The ecocentric ethic is 'grounded in the cosmos' (p. 74), whatever that may mean.

In between she places 'homocentrism', based on utilitarian philosophy and Marxism among other ideologies, and drawing on both mechanism and organicism. This does prioritise human values and desires, but its humanism does not lead to the destructive and short-sighted view of nature associated with egocentrism's aggressive and competitive individualism. Including social ecology and eco-socialism and most animal rights movements (which extend human ethics towards higher animals), homocentrism would steward nature in the attempt to maximise the sum of human happiness and welfare. It would include much of the 'communalist' element in O'Riordan's classification.

Green politics

Greens often hold it as a point of principle that they are neither left nor right, but 'forward', or 'above the old politics': 'The basic political choice today is not between Right, Left or Centre, but between conventional grey politicians and the Green Party', said that party's 1992 manifesto.

Yet this very rejection of traditional politics and politicians can itself be thought of as a fundamentally conservative sentiment. And when they discuss what we should all *do* about eco-crisis, greens certainly do invoke the 'old' politics. Table 1.9, which maps green politics against what more traditional political ideologies have to say about ecology, suggests that mainstream greens mainly straddle the categories of (a) welfare liberalism and (b) democratic socialism.

Thus greens say that social change must proceed from individuals (a) but change is also needed in the economic structures of society (b). They do not totally reject capitalism – indeed are enthusiastic for at least small-scale

Table 1.8 Threefold classification of environmentalism based on grounds for environmental ethics and showing antecedents in Western thought

Self: egocentric		Society: homocentric		Cosmos: ecocentric	
Self-interest	Religious	Utilitarian	Religious	Eco-scientific	Eco-religious
Thomas Hobbes John Locke Adam Smith Thomas Malthus Garrett Hardin	Judeo-Christian ethic Arminian 'heresy'	J. S. Mill Jeremy Bentham Gifford Pinchot Peter Singer Barry Commoner Murray Bookchin Social ecofeminists Left greens	John Ray William Derham René Dubos Robin Attfield	Aldo Leopold Rachel Carson Deep ecologists Restoration ecologists Biological control Sustainable agriculture	American Indian Buddhism Spiritual feminists Spiritual greens Process philosophers
Grounds for obligation					
Maximization of individual self-interest: what is good for each individual will benefit society as a whole Mutual coercion mutually agreed upon	Authority of God Genesis 1 Protestant ethic Individual salvation	Greatest good for the greatest number of people Social justice Duty to other humans	Stewardship by humans as God's caretakers Golden Rule Genesis 2	Rational, scientific belief system based on laws of ecology Unity, stability, diversity, harmony of ecosystem Balance of nature or chaotic systems approach	Faith that all living and non-living things have value Duty to whole environment Human and cosmic survival
Metaphysics					
Mechanism 1 Matter is composed of atomic parts 2 The whole is equal to the sum of the parts (law of identity) 3 Knowledge is context-independent 4 Change occurs by the rearrangement of parts 5 Dualism of mind and body, matter and spirit		Both mechanistic and holistic		Organicism (Holism) 1 Everything is connected to everything else 2 The whole is greater than the sum of the parts 3 Knowledge is context-dependent 4 The primacy of process over parts 5 The unity of humans and non-human nature	

Source: Merchant 1992

Table 1.9 Political philosophies and environmentalism

Traditional Conservatives (radical)	Are limits to growth, and enlightened private ownership is the best way to protect nature and environment from over-exploitation. Protect traditional landscapes, buildings, as part of our heritage.
	Anti-industrialism: human societies should model themselves on natural ecosystems, e.g. should be stable, and change slowly and organically. Need for diversity, but hierarchical structure: bound together by commonly held beliefs. Everyone to be content with their position (niche) in society. The family (perhaps extended) is the most important social unit. Admire tribal societies. Romantic: yearn for the past.
Market Liberals (reformist)	The free market, plus science and technology, will solve resource shortages and pollution problems. If resources get scarce, people will supply substitutes – if there's a market for them.
	Don't believe in 'overpopulation'; people are a resource.
	Capitalism can accommodate and thrive on protecting the environment.
	Consumer pressure for environment-friendly products will play a big part, capital will respond to this market.
Welfare Liberals (reformist)	Market economy, with private ownership, but managed. Reform laws, planning and taxation for environmental protection.
	Enlightened self-interest, tailored to the communal good, will solve the problems.
	Consumer pressure for environment-friendly products will play a large part. Pressure group campaigns, in a pluralist, parliamentary democracy will lead to appropriate legislation.
Democratic Socialist (reformist)	Decentralised socialism; local democracy; town-hall socialism.
	Mixed economy and parliamentary democracy – with strict controls on capitalism. Emphasises the role of labour and trade unions. A big role for the state (especially locally). Mixture of private and common ownership of resources. Emphasis on improving the urban environment. Production for social need. Big coops sector. State subsidises environment protection (e.g. public transport).
Revolutionary Socialist (radical)	Environment ills are specific to capitalism, so capitalism must be abolished: requiring some revolutionary change, perhaps brought on by environmental crises.
	Rejects the state ultimately, but perhaps needed in the transition to a communal (communist) economy (less conflict vital in social change to a green and socially just world – reject parliamentary reform
	Poverty, social injustice, squalid urban environments, all seen as part of the environmental crises.
	Similar visions of future to anarchists, but emphasise collective political action, and the state initially.

'Radical' = wanting to go back to the roots of society and change it fundamentally in some
ways, and quite rapidly.
'Reformist' = The present economic system is accepted: but it must be revised – in the
direction of either less or more interference in and management of the economy
– gradually and through parliamentary democracy.

Table 1.9 continued

*Mainstream Greens (radical aims, but reformist methods)
(inc. British Green Party: Friends of the Earth and other pressure groups)

A mix of welfare liberal and democratic socialist prescriptions but say they reject politics of left and right. Emphasise the importance of the *individual* and his or her need to revise values, lifestyles and consumer habits. Bioethic, limits to growth, utopianism.

Advocate a lifestyle of voluntary simplicity. Also, need to change social-economic structures, including putting an end to the 'industrial society'. Favour small-scale capitalism, but with profit motive secondary to production for social and environmental need. Also coops and communes. State has a role – especially locally. Romantic view of nature – spiritually important, especially in deep ecology and New Ageism, which all mainstream greens have tendencies towards. New Age irrationalism, mysticism, rejection of 'politics' and industrialism gives it a reactionary, conservative element.

* Green Anarchists and Eco-feminists (radical aims and methods)

Reject the state, class politics, parliamentary democracy and capitalism. People to organise themselves: have responsibility and power over their own lives. The *individual* very important, but the individual gets fulfilment in relation to the community. Decentralised economy and politics: common ownership of means of production, and distribution according to needs (income-sharing communes). Spontaneous and organically evolving society. Non-hierarchical direct democracy. Rural and urban communes and cooperatives. Bioregionalism.

- -

**These two together represent 'ecologism' (ecocentrism), which starts, unlike others, from the *ecological* imperative and the bioethic (nature as important as human society). But in their social prescriptions they mainly straddle welfare liberalism and socialism (with one or two elements of conservatism and revolutionary socialism).

versions of it (a), but they see social need and environmental quality as criteria to be elevated above the profit motive (b). The state does have a benign role (b), in facilitating the development of individual responsibility (a). This grudging acceptance of the state (and of parliamentary politics by Green Party supporters) constitutes a significant distinguishing feature marking off mainstream greens from eco-anarchism. But the elevation of natural laws and ecological principles also marks them off from 'straight' liberals and socialists, as does their sometimes expressed desire for more urgent and radical social change. Nature may be the source of social laws, but to many, principles of social justice are as important. However *eco-centrism* purports to make social justice part of a wider justice required for all life forms. Technology is not rejected, but it must be appropriate and democratic, as well as 'soft' on nature. Rationalism (a and b) must be balanced by elevating emotional and intuitional knowledge. Democracy and individual freedom (a) are cornerstones of mainstream green ideology – and that democracy is to be extended to all nature's creatures (animal rights, vegetarianism, veganism). But the importance of the community is stressed too (b).

Besides all this there is also that suspicion of the 'industrial way of life', and 'the old politics', coupled with a tendency towards irrationalism and mysticism. The innate conservatism of this is clear. It cannot be denied that despite the *emphasis* on left-liberalism in ecologism there is also a persistent strand of conservatism. It may be a minority strand, which is why the Table 1.9 does not extend 'mainstream greens' over towards the conservative side of the diagram, but it is there. The most prominent British radical environmentalist associated with it is Edward Goldsmith, who (1988) argues for commonly held belief systems such as those enshrined in strong religions as stabilising forces to create that *social unity* that he considers to be the key to an ecologically sound society. For Goldsmith the common values must, above all, be derived from ecological laws, creating an ecosystems model of society – they are not relative and arguable, they are *absolute*. He also eulogises 'traditional' values, constantly referring to 'primitive' peoples and tribes in Africa, Australasia, etc. as models for us. And he sees the family as the essential unit of social organisation: whatever preserves this (such as the traditional stereotyped role for women) is to be encouraged Finally, as a conservative ecocentric, he rejects 'industrial' society – it is aberrant.

The existence of a link between conservatism and ecologism is confirmed by conservatives (especially traditional conservatives rather than members of the New Right) themselves:

there are many natural affinities between conservative philosophy and Green thought . . . In repudiating the fashionable heresies of neo-liberalism [the New Right], conservatives are merely returning to an older and sounder Tory tradition, which perceived the illusoriness of the sovereign, autonomous chooser of liberal theory, and so insisted on the primacy of the common life. The importance of Green thought for conservatives today is that it recalls them to their historic task of giving shelter to communities and reproducing them across the generations – in a context of finite resources which dictates stability, not growth, as the pre-eminent conservative value.

(Gray 1993, 173)

Most radical greens are influenced by anarchism (Table 1.10 and Pepper 1993). They include 'eco-anarchists', 'eco-feminists' and 'eco-pacifists', who all believe in the need for 'organic societies'. Some express anti-urban, anti-industrial, conservative sentiments. But most have liberal leanings. Indeed, in emphasising the sovereignty of the individual, anarchism can be seen as extreme liberalism (Bottomore 1985).

But then anarcho-communism and anarcho-syndicalism also are forms of socialism, and eco-anarchists are mainly anarcho-*communists*, looking particularly to Kropotkin for inspiration (Chapter 4.5). They reject capitalism, want common ownership of the means of production (resources), and distribution according to need.

The eco-anarchist's utopia may involve rural communes and the craft-

Table 1.10 Types of anarchism

Individualism
Each individual follows own inclinations, but may enter into a 'union of egoists' for convenience.

Mutualism
Work organised around mutual credits. Federations of communes and workers' coops based on social contracts.

Collectivism
Voluntary groups of people or institutions sharing some goods. Individual still has right to enjoy his or her products.

Anarchist-communism
Voluntary federations of communes, *collectively* owning property. Distribution according to need. From each according to means.

Anarcho-syndicalism
Associations based on the workplace. Revolutionary trade unions run all production and distribution, alongside community groups.

Anarchist-pacifism
Non-violent resistance and revolution. Libertarian communities as a peaceful version of 'propaganda by deed'.

Source: After Woodcock 1975

based socialism of William Morris' *News from Nowhere*. It may also be inspired by urban anarchists like Colin Ward, being based in urban communes and the squatter movement. However Anglo-European eco-anarchism is not as uncompromisingly urban-centred and rooted in trade unionism as the Australian version (Purchase 1993). The former stresses small-scale, collective, decentralised commune-ism, participatory democracy through town and community meetings, low-growth (or no-growth) economy, non-hierarchical living and consensus decisions. All this echoes American populism (Roszak 1979), and is celebrated in Callenbach's (1978) fiction *Ecotopia*.

Chapter 2

SOME FUNDAMENTAL ISSUES IN RADICAL ENVIRONMENTALISM

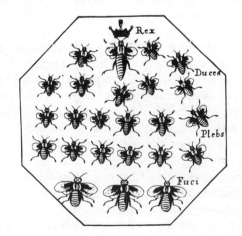

SOME FUNDAMENTAL ISSUES IN RADICAL ENVIRONMENTALISM

2.1 BIOCENTRISM AND INTRINSIC VALUE

The green theory of value

A theory of value concerns what creates, and measures, worth and good. Capitalist theories look to consumers to establish value (Table 2.1) by exchanging goods and services in a market place. When capitalist economics talks of value it therefore means *exchange* value. That which satisfies the 'wants' of most people has most value. What is most and least wanted is signalled (in theory) by price – high price denoting high demand in relation to supply. Marxist theory is producer-oriented, seeing the principal source of value as human labour – the more labour invested in products and services, the more valuable they tend to be – especially if they meet 'needs' rather than 'wants'. Here the latter are regarded as partly artificial: created by advertising, for instance, in consumer society. Socialist economics, then, do not focus on the process of exchange to establish value, but emphasise *social usefulness*. In capitalism, socially useful goods and services may not be provided to the poor because the poor cannot signal that they value them, by paying high prices.

All these theories, along with Christian perspectives regarding God as source and determinant of all value, are grounded within the tradition of anthropocentric humanism that has developed in Western Europe since the Middle Ages. Green theory, by contrast, sees nature as a prime source of worth (Goodin 1992).

This means more than just placing low values on goods and services that gobble up natural resources (though it does mean this). It also implies goodness and worth in nature *of itself,* apart from any usefulness to humans. Devall and Sessions (1985, 71) describe inherent value as 'independent of any awareness, interest or appreciation by a conscious being', that is, it is not merely in the eye of the conscious beholder: it is *objectively* there. They echo Heidegger, who was, claims Zimmerman (1983), in the 1930s and 1940s 'the first theoretician in the ecological struggle'. Heidegger pro-

Table 2.1 Theories of value

Neoclassical economics

Society is a collection of selfish, rational individuals, acting freely within markets to maximise opportunities to satisfy their wants. The source of value of goods and services is the 'subjective preference' of these individuals (i.e. the consumers, usually in aggregate). Hence increased demand in relation to supply confers increased value.

Neo-Ricardian school (Keynesian)

Markets must be managed by governments to maximise welfare, since totally free markets bring social disbenefits as well as individual benefits. Value is a function of costs of production, rather than consumers: partly determined by labour costs, partly by stages of technological development. These are a function of negotiations between different power groups in a pluralist democracy and government's job is to mediate in these negotiations for the general good.

Marxist

Value is a function largely of the labour power invested in goods and services. Private ownership of the means of production and the imperative of making and accumulating capital generates a class system whereby the people who produce goods and services are not paid the full value of them. Hence surplus value (amount paid to workers subtracted from market value of what is produced) is appropriated from the producers. This is the source of capital.

Greens

The value of goods and services and of things which are not goods and services is a function largely of the natural resources embodied in them. This implies not only the quantity of resources, but also their 'naturalness', i.e. the extent to which they have been created by natural processes rather than by 'artificial', i.e. human, processes. It follows that 'natural phenomena' which are completly untransformed by humans, and have never been, are most valuable of all.

Source: Cole *et al.* 1983, Goodin 1992

claimed that anthropocentric humanism took humans beyond their 'proper limits'. He proposed instead a non-anthropocentric relationship between humanity and nature. This must 'let beings be' – must be open to their being fulfilled in ways most appropriate for them, not humans. This would imply a considerable shift in how Western culture regards the society–nature relationship.

Since deep ecologists consider that humans are not outside or above nature, being just one of its constituencies, then humans cannot be the arbiters of the value of the rest of nature. Biological egalitarianism demands that humans must value and respect all other living and 'non-living' entities (in a sense everything is 'alive', being part of a whole ecological, self-repairing Gaian ecological community). From this it may follow that nature has *rights*:

> rocks, just like people, do have rights in and of themselves. It follows that it is in the rock's interests, not the human interested in the rock, that it is being protected.
>
> (Nash 1977, 10, cited in Merchant 1992, 76)

Such sentiments were responsible for a rash of legal actions in the USA in the 1970s 'on behalf of' threatened species and landscapes (see O'Riordan 1981, chapter 8), whereas much environmental law today (especially European) centres on the ill-effects to humans of environmental damage (see Hughes 1992).

Although such law aims to protect the environment, deep ecologists criticise it for being anthropocentric. 'Rights', in liberal society, form part of a social contract between humans, so extending rights to animals simply includes them in the human community: it imposes human concepts on the non-human world. This approach, says Fox (1990, 11–17), is undesirable because it is empirically incorrect – humans obviously are not at the universe's centre. And anthropocentrism has been disastrous in practice, resulting in industrialism, which destroys nature.

Objections to intrinsic value

O'Neill (1993a) considers that to have an environmental ethic you must believe that nature has intrinsic value. This is debateable, but deep ecologists certainly would agree. 'Intrinsic value', however, could have three meanings according to O'Neill (pp. 8–25). First, it is a synonym for non-instrumental value, so nature is not a means to an end but an end in itself. Second, it has value in terms of its own properties, not simply by virtue of its relationships to other entities, for example a forest might have value regardless of whether or not it was the only one of its kind. Third, it is a synonym for objective value, that is, value resides in nature independently of those who might also value it. In other words, if all human life ceased, the rest of nature would still have value and worth. O'Neill considers that ecocentrics unjustifiably conflate the three meanings, though in practice they do tend to overlap.

The problems of justifying 'intrinsic value' have attracted much critical attention. First, if, in O'Neill's third sense, you say that non-human nature has 'objective' value and worth, that is, not depending on human valuation, you might be implying that humans and nature are *separate* (in the Cartesian sense of humans as subject, separate from nature as object, see Chapter 3). This could contradict deep ecology's own conception of them as part of each other (Vogel 1988). Such a separation is also implied when 'ecosophy' castigates humans, alone among all the species, for behaving 'naturally' by changing their environment and using it for their own purposes. It makes humans

> separate from, different from, removed from or above nature . . . To avoid this special treatment of human beings as other than nature . . . we must stress that man's works (yes, including H-bombs and gas chambers) are as natural as those of bower birds and beavers.
>
> (Watson 1983, 252)

Second, the idea of objective value could be held to justify nature by itself, without reference to humans, so that if humans were removed from earth there would still be value in the rest of nature continuing. Yet the very ideas of *value, worth* and *rights* are *human* ideas: human concepts, assessments and valuations imposed onto nature. O'Neill argues that it is quite feasible for us to decide that a world without humans has value. Yet it is still *we* who are making this decision. We cannot, it seems, know whether other species value parts of nature other than themselves and that which gives them food and shelter. But it may be that the very concept of value has no meaning at all in the rest of nature: apart from what is associated with a very basic survival instinct. In Leopold's (1949) claim that something is right when it preserves the *integrity, beauty* and *stability* of the biotic community and wrong when it does otherwise, the words emphasised refer to essentially human-based qualities, which would be meaningless without humans to bestow them. Hence, as Merchant (1992, 78) says, 'At bottom, ecocentric ethics may have a homocentric justification.'

It might be objected that this kind of point is only too obvious, even trite. We have to use human language and concepts to describe everything. This does not mean that we have to be 'humanist' in the sense that 'only human states and achievements have value' (O'Neill 1994). We should acknowledge that there is a world 'out there', with its own objective qualities which, as we develop our own aesthetic sensibilities, we increasingly recognise.

Still, such recognition of nature 'for itself' is, really, for us, and it is important to accept this as legitimate, lest we fall into the trap of being anti-human. A 'response to the objects of the non-human world for their own qualities' is necessary, says O'Neill (1993b, 141), because it 'forms part of a life in which human capacities are developed. It is a component of human well-being.' Here we are clearly talking about a kind of human 'use' value and one that is ecologically acceptable. It satisfies the needs of 'humanism' when that word is defined in a second sense, of concern with human flourishing:

> the good life for humans involves among other things a recognition of the value of non-human beings in the natural world and a concern with the promotion of their well-being.
>
> (O'Neill 1994, 21)

This kind of use-value of nature might include what conventional economists call 'existence' value, where we like to think that, for instance, tropical rainforest exists, though we might see no immediate use for it and may never visit it. In this way, as Martell (1994) concedes, you can be a humanist and maintain a sound environmental ethic.

What we are doing in such an argument is deciding, subjectively and for us, to attribute 'intrinsic' (objective) value to nature. This may preserve ecosystems, but for Callicott (1985) and other deep ecologists it is still unsatisfactory. First, such value cannot be defended rationally, only asserted. Second, conceding that humans *do* bestow value, in a universe where they and the rest of nature are one, could lead us into saying that the worth of non-human nature ultimately *depends* on the consciousness of the observer, that is, human consciousness, which would be anthropocentric extremism (Sylvan 1985b).

Callicott, like Rolston (1989), believes that quantum theory may resolve the situation, by demonstrating that differences between 'objective' and 'subjective' qualities are anyway obsolete. This is a contentious interpretation of quantum theory (Chapter 5.3), but if correct it might imply that subjectively-bestowed value possesses essentially the same reality and status as that which we normally attribute to 'objective' facts. Equally, however, you could argue the opposite from quantum theory: that *all* qualities and attributes are really subjective – depending on the observer – and you are back to the extreme anthropocentrism where only humans bestow any meaning to the world.

We can try to sidestep the problem, like Skolimowski (1990), who prefers to say that qualities like value are *transpersonal* – going across and beyond the person or the individual entity, or species. It follows that if nature is continuous with the human self then obviously both the self and nature have value. If self-interested behaviour is rational then it is rational to act in nature's best interests. This defence sounds extremely pragmatic, but, again, this is no bad thing. As Weston (1985) says, people's instinctive feeling for nature, as revealed in the increasing popularity of outdoor pursuits like backpacking, constitutes an essential and perfectly adequate starting point for a defence of environmental values. It is a form of instrumentalism, albeit a broad one, not based on narrow material considerations.

Murray Bookchin is acutely aware of political dangers in insisting on a nature with 'objective' value. For we can begin to think of ourselves as helpless before nature's forces and laws, and strictly limited in our capacity to improve the social world by using nature. And we can think of ourselves as 'sinners' when we do use and change nature, defiling some supposed pristine purity. For Bookchin (1990, 44), organic evolution is a process whereby the planet directs itself towards greater consciousness and self reflectivity. It follows that the most conscious of life forms – humanity has particular responsibility to be the 'voice of a mute nature and to act to intelligently foster organic evolution'. If this evolutionary tendency and the particular importance in it of humans is denied then there is no reason why humanity should not, like other species, seek self-realisation merely at the expense of other species. Bookchin here attributes *special* value to humanity, as the 'brains' and custodians of the rest of nature.

This is a position which deep ecologists actually abhor. Yet their own alternative principle of biological egalitarianism is hugely problematic.

Merely because all parts of nature have value, this cannot, practically, imply *equal* value. For, as Sylvan (1985a) asks, are we then to infer that the life of the AIDS virus is as precious as that of a human? Such questions are particularly important in that special case of the biocentrism/intrinsic value issue which is the debate about animal rights.

Anthropocentrism and the case against animal rights

Ecocentrics are concerned about how animals are treated in Western 'industrial' societies, especially in factory farming. While not all are vegetarians, still less vegans, most ecocentrics insist that while they live animals should be treated 'humanely'. This very term implies anthropocentrically extending human rights and moral consideration to non-human nature. However, ecocentrics, as already noted, may reject anthropocentrism as an unfortunate product of the Enlightenment thinking which they generally criticise.

It included Descartes' (1596–1650) view that humans were distinguished from the rest of nature by having a soul, being self-reflective and able to think rationally and to contemplate the likely outcome of their actions. He saw the rest of nature, including animals, as machine-like. Animals reacted automatically to external stimuli, he thought, showing pain, but not as humans experience it – with anticipation, dread, sadness, nobility, etc. Hence if animals are machines we need have no scruples about how we treat them.

Modern versions of this argument (e.g. Francis and Norman 1978) accept that animals are not machines, but propose greater human consideration on account of human capacities for reason, understanding, emotion, and forming sophisticated socio-economic and familial relationships – as well as on account of sentience. Animals have only the last to any degree. And they do not have advanced language, which is the basis on which humans form a moral community – language is the starting point for an ethical code. Hence Frey (1980) argues that animals cannot have rights because they cannot have interests, because they cannot have desires or emotions, because they cannot have the thoughts required for them, because they cannot speak. This reasoning considers that language is thought and vice versa.

By extending the argument, animal rights critics echo the eighteenth-century philosopher David Hume's (1711–76) conception of 'rights' as a legal contract between mutually agreeing parties. Since only humans can assign or claim rights animals cannot have them. This is Passmore's (1980) position, and he adds that any community of humans, animals and the land is an ecological one only, not a moral one. Rose (1992) considers that because rights is a human concept, to extend them to animals is 'speciesist', i.e. prejudiced in favour of humans – the very thing the animal rights lobby abhors.

Most of these writers do not dismiss the need to treat animals well. But their reasons for good treatment lie in human, rather than animal, interests. Passmore reiterates the view that in abusing animals humans degrade *themselves*: 'He who is cruel to animals becomes hard also in his dealings with men' (Kant, cited in Midgley 1983, 51). Francis and Norman similarly advocate animal protection for the utilitarian reason that animals have sentimental, emotional value for *us*.

If it should come to a choice between animal and human interests, it is usually clear that we owe a species loyalty to humans. Rose asserts that to pretend that a struggle for animal rights is of the same order as that for human rights is perverse and anti-human. For the sake of humans it may be necessary to harm animals, in vivisection for instance. This is why biological egalitarianism cannot be practical, and why even injunctions not to cause *unnecessary* suffering are problematic.

For what is 'necessary' is relative – differing from culture to culture in time and space. Answering Peter Singer's (1983) absolute injunction never to eat meat, because rearing animals for food inevitably causes suffering, relativists point out that in some cultures vegetarianism is not an option, for religious reasons and/or because it is ecologically impractical (many areas have climates and soils that preclude arable farming). Furthermore a pure 'hands off living creatures' approach could lead to the absurdities of no defence against virus diseases, no photography and film, no stringed instruments or drums, and cloudy beer! If, as many animal liberationists do, you instead propose a hierarchy of creatures and moral consideration, you are really shifting the boundaries of hierarchical thinking rather than eliminating it – perhaps only a marginal improvement on a 'humans-only' position (Anton 1992).

For animals' interests

In defending animals, Johnson and Johnson (1992) do not propose that all life has great inherent value, for mowing a lawn is clearly not equivalent to wholesale slaughter. Yet, they say, it should still be possible to give animals rights on a non-utilitarian basis, so they are not just seen as human resources. Midgley thinks that there is no case for excluding animals from our community of rights, while Barbour (1980) reminds us of injunctions in Christianity commanding us not to do this. For some people animal rights can mean legal standing (Stone 1974), which is not jeopardised merely because animals themselves cannot sue. Humans can sue on behalf of animals, just as they can sue on behalf of babies (Warren 1983).

Whether this is accepted or not, however, it can be argued that all moral agents (i.e. humans who can assign and demand rights) have a *duty* not to harm other individuals, who may not be moral agents themselves. This is because all individuals have *interests*. Those interests add up to enjoyment of life to the maximum, and they should be protected if possible as if they were binding rights (Benson 1978).

For Singer (1985, 9) and Regan (1988), the most prominent advocates of this argument, the basis for ascribing interests to animals – particularly 'higher' animals – is that, like us, they feel: sentience is universal amongst all creatures with nervous systems. If you should want to exclude animals from the moral community on grounds that they have *no more* than sentience – they cannot assert their interests through speech and reason – then you would have also to exclude *human* non-moral agents: babies and mentally retarded people for instance. Singer adds that because animals, unlike humans, cannot understand what is *going* to happen to them their capacity for suffering could be more, not less, than humans: they cannot be reassured when they feel the pain, say, of an injection from the vet, that it will be for their good.

Regan claims more than mere sentience for mammals. Many have at least rudimentary perception, memory, emotion, desire, belief, self-consciousness, intention and a sense of the future. Therefore they have interests beyond mere survival; in living a full and pleasurable life. Johnson and Johnson assert that the mental powers of animals differ from humans only by degree: animals use abstract concepts like size, shape and hue to classify objects in the natural world. And Midgley rejects the extreme linguistic philosophy argument that language is the only source of conceptually ordering the world: merely because animals do not speak does not signify that they live in a disordered world. The burden of proof, says Regan, is on those who would deny that all this is so.

A resolution?

Is it possible to assert animal interests without being anthropomorphic on the one hand ('Disneyfying'), or regarding humans as merely another animal species on the other? Regan and Singer want to extend the 'circle' of moral concern, that is liberal-humanist values of equality and respect for the individual, across the species boundary, to higher animals. Benton (1993) echoes Rodman's (1977) unease at such anthropocentrism, and as a socialist, Benton particularly rejects the liberal-individualist view of rights which is being advanced.

He would be happier to extend *socialist* morality and insights across the species barrier. Through Marxist analysis this could show how animals, like humans, are turned into objects and commodities through capitalism, and how commodities produced from animals, like other commodities, manifest oppressive social relations between people and between people and animals. Partly by resurrecting and reconstructing earlier Marx (which saw nature and humans as part of each other in dialectical relationship), Benton advances a socialist position that recognises what is special about humans without regarding them as privileged by overemphasising what sets them apart from the rest of nature. It is a *naturalistic* position, accepting that humans are part of the order of nature, and that they share with animals needs such as health, physical security, nutrition and shelter. But, following Midgley (1979), this naturalistic understanding of humans is none the less compatible with the view that in some respects they are palpably different from animals.

It regards humans and animals (who do share common ancestry) as on a continuum, with no sharp divide between the nature of their being. Of course

> there are things (reading, writing, talking, composing symphonies, inventing weapons of mass destruction) which humans and only humans can do . . . [but they] are to be understood as rooted in the specifically human ways of doing things which other animals also do.
>
> (Benton 1993, 47–8)

That is, animals and humans have different ways of responding to characteristics and needs which they have universally in common – universals cutting across not only human cultures but also the human–animal barrier. These include the facts of birth and death, growth, development and decline and sexuality, and the need for social cooperation, stability of social order and the integration of social groups. Benton goes on to discuss how animals are intimately part of human society, ranging from sources of food, clothing, profit, companionship, to symbols and sources of metaphor for human relations. His naturalistic socialism would go far in meeting the suspicions which ecocentrics have about more orthodox Marxist socialism (see Pepper 1993).

2.2 TRAGEDY OF THE COMMONS

Hardin's parable

A central issue for ecocentrics is that of limits to the carrying capacity of earth's ecosystems. If animals or humans attempt to use the resources provided by ecosystems beyond those limits, then the ecosystems change (or 'degrade', as ecocentrics would say) to other sorts of ecosystems, which may be less productive.

This problem was graphically expressed in biologist Garrett Hardin's parable (1968), reviving a scenario proposed by mathematician William Lloyd in 1833. 'Picture a pasture open to all,' said Hardin. 'It is to be expected that each herdsman will try to keep as many cattle as possible on the commons'. Calculating rationally, each herder reckons to gain all the proceeds from selling each extra animal which they rear on the commons. (In economic terms, the profit will be 'internal' to each grazer.) Yet the costs of grazing each extra animal, in wear and tear on the soil, are shared *by all* (they are 'externalised' to the whole society). Apparently because the land is in common, herders either do not realise or care about the full costs of their individual actions. They treat the environment as a 'free' set of goods and services.

So carrying capacity becomes exhausted, soil poaches, and the grass dies off. Individual freedom in the commons has brought ruin to all. Hardin maintains that if there were just a few grazers the problem would not have been serious. But because so many have sought the benefits offered by the commons, they have destroyed the very thing they sought. Similarly, earth's commons (oceans, air, National Parks) become degraded because too many people (or nations) try to exploit them for their private gain, while everyone, unwittingly or unwillingly, shares the costs. The scenario is a 'tragedy' (i.e. remorseless), says Hardin, because people naturally tend to be selfish. Private self-seeking does not bring the communal good which Adam Smith's 'invisible hand' theory forecasts.

Implications

Hardin rejects both unfettered privatisation of the commons and state control as feasible solutions to this problem of people treating resources as 'free' when really they are scarce. Instead, he advocates mutual coercion, 'mutually agreed on', to stop people irresponsibly overbreeding and overusing the commons. This is not as authoritarian as it sounds, says Eckersley (1992), being really a version of liberal contract theory. Nonetheless Hardin's thesis does seem illiberal, since it reasons that people should not be allowed the freedom to be 'irresponsible'. They must be forced to behave properly or all will inevitably suffer:

> Injustice is preferable to total ruin . . . To couple the concept of freedom to breed [a UN human right] with the belief that everyone born has an equal right to the commons is to lock the world into a tragic course of action.
> (Hardin 1968, cited in O'Riordan and Turner 1983, 297, 294)

Mainstream economists, scientists and lawyers, as well as ecocentrics, have adopted Hardin's parable as a model for what is happening to the environment, though they may draw different implications from it. The mainstream generally favours as much privatisation as is practicable, together with state and meta-state solutions such as international management agencies (O'Riordan and Turner 1983).

Private property rights, it is argued, ensure that individuals will not overuse their resources for fear of devaluing them. However, common property rights over resources lead to overuse because no individual feels a specific interest in protecting them (Goodin 1992, 105–8). Economists of the free market Chicago School add that private property rights over environmental 'goods' make it easier to identify and sue the culprits in cases of abuse such as pollution (Mishan 1993).

Objections

Influential as it has been, the commons parable has also attracted many objectors. Some maintain that Hardin's assumption of finite carrying capacity is inaccurate because developments in technology and environmental design will extend this capacity. Others lack faith in coercive law, tied to private or state ownership. For it cannot be assumed that the law will be neutral, acting only for the general environmental good. Its concepts, statutes and practitioners will not be divorced from the vested interests of particular groups like the state or landowners.

Still others question the accuracy of Hardin's assumptions about people's actual attitudes to the commons. O'Riordan and Turner see about them today encouraging signs of public-spiritedness and willingness to try to formulate international environmental agreements. They contend that commons users never were oblivious of the common good. In the early commons in England 'a powerful sense of community obligation' caused users to talk with others before adding to their herds. Cox (1985) confirms that the traditional commons were mostly managed sustainably for mutual benefit. McEvoy (1987) describes how farmers met biannually to plan future production. Similarly, immigrant Californian fishermen, twentieth-century users of the ocean commons, tightly controlled their own allocation and harvesting of resources, to produce optimum yields for *the group*. Hence Hardin's unidimensional generalisation of humans as alienated, utility-maximising automata who do not communicate about common interests is inappropriate, although it may describe human behaviour in a particular context, for example in capitalist America. For McEvoy therefore (p. 300)

> Resource depletion may be more a social problem – evidence of a community's inability to integrate its social order in a self-sustaining way than it is a product of the alienating self-regarding profit motive that Hardin posits as simple human nature.

McEvoy points out how the existing economic system atomises society, encouraging individuals to think narrowly, in space and time, of their individual selves and short-term gain. It is rational to think in such terms in capitalism, says Stillman (1983). His radical solution is to change the herdsmen's rationality, to see the long-term community interest as their own. If we were to regard others as part of us, not external to us, we would automatically identify our individual interests with those of the wider community – 'externalising' social and environmental costs of what we do would be impossible. Many ecocentrics propose this solution, long-familiar to socialists and anarchists: a planned, cooperative society involving full democratic community participation, based on a deep sense of communality.

Gemeinschaft society

This call for greater communality of thought and action resounds throughout green and socialist literature:

> A green view of economic democracy coincides closely with that of William Morris, evident in a speech given in London on communism: 'The resources of Nature and the wealth used for the production of further wealth, the plant and stock, in short, should be communised'.
>
> (Kemp and Wall 1990, 81)

It goes further than the economic sphere. Community, says Maurice Ash (1980), is an everyday necessity, and future generations must recreate real communities.

For Goldsmith (1977, 139; 1988) we must recognise 'the need to subject what may appear to be our individual interests to those of the community and ecosystem'. As in 'traditional societies' the family and community, not the state, must be given responsibility and power to deal with what are their problems. They must not be allowed to export responsibility elsewhere for problems like waste generation or the size of their population. He proposes a society of self-regulating families and communities for pragmatic reasons. Such a society will behave 'responsibly' towards nature (observing laws like carrying capacity). It will be more willing to live with the lower material standards which underpin green society, since closer family and community ties will provide increased quality of life, compensating for smaller quantities of things consumed. It will also encourage lower consumption through communal sharing – of domestic appliances and housing for instance.

The sort of society which Goldsmith and other deep greens envisage is close to that called *Gemeinschaft* by Ferdinand Tönnies (1887) in his sociological work on forms of association (see Table 2.2). A *Gemeinschaft* society, thinks Dickens (1992), could be the basis of a 'biological society'; one conscious of its relationships with the natural world. Indeed Tönnies, like deep ecologists, saw people essentially as a particular kind of animal. His description of *Gesellschaft*, the liberal form of association which has largely replaced *Gemeinschaft* in modern capitalism, is 'not all that distant from the malaise often described by environmental writing in our own era' (Dickens 1992, 31). In *Gesellschaft* society we may be freed from the stifling impositions of near community and family, but we have lost close relationships, along with a close attachment to the land and nature.

Table 2.2 Forms of association

Gemeinschaft
Society is more than the sum of the individuals in it
The traditional order
People bound together in an intimately shared order
Unalienated, organic face-to-face relations
'Natural' associations, of kin, family, neighbours
Living and working together
House and home of special significance
Shared, known and loved territory
Close association with the land, a common heritage passed down
Shared values
Social cohesion via land ownership, evolved and inherited customs, religion,
    hierarchies and status inequality, but higher members have obligations to lower
Society perpetuated by a received wisdom handed down
Medieval society was this kind of organic totality

Gesellschaft
Society is the sum of the individuals making it up
Modern society
Atomistic relations, based on individual interests and rights
Relations based on division of labour
Contracts between discrete individuals serving their own interests (whole society
    benefits via 'invisible hand')
Association therefore a means to an end, law based on contract
Rational will supplants received wisdom
Alienation from close interpersonal and people–land relations

---

Source: After Tönnies 1887

The *Gemeinschaft* concept is also central to Gaia theory, according to Jones (1990). Whereas conventional sociology did not recognise a 'crisis in the industrial mode of production itself' (as Ivan Illich (1975, 11) put it), Tönnies was among a minority of sociologists (like Max Weber) who did. Hence *Gemeinschaft* society is Gaian, following

> collective sentiment rather than calculating and egotistical reason . . . governed by custom, folkways and religion. The social relations . . . are best expressed in the family, the village and the town, or the corporate organisation of guilds, colleges, churches and religious communities. Intimacy of scale is critical.
>
> (Kumar 1978, 80, cited in Jones 1990, 109)

Jones (p. 109) continues: 'The ethical and spiritual ties which lie at the basis of group unity are the social correlates of the [physical] regulatory mechanisms which ensure Gaian stability.' Here, Jones, like Goldsmith, echoes structural functionalist theory, which was popular in post-war sociology. This regarded society as an entity, all the parts of which function to maintain one another and the totality in equilibrium, like a physical system. Hence what each group does is best understood by reference to the functions which it has in relation to the whole. And if one part of the social system is disrupted this provokes readjustment among the others, to regain total stability. This is not only a way of describing how society does 'work'; it carries strong overtones of prescribing how society *should* work, if it is to maintain stability. Each individual and group should occupy (and presumably be satisfied with) a niche, in the manner of animals and plants in an ecosystem. Because it lends itself to stability and only slow change, such a perspective is inherently conservative (see Peet 1991, 22–8).

However, the kind of *Gemeinschaft* which might be appropriate for an ecological society, rather than conservatively idealising past societies and allowing for little change, should perhaps follow lines which socialists, like William Morris, and anarchists described as future communism. Being decentralised and democratic from the bottom up, communism is a cooperative, unhierarchical and secular *Gemeinschaft*, where property is social rather than private, labour has dignity, humans are equal and austerity, modesty and devotion to the public good are virtuous. The 'general will' is qualitatively different from the sum of individual wills, and the latter may have to be subordinate to the former, to express humanity's inherent social, communal *nature*. To be fully human is to live with others, and to be separated from this communal aspect of self, as in modern, liberal society, is to be alienated. This theme is strong in socialist *Gemeinschaft* (Kamenka 1982, 8–24), which is why many socialists (e.g. Grundmann 1991) consider that, by definition, a truly communist society must also be an ecologically sound one. Natural consideration for others would make the behaviour described in Hardin's commons unthinkable.

2.3 SOME FUNDAMENTAL ISSUES BEHIND GREEN ECONOMICS

Against neoclassicism and 'positive economics'

The world economy is in crisis . . . Orthodox economists, however, are powerless to assist . . . Economics . . . has developed enormously over the past decade, particularly in its mathematical sophistication, yet its understanding of the world is similar to that of the physical sciences in the Middle Ages . . . an intellectual orthodoxy has emerged, based on an idealised, mechanistic view of the world. Standard textbooks for economics degrees increasingly resemble engineering texts.

(Ormerod 1994)

Although this dissident view of orthodox – often called 'neoclassical' – economics does not come from a green, ecocentrics would most certainly endorse it. But while mainstream society criticises neoclassical economics for failing to solve problems of unemployment and poverty, ecocentrics also criticise orthodoxy's belief that the *value* of goods and services primarily comes from consumer preference. This conception is anthropocentric, because it derives value from *human* perception. It is incompatible with the green theory of intrinsic value in nature (Chapter 2.1), by which goods and services would be devalued in proportion to their environmental impact.

Neoclassical economics can be defined as

> the basis of the [conventionally adopted] view of how economic activity functions . . . in capitalist society. It represents the refinement and extension of ideas from the formative or classical phase of economics as an academic discipline.
>
> (Smith 1981)

Adam Smith's *Wealth of Nations* (1776) and John Stuart Mill's *Principles of Political Economy* (1848) define the classical period, which advocated free-market ('*laissez-faire*') economics. The aggregate effect of individuals seeking to maximise their personal welfare through market mechanisms was thought to bring greatest material benefit to society as a whole, as if by an 'invisible hand'.

In fact, classical economists did not necessarily lack a 'green' perspective. The law of diminishing marginal returns discussed by Malthus and Ricardo held that each extra unit input of labour and capital would yield progressively less in the way of agricultural products. Like green economics today, therefore, their economics started from assumptions about finite resources, limits to growth and potential scarcity (Dietz and Straaten 1993). They were also, like green economics, 'normative', that is openly advocating value positions, for example about the need to maximise human welfare.

By contrast, neoclassical economics claims to be 'positive', that is, value free – merely a statement about the way the world *is*. This claim is not valid, because neoclassical economics start from unchallenged but value-laden premises, for instance that markets are the most efficient way of allocating resources. These are the guiding assumptions of liberal capitalism's world view (Table 2.3).

Ecocentric economists generally challenge these assumptions particularly that which regards the 'natural' environment merely as a stock of assets providing *functions* for human activity. The first function is as resources for production; the second as a 'sink', neutralising and absorbing wastes; the third, giving 'environmental services', like life, health, amenity, spiritual and aesthetic welfare (Pearce and Turner 1990). Ecocentrics think this perspective is calculatingly utilitarian and instrumental: part of a linear view of development as increasing *quantities* of goods rather than improving ethics and quality of life (Norgaard 1992). It is self-centred, hedonistic, hard-nosed and crass towards higher things of life (Etzioni 1992). It sees

Table 2.3 Conventional and ecological economics compared

	'Conventional' economics	Ecological economics
Basic world view	Mechanistic, static, atomistic Individual tastes and preferences taken as given and the dominant force. The resource base viewed as essentially limitless due to technical progress and infinite substitutability	Dynamic, systems, evolutionary Human preferences, understanding, technology and organisation co-evolve to reflect broad ecological opportunities and constraints. Humans are responsible for understanding their role in the larger system and managing it sustainably
Time frame	Short 50 years max., 1–4 years usual	Multi-scale Days to eons, multi-scale synthesis
Space frame	Local to international Framework invariant at increasing spatial scale, basic units change from individuals to firms to countries	Local to global Hierarchy of scales
Species frame	Humans only Plants and animals only rarely included for contributory value	Whole ecosystem including humans Acknowledges interconnections between humans and rest of nature
Primary macro goal	Growth of national economy Max. profits (firms) Max. utility (individuals) All agents following micro goals leads to macro goal being fulfilled External costs and benefits given lip service but usually ignored	Ecological economic system sustainability Must be adjusted to reflect system goals Social organisation and cultural institutions at higher levels of the space/time hierarchy ameliorate conflicts produced by myopic pursuit of micro goals at lower levels
Assumptions about technical progress	Very optimistic	Prudently sceptical
Academic stance	Disciplinary Monistic, focus on mathematical tools	Transdisciplinary

Source: Costanza, Daly and Bartholomew 1991, 5, cited by Lutz 1992

nature as a state that can be copied by scientific product development, whereas ecocentrism sees nature as a fragile system constantly vulnerable to human production (Green and Yoxen 1993).

By contrast, says Etzioni, green economics consider that many human wants cannot be regulated neatly by price, and also that people often judge acts by how much they conform to principles and duties, not merely by utilitarian outcome. Hence the neoclassical assumption that people's market choices are rational and objective (taken by 'economic man': always maximising material satisfaction) is fundamentally flawed. In reality people do not function as isolated individuals: they seek and value communal good. Consequently they like to have nature *there* even if they do not want to use it: 'the pleasure of being of the earth and knowing its richness and variety' (Allison 1991, 161). Nature thus has 'option' or 'existence' value.

Such value is problematic for conventional economics, but green economics would take it into account. Greens would start from three assumptions: ecological sustainability is imperative; economic development is for human fulfilment: meaning all-round development and not just material things; and all must benefit from this development (Ekins 1992a). Such assumptions sharply differ from those of conventional economics (Table 2.3).

Nonetheless green economists may be ambiguous about whether they belong to or are separate from the neoclassicists. On the one hand, Paul Ekins (forthcoming), a prime mover in the 'New Economics Foundation' considers that

> Green economics does not reject the insights and methods of environment and resource economics but it seeks to incorporate them within a wide framework of analysis and ideas.

Defined thus, green economics easily becomes a subset of the main neoclassical school, in the hands of, say, Pearce *et al.* (1989).

By contrast, Lutz (1992) regards green economics as a development from humanist economics. This is peripheral to most Western political economy, and is not afraid to discuss spirit and conscience, moral purpose and the meaning of life, in sharp contrast to the amoral, individualistic and asocial neoclassical view. Its antecedents include Jean Sismondi (1773–1842), who focused on human welfare for all rather than Smith's wealth of nations, John Ruskin (Chapter 4.5), who opposed the idea of economic man and bemoaned the degradation and alienation of the artistic and aesthetic side of human nature, Richard Tawney (1880–1962), who socialistically defined equality as equality of consideration for all (rather than mere equality of opportunity) and opposed large-scale private property ownership, Mahatma Gandhi (1869–1948), with his rural *sarvodaya* (welfare of all) movement, and Fritz Schumacher (1973), who synthesised ideas associated with Gandhi, Lewis Mumford, Leopold Kohr and Ivan Illich into a classic green text of the 1970s, *Small is Beautiful: economics as if people really mattered*. The tension between this lineage and tendencies towards neoclassicism bedevils green economics, producing contradictions between its methods and policies and its underlying philosophy (see pp. 81–91).

Limits to growth

'The heart of the conflict [between production and the environment] lies in the finite carrying capacity of the environment,' says Roefie Hueting (1992, 62). More economic activity by more and more people, he continues, means that the 'natural' environment cannot meet existing demands, therefore we must increasingly choose what functions the environment is used for. This is the 1970s limits to growth theme, now the basic assumption (ecocentrics would say that it is scientifically proven) of green economics. It restates Malthus (Chapter 4.2), whose notion of inherent scarcity is also, paradoxically, fundamental to neoclassical economics. (More paradoxically, radical socialist economics, which greens often lump together with conventional economics as ecologically undesirable, do challenge Malthus (see Pepper 1993).)

Indeed, ecocentrism's message in the early 1970s was starkly neo-Malthusian. The original *Limits to Growth* report (Meadows *et al.* 1972) argued that if present growth trends in world population, industrialisation, pollution, food production and resource use continued, then the planet's carrying capacity would be exceeded within 100 years, bringing about a disastrous 'overshoot and collapse', leading to massive 'eco-catastrophe' (Ehrlich 1969), famines and wars.

The fundamental problem, according to *Limits*, is that global growth in resource use, industrial output, population and pollution is exponential, that is, increasing by a constant factor. Exponential growth displays a gentle and gradual curve for a long time, but rapidly shoots up in a very short period (Figure 2.1: multiply one by two on a calculator, then keep multiplying the result by two to see this effect). *Limits'* computer model of the world economy holds that exponential economic and population growth cause exponential resource decline. And, because resources are (deemed to be) finite, when population overshoots earth's carrying capacity, it collapses into mass starvation. If we observe signals of impending danger and alter behaviour accordingly, however, a sigmoid growth curve can bring stabilisation below carrying capacity (Figure 2.2).

Limits tests several scenarios, feeding different assumptions into its model. Trying to increase food production through technological intensification, for instance, delays but does not avoid overshoot, which is then induced by increased pollution and lack of money for investment and future growth. The message is, always, if you remove or raise one limit

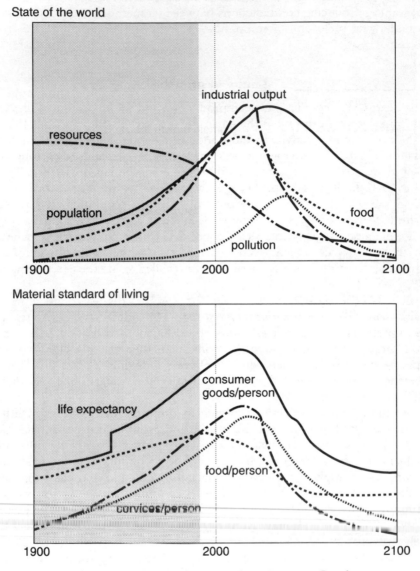

State of the world

Material standard of living

Figure 2.1 The 'standard run' from *Limits to Growth*
Source: Meadows *et al.* 1992

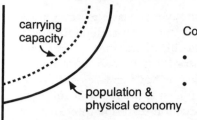

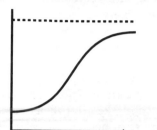

Continuous growth results if

- Physical limits are very far off, or

- Physical limits are themselves growing exponentially.

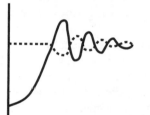

Sigmoid growth results if

- Signals from physical limits to growing economy are instant, accurate, and responded to immediately, or

- The population or economy limits itself without needing signals from external limits.

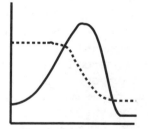

Overshoot and oscillation results if

- Signals or responses are delayed, and

- Limits are unerodable or are able to recover quickly from erosion.

Overshoot and collapse results if

- Signals or responses are delayed, and

- Limits are erodable (irreversibly degraded when exceeded).

Figure 2.2 Structural causes of the four possible behaviour modes of the world model
Source: Meadows *et al.* 1992

you eventually encounter another. Only by reducing and stabilising *all growth simultaneously* could ultimate collapse be avoided. Although the 1972 report said that growth trends could be altered to attain the desirable stable state economy, it regarded the chances of success as slimmer the longer that radical reform of society and economy was delayed.

67

Limits to growth theory applies thermodynamic laws to economics, after Georgescu-Roegen (1971). These imply that all production which uses materials and energy eventually transforms them into a more random, that is, chaotic or disordered, state. Disorder is termed 'entropy', and the second law of thermodynamics says that entropy increases with time. New energy from the sun slows disorder, but eventually this will run down and the solar system will die. Industrial production featuring intensive through-put of energy and materials hastens the decay: recycling delays it, but still uses much energy. Solar energy-based production with recycling delays it yet further, but the best approach of all is to reduce demand for resources. The Limits school questioned whether economic growth was anyway desirable: its costs in pollution and social dislocation might outweigh its benefits (Mishan 1967).

Limits theory was attacked from many directions. Hirsch (1976) pointed out that social rather than ecological limits were more immediate. As people got richer they demanded 'positional' goods which marked out their status and individuality, such as unfettered road transport or access to secluded nature. But the 'commons' principle implied that these advantages ceased to exist as more people got them. Hence demand for positional goods would be self-limiting.

Simon and Kahn (1984), free market 'cornucopians', questioned the accuracy of *Limits'* data and the usefulness of its methods. They argued, correctly, that economic activity everywhere was becoming more energy efficient, and that resource prices were falling. They regarded the latter as signifying that resources were becoming increasingly available over time, forgetting, however, that conventional markets take only a short term perspective, and prospects over, say, a century do not affect prices now.

Two decades later, says Ekins (1993), even resource optimists accept that there are limits of some immediate relevance, and that action must be taken to at least modify the workings of the economy. At the same time, resource pessimists are now more cautious in their pronouncements. For instance they see population densities and the displacement of poor people from their land as the ultimate problem (Ekins 1992a), rather than absolute population increases. And it is waste-generating growth that uses up resources which is unacceptable, not growth in, say, biomass or in human welfare (Ekins 1994).

Even the revived *Beyond the Limits* study (Meadows *et al.* 1992), is at pains to stress that its conclusions constitute merely a conditional warning. Those conclusions are that human use of many essential resources has already surpassed sustainable rates, and without significant reductions in material and energy flows in coming decades an uncontrollable decline in per capita food output and consumption and a rise in pollution will ensue.

But the 1992 *Limits* no longer suggests that resource exhaustion is imminent. It concedes that known reserves of non-renewable fuels have

68

increased over the twenty years since the original *Limits* suggested rapid depletion. However, it insists that this does not invalidate the general conclusions. More people consume fuel, as car ownership spreads. Even with catalytic converters and fuel economy this spells more pollution, therefore more capital for cleaning up and finding more resources, hence less for relieving poverty.

The new *Limits* is more politically sophisticated, recognising at least some of the political-economic ramifications of its analysis. Environmental impact, it argues, is a function of population numbers plus affluence levels, and the availability or otherwise of environmentally benign technology. In this formulation, reducing impact is therefore politically problematic for the South, West, and East respectively. *Limits* urges 'distributional and institutional change'. A sustainable economy is still technically possible, but only if more 'maturity, compassion and wisdom' is shown. Part of this wisdom is the ability to recognise accurately and in time the causes and symptoms of overshoot and to conserve the negative feedback mechanisms which would help stabilise the global ecosystem (rather than, for example, removing predators from a grassland, allowing the prey to overbreed, therefore overgraze, thereby causing soil erosion and destruction of the grassland). Recognition and evasive action depends on availability of accurate advance signals of impending trouble. *Limits* is sceptical about whether the current market-based economics can achieve effective mechanisms for recognising signals or acting on them.

Challenging convention: redefining wealth and needs

If green economics does not challenge the assumption of inherent scarcity, it does, however, challenge many other fundamentals of orthodox economics. Chief among these is the very definition of wealth. Money itself is not wealth, says Ekins (1992a), and economy is only one of four essential dimensions of the human condition: society, ethics and ecology are the others. Hence economic wealth really consists of what can be bought and sold *plus* satisfaction of wants not satisfied through markets, including belonging to community and family. 'Green economics is the economics of enough' (p. 31), which means distinguishing between needs and the wants that are artificially stimulated in a consumer society. Earth's resources can meet everyone's needs, but not all wants. Ekins is aware that this green redefinition of wealth restates William Morris (1885): 'Wealth is what nature gives us and what a reasonable man can make out of the gifts of nature for his reasonable use'.

Wall (1994, 14, 54) regards Morris as 'an eco-socialist in the 1880s', belonging to a minority group of Victorian critics of capitalist expansion, along with Carlyle, Marx and Ruskin (see Chapter 4.5). 'What do I need?' asked Morris. His answer was lack of hunger, good health, unashamed enjoyment of physical pleasure, an active mind and the kind of education to stimulate it, travel, useful work (not producing luxuries or war machines), enjoyment, sociality, fun, creativity and artistry in work, leisure (in which probably to do more good and enjoyable work), machines to help but not cheapen labour, pleasant and healthy workplaces and pleasant, generous and beautiful material surroundings. Most contemporary Westerners would endorse this list, and would probably admit to not having many of the items on it: despite their material possessions they might also concede having a vacuum in life caused by lack of other things. They might even agree with Thoreau (cited in Wall, 211):

> Most of the luxuries, and many of the so-called comforts of life, are not only not indispensable, but positive hindrances to the elevation of mankind.

But this view is of course disastrous for capitalism: hence capitalism's media and socialisation mechanisms generally work hard to marginalise it, branding people like Marshal Sahlins as 'cranks' or 'unrealistic' when they say:

> there are two possible courses to affluence. Wants may be 'easily satisfied' either by producing much or desiring little. The familiar conception, the Galbraithian way, makes assumptions peculiarly appropriate to market economies: that man's wants are great, not to say infinite, whereas his means are limited . . . But there is also a Zen road to affluence, departing from premises somewhat different from our own: that human material wants are finite and few, and technical means unchanging but on the whole adequate. Adopting the Zen strategy, a people can enjoy an unparalleled material plenty – with a low standard of living.
>
> (Sahlins 1972, cited in Wall 1994, 24–5)

Hence in green economics redefining wealth and the planet's capacity to produce enough crucially depends on the old problem of redefining needs. Max-Neef (1992) has comprehensively attempted this. Fundamental needs, he says, are similar in all cultures and include subsistence, protection, affection, understanding, participation, creation, leisure, identity and freedom. Other needs are false. The latter are met by 'pseudo satisfiers', while other satisfiers meet only one need. Clearly the satisfiers to strive for are those which meet multiple needs (Table 2.4). Poverty can be redefined as failure to meet any sort of need: there are lots of different poverties. Satisfiers are not just economic goods but *everything* which meets a need.

Expanding horizons: measuring wealth and welfare

Green dissatisfaction with Western concepts of wealth extends to how we measure it. GNP (gross national product) has become equated with wealth and welfare, although it really measures economic activity in only a narrow sense. If you spent an hour lobbing hand grenades into a busy shopping centre you would be contributing substantially to GNP. Firemen, police, doctors, nurses, ambulance drivers, funeral directors and many more paid employees would all be active as a consequence of your action. The fact that it is undesirable is not relevant. GNP is ostensibly a 'positive' (i.e. value-free) indicator, and it rises when any economic activity rises. But if instead you spent the same hour making love, doing housework or looking after a disabled relative, this would not increase GNP, even though contributing substantially to welfare and quality of life. Clearly GNP is not as value free as it seems since it rules out a vast amount of human (unpaid) activity, even though it is good.

GNP deals with flows and not stocks, hence it does not register the negative effects of resource depletion. It deals with money transactions only, hence implicitly accepting market values, and it takes no account of economic inequality. Miles (1992) points out that although often regarded as a welfare concept, paradoxically GNP and quality of life are often inversely related. What is needed is a modified GNP measuring true welfare, or preferably a whole set of alternative economic indicators, since no one indicator is totally satisfactory.

Ekins (1992a) proposes an Adjusted National Product, indicating *sustainable* income. This would indirectly derive from Gross National Product by subtracting any effects detrimental to 'human capital', that reduce, say, health, knowledge and skills or motivation. ANP would also subtract from GNP anything which (i) depreciated natural capital (e.g. consumed non-renewable resources); (ii) constituted 'defensive' expenditure (e.g. making good any deterioration of environment, health, civil security); and (iii) constituted loss of sustainability (e.g. extinction of species).

To make such modifications would require painstaking ecological accounting, where all industry and commerce would be audited for environmental impact. Some sophisticated auditing systems have already been developed. Norway uses natural resource accounting, trying to measure resource quantities and link resource stocks and flows and balances of energy and materials to economic valuations of development. And Norway attempts environmental quality accounting: measuring the state of resources like air, water and land, and drawing implications for human welfare (Lone 1992).

Such accounting is problematic. It needs not only adequate indicators, about which there will be disagreement, but also accurate statistics, which can be reduced to common units (e.g. money or energy). Additionally, critics will protest that some aspects of welfare cannot be measured at all. None the less work has been done on alternative economic indicators, reasoning that any such attempts must produce something better than raw GNP. Anderson (1991), for instance, proposes social indicators such as

71

Table 2.4 Needs, satisfiers and false satisfiers

Singular satisfiers*

Satisfier	Need which it satisfies
1 Programmes to provide food	Subsistence
2 Welfare programmes to provide dwelling	Subsistence
3 Curative medicine	Subsistence
4 Insurance systems	Protection
5 Professional armies	Protection
6 Ballot	Participation
7 Sports spectacles	Leisure
8 Nationality	Identity
9 Guided tours	Leisure
10 Gifts	Affection

* Singular satisfiers are those which aim at the satisfaction of a single need and are, therefore, neutral as regards the satisfaction of other needs. They are very characteristic of development and cooperation schemes and programmes.

Synergic satisfiers*

Satisfier	Need	Needs, whose satisfaction it stimulates
1 Breast-feeding	Subsistence	Protection, Affection, Identity
2 Self-managed production	Subsistence	Understanding, Participation, Creation, Identity, Freedom
3 Popular education	Understanding	Protection, Participation, Creation, Identity, Freedom
4 Democratic community organisations	Participation	Protection, Affection, Leisure, Creation, Freedom,
5 Barefoot medicine	Protection	Subsistence, Understanding, Participation
6 Barefoot banking	Protection	Subsistence, Participation, Creation, Freedom
7 Democratic trade unions	Protection	Understanding, Participation, Indentity
8 Direct democracy	Participation	Protection, Understanding, Identity, Freedom
9 Educational games	Leisure	Understanding, Creation
10 Self-managed house-building programmes	Subsistence	Understanding, Participation
11 Preventive medicine	Protection	Understanding, Participation, Subsistence
12 Meditation	Understanding	Leisure, Creation, Identity
13 Cultural television	Leisure	Understanding

* Synergic satisfiers are those which, by the way in which they satisfy a given need, stimulate and contribute to the simultaneous satisfaction of other needs.

Table 2.4 continued

Pseudo-satisfiers*

Satisfier	Need which it seemingly satisfies
1 Mechanistic medicine: 'A pill for every ill'	Protection
2 Over-exploitation of natural resources	Subsistence
3 Chauvinistic nationalism	Identity
4 Formal democracy	Participation
5 Stereotypes	Understanding
6 Aggregate economic indicators	Understanding
7 Cultural control	Creation
8 Prostitution	Affection
9 Status symbols	Identity
10 Obsessive productivity with a bias to efficiency	Subsistence
11 Indoctrination	Understanding
12 Charity	Subsistence
13 Fashions and fads	Identity

*Pseudo-satisfiers are elements which stimulate a false sensation of satisfying a given need. Though they lack the aggressiveness of violators, they may, on occasion, annul, in the medium term, the possibility of satisfying the need they were originally aimed at.
Source: Max-Neef 1992

education, literacy, unemployment, unpaid work, wealth consumption (food, water, energy, telephones) *and distribution*. He also suggests environmental indicators such as tropical deforestation, species extinction, greenhouse gas build up, desertification, and rates of energy use and of recycling. His compilation of such indices shows up sharp South–North differences, and general improvements in social welfare being offset by environmental deterioration.

Expanding horizons: redefining capital and time scales

Capital is central to orthodox economics. Neoclassicists define it as a stock of all sorts of resources, grouped under the headings land, labour and manufactured capital (tools, machines, buildings, technologies, infrastructure). Ekins (1992a) proposes replacing 'land' by a more holistic category of 'ecological capital', which would embrace all the functions provided by the 'natural' environment. And he would add a fourth – 'social/organisational' – capital which recognises human welfare and cooperation as positive assets in wealth creation. Although he recognises a risk in regarding nature and humans as 'capital' and therefore just another factor of production, he promotes his four-capital model as richer and more holistic than the neoclassical one.

More holistic, too, is the conception of social justice in green economics, which requires justice for future generations. This ethic of *futurity* starkly contrasts with the standard economic practice of *discounting* the future. The

73

discounting philosophy says that any benefit to be gained or cost to be forfeited through economic activity counts for more now than it does if deferred into the future. Thus, leaving aside inflation, and adopting a 5 per cent discount rate, what is worth £1m. today is worth only £952,400 next year, or merely £761 in one hundred years. Essentially, a bird in the hand is worth two in the bush; it is better to enjoy now what we may not be here to enjoy tomorrow, and deferred gratification is worse than instant gratification.

Discounting also implies that environmental loss is not likely to be regarded by future generations as so serious or so great as by us – partly revealing an implicit faith that future technologies will cope with or mitigate loss. As Goodin (1992, 65–73) points out, discounting is supposed to be economically rational, but it is neither rational nor ethical. Technology does not inevitably have to improve, neither does it improve at a constant rate, as discounting implies. The compound interest principle, of which discounting is a variation, assumes not only that in order to buy a good in the future rather than now progressively more money is needed each year, *but also* that this good will continue to be available. Yet it is illogical to apply this principle to resources, where using them now means that future generations do not have the *choice* to use them. This is where the immorality lies, for in liberal justice theory (spelled out by Rawls 1971) equal opportunity must be given to all regardless of arbitrary factors like colour, class and *time* of birth.

Sustainable development

From this moral principle comes the green insistence on sustainable development, defined by the Brundtland Report (UN 1987) as meeting the needs of the present without compromising the ability of future generations to meet their own needs. Such a definition is often accepted as relatively unproblematical by mainstream economists, including techno-centric 'greens' such as Pearce *et al.* (1989), who see sustainable development as compatible with conventional economic growth. Porritt (1992), however, typifies ecocentric cynicism at the ambiguity of Brundtland's definition.

> it allows politicians and economists to prattle on about 'sustainable growth' even though current patterns of economic growth and genuine sustainability are wholly contradictory concepts.

Ecocentric sustainable development, says Porritt, emphasises not growth but 'improving the quality of human life while living within the carrying capacity of supporting ecosystems' and preserving biological and cultural diversity. Thus renewable resource use would not exceed the rate of regeneration, non-renewable use would not exceed the rate of development of sustainable substitutes, and pollution emissions would not exceed the assimilative capacity of the environment (Daley 1991).

There is, then, no consensus over whether sustainable development allows economic growth, as Ekins (1993) reminds us. But, he says, all sides agree that negative environmental externalities (e.g. pollution) should be either internalised or reduced via technology, and there should be a non-declining stock of ecological and social/organisational capital.

This involves many politically contentious conditions for sustainability , like forgoing high-risk (nuclear) technology or absolutely protecting important ecosystems (Ekins 1994), or applying environmental standards not (a) to individual factories and enterprises but (b) as general 'envelopes' that whole areas of activity must fall within (Jacobs 1991). In this last, the implication of (a) is that as long as, say, no individual power generator exceeded x tonnes of output of sulphur there could be any number of generators, thus no limit on total pollution. In (b) however, there is a total limit, no matter how many generators. Thus (a) would have us all fit catalytic converters to our cars but, as now, would not restrict our use or ownership. This secures no general emission reduction, because although each car pollutes less there is unrelenting growth in cars. Strategy (b) would curtail car use.

When definitions of sustainable development are extended to include 'human capital' they become even more politically contentious. Engel (1990), for instance, insists that ethical sustainable development involves people as well as plants and animals. It means commitment to human solidarity and distributional justice (including future generations), a good life for all, a shared world-wide morality, spiritually nourishing as well as physically prolonged lives, moral and religious rearmament, reaffirmation of the individual-in-community and of the earth as a mosaic of diverse, co-evolving, self-governing, self-planning communities (hence individualistic societies and nation states are inadmissible). This all sounds like socialism (or as Dobson 1990, 84 has it the 'modified anarchist' view), and even Pearce *et al.* acknowledge that sustainable development requires equity within and between generations.

Internalising externalities: creating property rights

All green economics has to deal with the tragedy of the commons issue. The root of this is the grazers' failure to internalise all the environmental costs of their actions, either because they do not accurately perceive what these costs are or because they do not care. There are various possible solutions to this. That of a centralised authoritarian 'command and control' economy is almost universally rejected. At the other end of the spectrum are radically democratic solutions, involving common ownership in small-scale societies with restricted or no use for money. These green anarchist visions are described in Chapter 6. Most of the arguments amongst and between neoclassical and mainstream green economists are about how much state intervention there should be in an essentially market-regulated private property system.

Neoclassicists often cite Coase's (1960) argument that creating private property rights over environmental 'goods and services' would internalise pollution, making it worthwhile for firms to eliminate them. This argument for property rights is that private owners will care for their environment because they know that they will reap the benefits of doing so (or suffer the consequences of failing to). Conversely, common rather than private property rights lead to abuse and overuse because no individuals feel any special interest in protecting the property.

However, it is difficult to establish exclusive rights over much of the commons, such as air or oceans. For property rights to work certain conditions must apply, and it seems doubtful that in most cases they would. For instance, areas and dimensions of environmental property must be capable of clear demarcation, and for air or water this is difficult or impossible. Neither is it always feasible to discover the source of pollutants, enabling individuals to take legal action against polluters. Furthermore the costs of enclosing property and monitoring and prosecuting against infringements must be less than the value of the benefits to be gained by ownership to make it worthwhile, and this is not usually so, for example in cases of noise pollution or wildlife preservation. Neither are property rights relevant in some environmental issues, like population growth. Mishan (1993) therefore concludes that the claims made for markets are overstated. Existing property rights have clearly done little to mitigate agricultural pollution: they have usually allowed the rich to pollute and develop, while conservation groups have not been able to afford to buy the land that they wanted to protect.

Free market environmentalists such as Anderson and Leal (1991) try to get round such objections by proposing tradeable property rights in environmental 'goods' and 'bads', that is, the right to pollute air or water, or to exploit or preserve a forest or mineral reserve. Such rights could be bought and sold in a market, and the more expensive exploitation or pollution rights were, the more incentive there would be not to exploit or pollute. This approach would allow 'envelopes' – limiting total pollution to a ceiling decided by government and attained by restricting, even progressively reducing, the number of permits issued. As Jacobs (1993) underlines, this is not a very 'free' market since the state decides on how much pollution is generated as a whole – it is as if government decided on how many cars would be produced then sold rights to produce quotas within this total. However, tradeable pollution would give firms some choice; about whether they polluted or not.

Tradeable pollution permits operate in the USA. The government issued the first sulphur dioxide permits in 1992, costing $US250–300/ton emitted. The first sale was by a Wisconsin Power Utility to the Tennessee Valley Authority (Ingham 1993). The former, for whom pollution abatement equipment was relatively cheap, sold its rights to pollute to the latter, who found it cheaper to buy rights than to stop polluting. Thus, business and industry responded flexibly to ceilings on total pollution levels. Credits could even be banked for future use. And total pollution could be brought down over time by government taking (or buying) back permits.

However, the system's effectiveness has been limited. Midwest industries have bought permits from Eastern industries, but the pollution created by the former has killed trees in the East because prevailing winds blow west–east. Some attempts at trading have failed, because of high transaction costs, or high prices, or concerns that holding a stock of permits could be risky because government might just cancel them in future. And buying permits has been bad public relations for some firms, signalling that they are about to increase pollution. Environmentalists have generally opposed the system on fundamental grounds: that it sanctions pollution.

Eckersley (1993) considers that free market environmentalism is driven more by ideological commitment against state intervention than by sense. She regards it as anthropocentric, technocentric and socially unjust.

Internalising externalities: market-based incentives and environmental evaluation

Market-based incentives (MBIs) modify the market's tendency to treat the environment as a free good by having the state place prices on environmental goods and bads. These are then built in, via taxes or other means, to market prices of goods and services, which are traded normally. In theory, this makes goods and services which are particularly environmentally demanding more expensive, so that demand for them falls, discouraging their continued production. Heavier taxes on leaded than unleaded petrol, for example, have greatly encouraged car drivers to use the latter in Britain.

MBIs encourage environmental protection, while apparently preserving the liberal principle of freedom of choice for potential polluters. Initially the state sets an environmental standard (e.g. so many parts per million of a substance in the air represent an 'acceptable' level in terms of health and environmental protection – see, for instance, Elsom 1992, 15–20). Producers can then chose to install equipment to remove pollutants above this level from their emissions, *or* they can continue emitting above the level but pay a tax, fine, or other levy.

Pearce *et al.* (1989, 161–2) consider this choice to be a great virtue of MBIs: it 'introduces flexibility into the compliance mechanism'. Polluters with high abatement costs will prefer to pay up, those with low costs will install cleansing equipment: this produces lower compliance costs. To most environmentalists the logic of this argument seems odd, since it allows producers to continue to pollute if they can afford it. Hence, as Martell (1994, 71–2) observes, maintaining the producer's freedom thus contradicts other people's freedom, to breathe clean air. That freedom could be reinforced by making environmental taxes very heavy, but this would effectively force environmental responsibility on to firms 'just as much as state coercion'.

The market is not the only decision-making arena with implications for environmental quality. There are more planned development processes which also need to have environmental costs and benefits built into them to help decide whether and in what forms development should proceed. Here an array of techniques has been developed, such as cost-benefit analysis (CBA), comparative risk analysis, environmental impact analysis and multi-criteria analysis (Jacobs 1991).

Internalising environmental effects through such techniques involves translating them into money terms. Whether or not this is desirable or possible is contentious: what price could one put on a favourite view? Green economists hesitate to try, but often come to believe that attempting monetisation is unavoidable.

There are different approaches to calculating money value of environmental conservation or degradation. It can be done empirically, so a National Park's value, for instance, is calculated as the aggregate expenditure on travel there by visitors. Or the costs of mitigating environmental damage (e.g. putting lime on soil to counter acid rain) or abating it (flue gas desulphurisation) can be directly assessed (Stirling 1993). More difficult approaches assess damage costs, for example some kind of compensation to people who lose environmental amenity or function (peace and quiet), or create surrogate markets – assessing how much people would theoretically be prepared to pay to conserve an amenity.

Valuing costs and benefits and building them into market-based incentives and disincentives constitutes the most widely accepted approach to the economics of environment, yet it is full of problems. First, the environmental effects of economic activity have to be accurately identified, their causes determined and traced to specific sources. Then standards for effects on biota, and ultimately for emissions and other environmental effects, must be set. Equipment meeting these standards must be devised and its use encouraged via the market or enforced if the market fails.

And the complex and multidimensional nature of environmental effects means that producing a single money index of value omits vital contextual information. It is like, says Stirling, trying to describe a three-dimensional object by reference only to its length. For instance

Microeconomic techniques . . . are not able realistically to assess the economic costs of displacing millions of people from low-lying coastal

78

areas (global warming); of hundreds of thousands of extra eye-cataracts and skin cancers (ozone depletion)

says Ekins (1994), concluding therefore that the *implicit* costs of unsustainability approach infinity.

While certain forms of value are beyond price, neoclassicists like Pearce say that even human life has a limited price since we are not prepared to pay infinite amounts on health care or life insurance. Basically:

> The very idea of monetary valuation implies tradeability and purchase. If the value of a life is £9000, this carries the clear implication that anyone with £9000 can purchase a life.
>
> (Mulberg 1993, 110)

This is why socialist, anarchist and deep greens object to monetising the environment. They consider that it should not be for sale.

Actual attempts at monetisation bear out misgivings. Despite the misleading precision with which the authors of valuation studies express their results, says Stirling, those results are usually inaccurate, or useless – for instance various studies of the external costs of coal-fired electricity produced a range of results varying by a factor of fifty thousand.

It is relatively easy to construct a 'supply curve' for sustainability, that is, to calculate the costs of measures which will mitigate, abate or eliminate environmental damage. Demand curves are much more problematic, since they require some assessment of what people would pay for sustainability. *Contingent valuation* (CV) tries to achieve this. People are simply asked what they would be prepared to pay (e.g. in taxes) to secure, say, a stable and diverse moorland environment, or clean water. Alternatively they may be asked how much compensation would make up for the loss of such environmental 'services'.

CV is highly questionable. For instance, people are usually willing to pay (to avoid an ill-effect) much lower amounts than they would demand in compensation (once the ill-effect has occurred) (Stirling 1993). Additionally, what they say in a questionnaire they would hypothetically do is often different from what they actually do, as voter intention polls often show. Respondents' answers are also circumscribed by the form in which questions are put, and by incomplete information about environmental risks and effects. And many just refuse to participate in such exercises, feeling that someone else (e.g. the polluter) should pay, or being averse to monetisation (see Hueting 1992).

CV produces some bizarre results, for instance that Americans would pay $40/annum to save humpback whales, but only $9.3 and $1.2 for blue whales and whooping cranes respectively. And their initial willingness to stump up $85/annum to protect Prince William Sound from disasters like Exxon Valdez dropped to 29 cents/annum when researchers reminded them of other spending targets like schools and hospitals (BBC 1993). Pearce (1993) meanwhile, reports 400-fold variations in people's willingness to pay for tropical rainforest conservation. Yet despite such farcical outcomes the British Department of the Environment is required to use CV before advising ministers.

Perhaps the most fundamental objection to all monetised CBA is that it tends to discount for future costs, effectively devaluing the votes of coming generations. Meanwhile, MBIs certainly do increase consumer prices as firms pass on costs of compliance with environmental standards, raising the further principle of present equity. Should the poor be pushed onto inferior public transport by, for instance, not being able to afford to pay environmental taxes on their cars? Reformist advocates of 'sustainable development' would accept this; radicals would not. Other MBIs have gained wide public disapproval for being socially unjust. The domestic fuel consumption tax introduced in Britain in 1994 is what greens generally advocate (e.g. Loske 1991). Politically, however, it proved disastrous. In December 1994 the Conservative government was defeated in a House of Commons vote which would have increased the tax further, some of its own members voting against it.

The local state and local democracy

Despite a persistent anarchist streak in ecocentrism, green economics generally has few qualms about state intervention to set and enforce environmental standards and ceilings on pollution: 'Markets can quite happily coexist with planning, so long as the conditions prevailing in the markets are adjusted to generate the targeted outcome,' says Jacobs (1991, 125). He rejects complete centralised planning, considering dismissively that the Soviet experience proved that direct administration of environmental impacts is extremely inefficient. Instead, he suggests, planning should 'administer' the 'macroeconomic outcomes' of market forces, while markets themselves constitute the 'microeconomic' methods by which those outcomes are reached.

Left alone, Jacobs argues, markets do not produce an 'invisible hand' so much as an 'invisible elbow', either deliberately pushing nature aside or inadvertently causing it damage through clumsiness. What he wants is to influence market behaviour towards particular levels of environmental impact. For even if in a green decentralised society all values were ecocentric there would still be some need for overall planning since local communities could not act in full knowledge of what all other communities were deciding. Individual choice must therefore be 'informed . . . by a centralised state authority setting targets'.

Having thus affirmed the state, Jacobs seeks to allay green fears about it by emphasising its *local*, town-hall form. Indeed, he asserts, with Levett and

Stott (Jacobs *et al.* 1993), that the Economic Development Units of most existing British city and county councils, for example Oxfordshire and Sheffield, are already encouraging firms onto an environmental path, fostering local sustainable development.

This is becoming a familiar line in green economics: the benign local state constitutes a bridge between suspicious anarchist and liberal greens and those who are more state socialistically inclined. The now-extinct Greater London Council is often cited: it was neither the minimal state of the new right nor the depersonalising bureaucratic state of 'Fordism', but an enabler of local communities, groups and initiatives (Mulgan and Wilkinson 1992). 'The best thing the GLC did was to give people the resources and encouragement to think for themselves', working with women's groups, ethnic minorities and environmental groups, especially in alternative energy. It extended democracy beyond the ballot, after William Morris and G. D. H. Cole, to whom socialist economic policies were to be based on the ideas and organisation of working class people (Mackintosh and Wainwright 1992). This kind of state, then, goes beyond welfare, which now has connotations of creating dependency. Instead it is democratic, empowering people by giving them access to the mainstream of society, however disadvantaged they may be, with a say in support services securing their civil rights (Beresford and Croft 1992). It can also help to facilitate local economic self-empowerment in the Third World, which greens increasingly regard as the key to arresting problems of desertification and deforestation (see Wignaraja 1992, Agarwal and Narain 1990 and Ekins 1992c).

The contradictions of 'progressive markets'

the market can also be organised and harnessed to exert a powerful progressive influence . . . this is fully in accordance with standard neoclassical theory.

(Ekins 1992b, 322)

Ekins' 'progressive market' consists of 'progressive' business, investors and consumers. The first thinks not just about its shareholders but its *stakeholders*, who include customers, employees, local community and environment. NCR, Johnson Wax and IBM are among big companies who have at least voiced good intentions towards stakeholders – they are 'green capitalists' (Elkington and Burke 1987).

There are also thousands of producer coops and share owner-employee companies, as well as alternative trading companies (Traidcraft, Café Direct) who share this wider conception of their obligations. And progressive investors in the USA, Ekins tells us, put $625 billion into 'ethical' enterprises, including ethical banks. Progressive consumers, too, are 'stirring from the deep sleep of commodity fetishism' (p. 325) to face their responsibilities: critical and environmental awareness about what they buy, social concern, action and solidarity.

Yet Ekins, who is often enthusiastic about the market, also conveys that contradictory sense of unease about it which most mainstream green economists seem to share. The progressive market is in accordance with standard classical theory, Ekins says, but he also admits that progressiveness rests on a conception of markets motivated by ethical, social and ecological criteria: at variance with the neoclassical 'economic man' motivated by narrow self-interest, individual gain and profit maximisation.

Elsewhere Ekins (1993) lists some requirements of sustainability. These include that businesses should not prioritise making money for shareholders, but also aim to generate satisfying livelihoods for employees and society. For this to happen, says the unabashed Ekins, the owners of business should be the workers, not the lenders of capital, and managers should be accountable to all stakeholders, including the wider community. Company law should accordingly be reformed, while the present banking system should be replaced by radical alternatives such as people's credit schemes. These are mighty tall orders of capitalism, as socialists have discovered over two or more centuries.

One of the root problems about green demands for progressive business and corporate responsibility is that these reformist demands are underlain by a critique of technocracy which 'strikes at the heart of the corporate power base' (Smith 1993). Being green and ethical makes for extra expense and lower profits, undercutting competitive advantage, unless there is 'a level playing field'. The more that trade becomes internationalised, the less possible this seems: transnational corporations can and do avoid meeting strict environmental standards and wider social duties by locating in countries that do not fuss about such things. Even within the unified European Union trade area Britain's business-oriented government insists on its rights of 'subsidiarity'. Ironically for environmentalists, who support subsidiarity everywhere, this allows government to lower environmental standards, helping businesses to increase profits, without fear of European retribution.

The transnational corporations, which produce 29 per cent of world manufactures, have fantastic power. They can stimulate or undermine local industry, can generate revenue but avoid taxes, can monopolise technology, can catalyse modernisation throughout the world – simultaneously reinforcing inequalities and promoting Western cultural hegemony, can create low standards for employees (child labour, temporary low-paid work, alienating workplaces) and can critically alter the natural environment (*New Consumer* 1993).

To maintain this power, most corporations fight against state regulation and nationalisation. Hence in 1990 the EC announced that it was preparing to direct companies to make regular cradle-to-grave evaluations of their environmental impacts, for public scrutiny. But by 1992 this promised directive had 'been so watered down as to be almost useless' (Pearce 1992). For two decades UK (Blowers 1987) and US (Faber and O'Connor 1989) governments have sympathetically responded to constant pressures by business and industry to reduce environmental standards and regulations. Business argues that only inducement, not directives, can yield corporate environmental responsibility. Inducement might come from MBIs and from investors and consumer pressure.

But is such pressure effective? Denis Smith, a Business Professor, concedes (1993) that the idea of 'democratic shareholder control' of companies does not work, because company owners or owner-managers own most of the shares. And 'ethical' and 'green' investment is full of ambiguity. The Merlin Green Fund's approach, for instance

> is not so much to see one sector as green and to avoid more contentious areas like chemicals or mining, but rather to seek out the most environmentally responsible companies within each sector.

It therefore has in its portfolio a goldmine, a Christmas tree producer, Tesco supermarkets (who, despite their 'green' PR, take every opportunity to press for greenfield out-of-town hypermarkets to which people must drive), glass recycling, fish farmers, gas, oil and coal producers and explorers and a landfill company. Radical greens would question the acceptability of all of these.

So, too, would they reject the idea of the environment as 'a potential marketing tool for retailers and manufacturers' (McCloskey et al. 1993). Yet this is what green consumerism makes it, sometimes nakedly (see Figures 2.3, 2.4). Market research reveals that 62 per cent of US consumers want and read product information on the environment, while 41 per cent say that such information influences their buying behaviour, according to McCloskey et al., who consider that the boom in environmentally friendly products has largely been consumer led. Not just a passing fad, this is 'a fundamental shift in consumer behaviour'.

All of this is highly questionable, however. It looks as if, in Britain, 'green' products are no longer very popular – major supermarkets stock fewer of them. And the idea that green consumers can greatly influence decisions about what is made, and how, is a re-emergence of the myth of consumer autonomy, exposed by Galbraith (1958), who showed how demand is stimulated and led *by producers*. Indeed

> Few producers of consumer goods would care to leave the purchases of their products to the spontaneous and hence unmanaged responses of the public.
>
> (J. K. Galbraith, cited in Young 1990, 17)

Figures 2.3 'Acid rain: save your car'

Figure 2.4 'Do you care about the health of your Planet?'

And while free and accurate product information is the basis of green consumer theory, in practice UK companies have responded little to the emerging green agenda by making information available to the public (Owen 1993).

Irvine (1989) has exposed many shortcomings of green consumerism. Not only does it barely work in its own terms, he points out, it is a deeply contradictory part of 'green' economics, because such economics should be about affluent Westerners *consuming less.* Thus the *New Consumer* slogan, 'let's all buy for a better world' belies the green intentions behind that consumer body's existence.

Green consumerism is part of the problem, not the solution, because it incorporates so many of the assumptions of free market liberalism. Anita Roddick (1988) of the Body Shop faithfully repeats one of liberalism's main economic shibboleths, laid down by Hayek (of the Austrian economic school, which bred monetarism in the 1980s). She says

> Don't just grin and bear it. As consumers we have real power to effect change . . . we can use our ultimate power, voting with our feet and wallets – in either buying a product somewhere else or not buying it at all.

The idea that, through the market, money can be a vote for desirable change is flawed from an ecocentric point of view, because many of the constituents represented by ecocentrism – the poor, indigenous people, future generations, animals – have little or no money, therefore no effective vote.

Really, green consumerism is reactionary. In elevating the power of the individual, as consumer, to effect social change through personal lifestyle, it encourages us to forget our potentially greater collective power as producers, able to withdraw labour. So it is politically anaesthetising (Luke 1993). Furthermore its strategy depends on voluntarism rather than regulation, which is usually a recipe for minimal change. This is old-fashioned economics. But, say many greens, it is surely better than doing nothing.

Internal contradictions of green economics

Herein lies a central dilemma for mainstream green economics. Its critique of conventional economics is often radical, but, partly for pragmatic reasons, its analysis of why they are as they are, or what to do about it is not radical. Indeed prescriptions may be quite anodyne: essentially a different version of the mixed economy (markets with state ownership and planning) which most European countries have practised for decades, with limited success.

Most greens have something good to say about markets. Achterberg (1993, 91), for instance, echoes neoclassicists in regarding them as a 'triumphant expression of individual rights and civil liberties', stimulating efficiency and inventiveness. Ekins (1993, 272) thinks they are 'a marvellous institution for effecting individual preferences' and 'a wonderfully democratic mechanism if the conditions for the "invisible hand" to work are all met'. Yet he is also aware (1992a, 24, 26) of 'chilling confirmation . . . of Marx's century old-perception that capitalism's principal tendency is the concentration of wealth and power', that 'making money out of money independently of productive activity or real wealth creation has become a debilitating virus in the economic system', and that 'Most of the ravaging by the industrial economy of the planet, its people and communities is an unexceptional part of the system's operation'. All for what?, Ekins asks. For the vacuousness and unhappiness of the American middle-class way of life. TV, shopping, lack of individualism, and wanting to be richer than others

The ambiguity of this critique is matched by that of green solutions. Greens, says Ekins, should 'reform' the financial system, giving rich and poor access to money via banks and each of the four capitals. Money should reflect value created by real goods and services, not by speculation or paper transactions. He proposes a green mixed economy with capital–labour partnerships, not specifying how power is to become evenly shared.

Many of the features of that economy (Table 2.5) are unexceptional and are or could be accommodated by current mixed economies, for instance the 'flexible' employment pattern, preventive medicine, commitment to

Table 2.5 Features of a green mixed economy

- Universal small-scale ownership of capital enough for self-reliance
- Collective organisation of capital, not leaving it to the market
- Land reform, and access to all capitals
- Commitment to monetary stability
- Local credit, and, within limits, monetary autonomy
- Cooperative working: either profit-sharing or full worker ownership
- A vigorously decentralised progressive market, with informed consumers
- An enabling, and enforcing state
- More renewable, less non-renewable, no nuclear, energy
- Energy efficient production and extensive heat insulation
- Green production, with annual environmental audits
- Circular, not linear, resource flows
- A zero pollution goal, with pollution taxes
- Product lifecycle analysis
- De-emphasising road transport
- Taxing personal and freight travel
- Local/national food self-reliance
- Labour intensive, organic farming
- Maximised human capital (social justice, education, enhanced spirituality, ending women's oppression)
- Stable population levels
- Health as wealth: preventive healthcare
- Education as wealth (education for personal fulfilment and cooperation)
- Green employment: shorter hours, part-time and job-sharing, self-employment, ownwork
- Basic income scheme
- Strengthening of families, communities and networks
- Democratic, local politics
- Peace through the UN
- Alternative, self-reliant, development models
- Import tariffs, including environmental taxes

Source: Ekins 1992a and Kemball-Cook *et al.* 1991

monetary stability, energy efficiency, some decentralisation of power. Others, like tariff barriers, land reform or genuine collective organisation of capital, are deeply at odds with prevailing economic arrangements.

Limits, with similar ambiguity, firmly asserts that markets and technological innovation can only delay, not avoid, overshoot and collapse. Markets operate through feedback mechanisms which themselves involve environmental costs, and are further impaired and distorted by imperfect information. And

> The market is blind to the long term and pays no attention to ultimate sources and sinks, until they are nearly exhausted, when it is too late to act.
>
> (Meadows *et al.* 1992, 184)

That is, there is no corrective feedback to keep competitors from over-exploiting the commons. In the fishing industry, for instance, high prices do not signal scarcity, deterring fishing. On the contrary, they encourage more fishing. Thus the whaling industry is

> a huge quantity of [financial] capital attempting to earn the highest possible return. If it can exterminate whales in ten years and make a fifteen per cent profit, but it could only make ten per cent with a sustainable harvest, then it will exterminate them in ten years. After that, the money will be moved to exterminating some other resource.
>
> (a whaling journalist, reported by Ehrlich and Hoage 1985, cited in Meadows *et al.* 1992, 188)

This trenchant indictment leads *Limits* (p. 191) to conclude that we must

> step back and acknowledge that the human socioeconomic system as currently structured is unmanageable, has overshot its limits, and is headed for collapse . . . and . . . change the structure of the system.

But this is immediately contradicted (p. 192) by the assertion that the same combination of people, institutions and physical structures 'can behave completely differently if its actors can see a good reason for doing so . . . no-one need engage in sacrifice or strong-arming'. The less-than-revolutionary conclusion to *Limits* (pp. 222–3) is that information is the key to transformation, that 'only individuals, by perceiving the need for new information, rules and goals, communicating about them, and trying them out, can make the changes that transform systems', and that we need visioning (by visionaries), networking (the green panacea), truth-telling, learning and loving.

Other greens repeat such glibness. Daley and Cobb (1990) write of capitalism's inherent self-destructive contradictions, but then argue that markets and profits are needed for diversity and should merely be tempered by governments and communities, in a climate of religious replenishment. Norgaard (1992) wants no less than a complete paradigm change from the linear progressive model of neoclassical economics, but thinks this can come just through changes in management, technologies and institutions. Even Commoner (1990), one of the most left wing American environmentalists, having blamed environmental 'crisis' on leading corporations and their production techniques and arrangements, lamely seeks merely to curb corporate power by introducing social governance into economic decision-making.

Internal contradictions of capitalism

Most social ecologists and all eco-socialists are informed by a Marxist perspective on capitalism. They would argue that green economics usually produces reformist, mixed-economy prescriptions because it lacks that perspective. Green economists think that it is possible to somehow reconcile the destructive tendencies in neoclassical, capitalist economics with radical sustainable development. Marxism, however, reveals that capitalism is fundamentally *not sustainable*. It has within it internal contradictions: features which work to destroy itself and the natural environment which constitutes one of its means of production.

There is the contradiction of overproduction. Capital accumulation depends on selling goods and services in the marketplace for profit, that is, the surplus over what it costs to produce them. A principal item of cost is labour. It follows, therefore, that the labour force, *by definition*, cannot afford to buy all of the things that it produces, since it is not paid their full market value. Hence there are 'too many' goods and services for the immediate market (the labour force) to buy, so there must be a constant effort to bring more people into that market by geographical expansion. The purpose of capitalism, by definition, is to accumulate capital. Therefore, when profits have been made they must be reinvested in more production, to keep the process going. More production implies more consumption. This of itself tends to be wasteful in a consumer society, and sometimes there is very visible waste, such as destroying 'surplus' food mountains rather than selling them cheaply or giving them away.

Capitalist productive relations are competitive. To make products more cheaply than competitors, industry must be 'efficient' so labour must be ever more 'productive'. This means minimising costs in relation to selling price. Here is the source of inexorable pressure to externalise social and environmental costs. Firms create unemployment by automating production, and they create wastes which pollute unless recycled. By definition, it is good business practice to get wider society (e.g. the state) to pick up the bill for these externalities, rather than internalising them, especially during the frequent recessions which characterise the system.

Perhaps most insidious of all the inherent anti-ecological and anti-human tendencies of capitalism is the fact that the very *purpose* of production is to realise profit. This, the 'cash nexus', *must* be the main criterion determining what is and is not produced. Hence much social need is unmet in 'affluent' societies because to meet it is not profitable.

Profit comes by putting goods and services, as 'commodities', onto an anonymous market. This encourages consumers not to think about the 'relations of production' which commodities encapsulate. These are the relations between people, and between people and nature, which result from the way we organise to produce things. We are alienated – separated – from true understanding of our relations with others, and with nature, by this anonymous market. We regard our digital watches or music systems as *things*, not as outcomes of Asian sweatshop labour, for instance. We think about *objects* of furniture, not hardwood trees felled from tropical forests.

For all their shortcomings, green consumer movements have done much to counter this objectification of relations of production in capitalism, encouraging us to look beneath the superficial appearance of commodities as mere depersonalised things.

A more thorough analysis of the 'ecological contradictions' of capitalism can be found in Schnaiberg's (1980) and Johnston's (1989) Marxist accounts. Wall (1990) and Singh (1989) are among a minority whose green economics do take a Marxist perspective. The former (p. 80–1) says

> It is very difficult to see how Green economics could function in theory or practice . . . A Green economy would be built on assumptions so different from our present society that it is difficult to see how the transition could be made smoothly, if at all . . . Neither workers' control nor the growth of community businesses outlined by Guy Dauncey (1988) and others would solve the growth problem. Neither the idealized 'workers' nor the 'small businessmen' would have an incentive to go for negative or zero growth . . . *Ecology is incompatible with the market.*

The latter (Singh pp. 28–9) cites Marx's *Capital*, in dismissing the idea of a restricted- or non-growing capitalism. The capitalist

> shares with the miser the passion for wealth as wealth. But that which in the miser is a mere idiosyncrasy is, in the capitalist, the effect of the social mechanism, of which he is but one of the wheels.

In other words, the system itself constrains people to behave in certain ways. Whether greens fully realise this is arguable. One might well hear greens attributing the following conversation to 'ignorance', 'greed' or some other malice in the nature of the individual person being interviewed. A Marxist, by contrast, would regard such brutalisation and alienation from humanity as *necessary qualities* (though not usually displayed so nakedly) in leading functionaries of the system.

Question. Did it ever bother you personally that this British equipment was causing such mayhem and human suffering? [Hawk aircraft, sold by the British Government to Indonesia, were being used to commit genocide against East Timor]

Answer. No, not in the slightest. Never entered my head.

Question. The fact that we supply highly effective equipment to a regime like that, then, is not a consideration as far as you're concerned?

Answer. No.

Question. I ask because I read that you are vegetarian and seriously concerned about how animals are killed. Doesn't that concern extend to the way that humans, albeit foreigners, are killed?

Answer. Curiously, no.

This conversation was between journalist John Pilger and former British defence minister Alan Clark (Central TV 1994). In his post Clark was anxious to help the British 'defence' industry to make profits in order to underwrite weapons production for British forces. Clark was also quoted as having said: 'Does anyone know where East Timor is? I don't really fill my mind much with what one set of foreigners is doing to another'.

The kind of radical green economy which might replace capitalism and heal such desperate alienation from our own humanity, is described in Chapter 6.2 and 6.3.

2.4 TECHNOLOGY AND THE ECOLOGICAL SOCIETY

Technological and social determinism

Ecocentrics usually oppose 'high' technology, such as nuclear weapons or power, green revolution agriculture or genetic engineering, for reasons outlined in Chapter 1.1. But they are not against all technology, favouring 'alternative' forms, known variously as 'appropriate', 'intermediate', 'soft', or, less commonly, 'radical' or 'utopian'. The alternative technology (AT) movement became prominent in the early 1970s, and its icons – windmills, solar panels, organic vegetables, etc. – were particularly associated with ecocentrism's back-to-the-land communes-founding phase: part of the 'cranky' would-be ecotopian decentralised, small-scale society and land-scape. Now, however, these icons are part of mainstream 'common sense' in the West, and are also seen as important in Third World development. But with acceptance by mainstream society, part of the AT movement has fallen into what more radical ecocentrics would regard as 'traps'.

The trap of technological determinism holds that technology is what largely shapes society (as in many conventional explanations of the history of industrial society in Western Europe). It follows that a different, (ecological) society can be got by introducing different (ecologically-sound or 'soft') technologies – the technology itself is seen as a route to social change. This position soon becomes technocentric, since it essentially puts faith in AT to secure social change as it spreads.

Futurologists, particularly, do this, arguing for instance that information technology (IT) will bring the 'electronic cottage', thus people will not have to travel to work, resulting in a decentralised, democratic (two-way videos allow each citizen to vote on everything in a televised parliament) society with far less energy consumption than at present (J. Martin 1978, Tofler 1980). Webster and Robbins (1986) expose the absurdities of such 'post-industrial' scenarios. The reality is that in an abidingly capitalist society, IT largely serves the interests of capital, assisting more centralisation of political and economic power, helping to throw millions out of work and de-skilling jobs.

This suggests that Winner (1986, 64–6) is right to warn the AT movement against mere 'sociological tinkering' rather than confronting the basic power relations in society. His review of 'new age' writers who focused on technology, such as Marcuse, Mumford, Roszak, Goodman and Ellul, suggests that they fell into the parallel trap of idealism. That is, they attributed problems associated with high technology merely to the wrong 'ideas' of the developers and users of that technology – social values which were aggressive, excessively rational or just plain greedy or arrogant (Schumacher 1973 also argued this). They therefore underestimated the importance of material conditions – what people do economically – in shaping technology and socio-political relationships. These must change to create an ecologically sound society.

A similar trap involves assessing technology simply 'technically', that is, in terms of risk- or cost-benefit analyses, as do standard, technocentric environmental management approaches. Such approaches, as Goldman and O'Connor (1988, 92) reveal, again slight or ignore 'the problem of technology as the content and context of social domination, exploitation of labour and accumulation of capital'.

Some ecocentrics and politically left analysts (such as Tony Benn in the British Labour Party) may seek to recognise this issue of the social context of technology by arguing that it is not so much the technology itself as its *ownership* which is crucial. But this would make collectively owned nuclear power stations acceptable, a position which few greens would entertain.

Clearly ownership is not the only factor that makes wind power, for instance, an environmental *threat* in the 1990s in Britain. There is growing protest against the noise and visual intrusion of the 400 turbines (each about 100 feet high) already operating on upland Britain's hill farms. And groups are apprehensive about the 1200 proposals pending from 'get-rich-quick' farmers anxious to snap up subsidies as part of a government programme to produce 3 per cent of electricity from renewables by AD 2000 (Engel 1994). However, if the land were owned by many greens it would still be subjected to the same treatment, for longtime environmental campaigners from the Green Party and Friends of the Earth 'have transformed themselves into developers' actively supporting the windfarms (Friends of the Earth 1994).

Greens might typically argue that scale, not ownership, is the problem. Environmentally benign technologies are transformed into an environmental threat when they become excessively large and extensive. But here the analysis may stop, without making essential connections between scale and economic relations of production, so that campaigners may end up pleading for what is essentially small-scale capitalism: a contradiction in terms.

None of these analyses, then, are adequate in themselves. The whole panoply of social and cultural relations of production needs to be considered to explain the relationship between technology and social change. In trying to do this, however, we should also steer clear of the trap of *social* determinism; the opposite of technological determinism. For this would place *all* the onus onto society – its culture, its economic system and the social (including society–nature) relations they give rise to – to explain what technologies we have and their effects. Albury and Schwartz (1982) tend to do this, arguing that technologies as disparate as the Davy miners' lamp, green revolution agriculture and IT were developed by capitalist business and industry specifically to serve the interests of capital accumulation: as if they would not have been developed in a non-capitalist society. This may or may not be true. In fact the evolution of these technologies could probably be set within a context of social or technological determinism to give two equally convincing accounts. As Pacey (1983, 25) says, most inventions are made with a social purpose in mind, but many also have social influences and consequences not anticipated or intended.

Radical, utopian technology

Perhaps that arm of the AT movement which most appreciates this importance of social context but at the same time sees technology as partly a force for social change is what Boyle and Harper (1976) called 'radical' technology. Dickson (1974) called it 'utopian', because it could not be widespread in any society which had politics, economics and social relationships similar to those of existing society. Dickson insisted that technological development is essentially a political process. Technology is not neutral, since, both symbolically and practically, the dominant technologies support and promote the interests of dominant social groups. In capitalism their prime purpose is to help in accumulating capital, so

> The dominant modes [in capitalism] of hierarchical organisation and authoritarian control . . . become incorporated in . . . the technology that is developed in capitalist societies.

An example is nuclear power, which is so complex and dangerous that it *has* to be surrounded by secrecy and private police. It cannot be owned, controlled and understood by ordinary people; only by a hierarchy of experts and managers backed by the massive resources at the command of the state and big corporations.

By contrast, utopian technology

> would embrace the tools, machines and techniques necessary to reflect and maintain non-oppressive and non-manipulative modes of social production and a non exploitative relationship to the natural environment.

> (Dickson, p. 11, see Table 2.6, contrasting 'soft'
> utopian with 'hard' capitalist technology)

93

Table 2.6 Hard and soft technology societies

'Hard' technology society	'Soft' technology society
1 Ecologically unsound	Ecologically sound
2 Large energy input	Small energy input
3 High pollution rate	Low or no pollution rate
4 Non-reversible use of materials and energy sources	Reversible materials and energy sources only
5 Functional for limited time only	Functional for all time
6 Mass production	Craft industry
7 High specialisation	Low specialisation
8 Nuclear family	Communal units
9 City emphasis	Village emphasis
10 Alienation from nature	Integration with nature
11 Consensus politics	Democratic politics
12 Technical boundaries set by wealth	Technical boundaries set by nature
13 World-wide trade	Local bartering
14 Destructive of local culture	Compatible with local culture
15 Technology liable to misuse	Safeguards against misuse
16 Highly destructive to other species	Dependent on well-being of other species
17 Innovation regulated by profit and war	Innovation regulated by need
18 Growth-oriented economy	Steady-state economy
19 Capital intensive	Labour intensive
20 Alienates young and old	Integrates young and old
21 Centralist	Decentralist
22 General efficiency increases with size	General efficiency increases with smallness
23 Operating modes too complicated for general comprehension	Operating modes understandable by all
24 Technological accidents frequent and serious	Technological accidents few and unimportant
25 Singular solutions to technical and social problems	Diverse solutions to technical and social problems
26 Agricultural emphasis on mono-culture	Agricultural emphasis on diversity
27 Quantity criteria highly valued	Quality criteria highly valued
28 Food production specialised industry	Food production shared by all
29 Work undertaken primarily for income	Work undertaken primarily for satisfaction
30 Small units totally dependent on others	Small units self-sufficient
31 Science and technology alienated from culture	Science and technology integrated with culture
32 Science and technology performed by specialist élites	Science and technology performed by all
33 Strong work/leisure distinction	Weak or non-existent work/leisure distinction
34 High unemployment	(concept not valid)
35 Technical goals valid for only a small proportion of the globe for a finite time	Technical goals valid 'for all men for all time'

Source: Dickson 1974

Boyle and Harper (p. 5) argue similarly that radical technologies would

> help create a less oppressive and more fulfilling society . . . [involving] small scale techniques suitable for use by individuals and communities, in a wider social context of humanised production under workers' and consumers' control.

In other words it is not only specific technologies, it is those technologies set in a non-capitalist, indeed socialist, society. Accurately, Boyle and Harper refer to the (politically ambiguous) ecology movement as only an 'aunt' of radical technology, which has many other influences. They include critics of industrial society like Huxley, Illich, Mumford, Marcuse and Roszak, the 1960s counterculture, anarchism and utopian socialism, the intermediate technology movement – and China in the 1960s, apparently seeking social justice and minimum living standards for all through a judicious combination of big and small industry, decentralisation and ruralisation: 'We have more to learn from the Chinese than from any other

Table 2.7 Radical technology and its programme

Some examples of 'radical technology'

Organic agriculture and gardening, 'biodynamic' agriculture and gardening (after Rudolph Steiner, 'muck and magic'), vegetarianism, hydroponics, soft energy (solar, wind, etc.), insulation, low-cost housing, tree houses, shanty houses (shacks, chalets), 'folk-built' houses using traditional methods, houses built from subsoil, self-build houses, housing associations, solar dwellings, domestic paper-making, carpentry, scrap reclamation, printing, community and pirate radio, collectivised gardens, collective workshops for clothesmaking, shoe repair, pottery, household decoration and repairs, autonomous housing estates, autonomous rural villages and urban streets and neighbourhoods.

The radical technology programme

To include the following:

- Self-organised community projects
- Industrial ownership by workers' coops
- Alternative work for scientists and engineers, serving the above
- Land reform featuring collective ownership
- Repopulation of the countryside
- New rural villages
- Labour intensive husbandry
- Alternative financial institutions
- Basic income scheme
- Low-energy production
- Production for social use, not primarily as commodities for exchange and profit
- Sharing of some goods by neighbourhoods and communities (e.g. TVs, washing machines, cars)

Source: Boyle and Harper 1976

source save our own imagination' (p. 231). This idealisation of 'communist' China as somehow an ecotopian prototype was common in the 1970s (see Sandbach 1980), and seems very misplaced in the wake of post-Tiananmen Square capitalist China.

Boyle and Harper detail radical technologies, outlining social programmes which they would underpin (see Table 2.7). From them and Dickson we can also draw the following basic tenets of radical/utopian technology.

It acknowledges physical/biological limits and constraints on human activity hence it is ecologically sustainable. We can understand, own and control it, not vice versa, such control being democratic from the bottom up. It helps communities to be autonomous, depending minimally on outside money, materials or expertise, although this is not practicable in some areas (social services, many domestic goods, educational provision) where large-scale intra-community production should continue. It is associated with material simplicity, even frugality, compensated for by good quality of life which partly comes from enjoyment in production rather than through buying commodities. Hence a prime purpose behind it is to provide fulfilling work. Labour is judiciously substituted for machines and craft and creativity are maximised while repetition and de-skilling through division of labour are minimised.

Again, for some products near fully-automated large-scale factory production would be appropriate, but this would be under collective worker-ownership and control. Principles of good design would ensure that more was obtained from less materials and energy, so available, local, cheap, recycled materials would be common, insulation would replace much energy generation and so on.

Radical/utopian technology meets direct social need, not production for profit, and is non-alienating. It minimises hierarchical relationships, being owned and understood by many, including the poorest, not just an élite minority of 'expert' men politically controlled by an élite class.

Since this clearly is a socialist agenda, it is unsurprising that it is mainly the technical rather than the radical side of AT which persists into the 1990s, when leading oil companies are among the corporate interests developing 'soft' energy. Even in the 'intermediate technology' movement, which has focused on Third World peasant societies where the case for socialism was overwhelming, the all-important perception of context has changed. In the 1970s many saw intermediate technology as part of a strategy of land reform and small-scale 'alternative' rural, decentralised development, subject to democratic community control and for self-reliance (Omo-Fadaka 1976). It aimed to provide work with minimal capital input from outside, through labour-intensive techniques using local materials in cheap and simple designs, mainly to satisfy local needs (e.g. low-cost tools for blacksmithing, small-engine wire-drawn mechanical ploughs, charcoal-fired metal foundries, egg-carton machines using waste paper; see Schumacher 1973 and McRobie 1982).

Today, however, the Intermediate Technology Development Group, founded by Schumacher in 1966, may well be realising the worst fears of some ecocentrics. For, as Dickson foresaw, it often operates mainly in a context where this kind of development merely complements rather than replaces, large-scale, capital-intensive development funded from outside and subject to Western influence and control. This latter produces unemployment, rural depopulation, dispossession from the land, an urban proletariat and ecological degradation. The intermediate technology component, then, is not part of radically different communitarian social relations, but mostly provides the seed bed for complementary small-scale capitalism, from which large-scale capitalism tends to grow: part of the problem rather than the cure.

2.5 THE GLOBAL DIMENSION

Early concerns and the lifeboat ethic

From its beginnings in the 1960s, environmentalism's concerns have been global. Themes such as ecosystems, interconnectedness, Gaia and the personal-is-political underlined a realisation that Westerners (the 'First World') could not assess their environmental problems and their solutions in isolation from 'Second' (former 'communist') and 'Third' worlds. The *Limits to Growth* report followed Malthus by discussing population and resources in terms of universal, global aggregates and principles. Unfortunately, many of the implications about who was most to blame for supposed 'overpopulation', and what should be done about it, were neo-Malthusian too, in the worst sense, that is, they could be read as essentially putting the onus onto those who suffered most from poverty and starvation (Chapter 5.6).

Most infamously, Hardin (1974) proposed a 'lifeboat ethic'. If ten men were adrift in a lifeboat with exactly enough supplies for ten, any compassionate attempts to try to save drowning people by dragging them into the boat were doomed: they would merely overtax the boat's carrying capacity and all would perish. Hence not all Third World countries could, or should, be saved from famine by the Western 'lifeboat'. Only those with a vigorous population policy should be aided. This conclusion, and the zero population growth movement associated with it, was roundly attacked as 'ecofascism' and 'scientific racism' – using supposedly 'objective' scientific (ecological) principles to advance an anti-Third-World position (Buchanan 1973, Bookchin 1979, Chase 1980).

Today, much radical environmentalism adopts almost an exactly converse position: that environmental problems result from an exploitative *Western* economic system which particularly victimises the Third World. Despite the popular myth of aid flowing from North to South, in fact there is a large net *outflow* of wealth in the reverse direction, when debt, trade and aid are aggregated.

Partly following the Brundtland Report (UN 1987), ecocentrics increasingly see global economic and social justice, eliminating Third World poverty, as the key to ecological sustainability. To do this entails adopting a development model for both First and Third Worlds which is very different from that of conventional wisdom.

Development theories

Conventional wisdom equates 'development' simply with *modernisation* of the whole globe after the West's example. Ecocentrism prefers, however, a 'sustainable development' (SD) model of the more radical type (Chapter 2.3). As Peet (1991) describes, both of these have their roots in nineteenth-century ideas about social development, such as structural-functionalism and environmental determinism (see also Pepper 1993, 23–7).

Modernisation theory holds that the more structurally specialised and differentiated a society is, the more *modern*, and therefore 'developed' and 'progressive' it is. Modernisation involves technological sophistication, urbanisation, the spread of markets, 'democracy', social and economic mobility and the weakening of traditional élites, collectivities and kinships. Individualism and self-advancement are equated with the notion of general social *progress* (mainly a nineteenth-century idea) – the two being related by an 'invisible hand' after Adam Smith. 'Modern' education displaces older forms of knowledge with concepts claiming to be superior and universally valid, such as those of classical science (Chapter 3.2), of the commoditisation of relationships (impersonal buyer–seller, employer–employee), and of materialism and consumerism (Saurin 1993).

Modernisation theory is epitomised in Rostow's (1960) influential model of 'stages of economic growth'. These described how 'traditional' societies (with 'primitive' technologies and spiritual attitudes to nature), 'develop' to 'pre conditions for economic take off' (like that experienced in seventeenth- and eighteenth-century Western Europe). 'Take off' follows, where new industries and entrepreneurial classes emerge. In 'maturity' steady economic growth outstrips population growth, then a 'final stage of high mass consumption' allows the emergence of social welfare.

Obviously, this model is Eurocentric and imperialist, and it advances capitalism as 'progress' in itself. Third world societies, particularly, and all communities (actual or potential) basing their economic and social relations on localism and kinship are, by definition, 'backward'. Their 'development' must come by opening up borders to the influence of Western economic interests, via arrangements like GATT (the General Agreement on Tariffs and Trade).

In rejecting this modernisation model ecocentrics find common ground with underdevelopment and dependent development models, whose origins partly lie in Marxist analysis. In their Communist Party manifesto, Marx and Engels described capitalism's growth dynamic and the contradiction whereby more goods were produced than could be bought by those who produced them. Needing therefore 'a constantly expanding market for its products', capitalism, inherently, had to roll 'around the world looking for the best place to grow', creating new markets and

> constantly revolutionising the instruments of production [in the interests of 'efficiency', i.e. increased profit margins] and thereby the relations of production . . . In place of the old wants, satisfied by the productions of the county, we find new wants, requiring for their satisfaction the products of distant lands and climes. In place of the old local and national seclusion and self-sufficiency, we have intercourse in every direction, universal interdependence of nations . . . The bourgeoisie . . . compels all nations, on pain of extinction, to adopt the bourgeois mode of production; it compels them to introduce what it calls civilisation into their midst.
>
> (pp. 52–3, 1872 edition)

Capitalism therefore forms a world system, with the older industrialised nations a 'core' region bleeding wealth from the dependent Third World periphery, as resources, repatriated profits, cheap labour and new markets (see Figure 2.5 and Wallerstein 1974). Hence capitalist development *produces* Third World underdevelopment rather than eliminating it.

Processes produced by capitalist industrialisation in nineteenth-century Europe are today replicated in the Third World, for instance rural depopulation and the creation of an urban proletariat, centralised production, concentration of wealth and political power among élites, exploitative, cheap, sweatshop labour, and a cash nexus overriding other bases for relationships like religion, spirituality, family ties or loyalty.

More sophisticated versions of this idea might see a hierarchy of cores and peripheries ('metropoles' and 'satellites' according to Frank 1989) whereby Western core nations extract wealth from Third World cores, which in turn extract wealth from their poorer peripheries. Indeed, for most contemporary Marxists dependency models are too simplistic, neglecting the observable fact that 'peripheries' can show higher growth rates than cores (compare Western Europe with some Far East countries). Like Peet (pp. 73–7), they see the true development picture as multilineal, not unilinear, displaying several, often conflicting, tendencies. Nonetheless, in most modernisation ruling élites must extract surpluses, hence SD is not everywhere achievable in a world capitalist system.

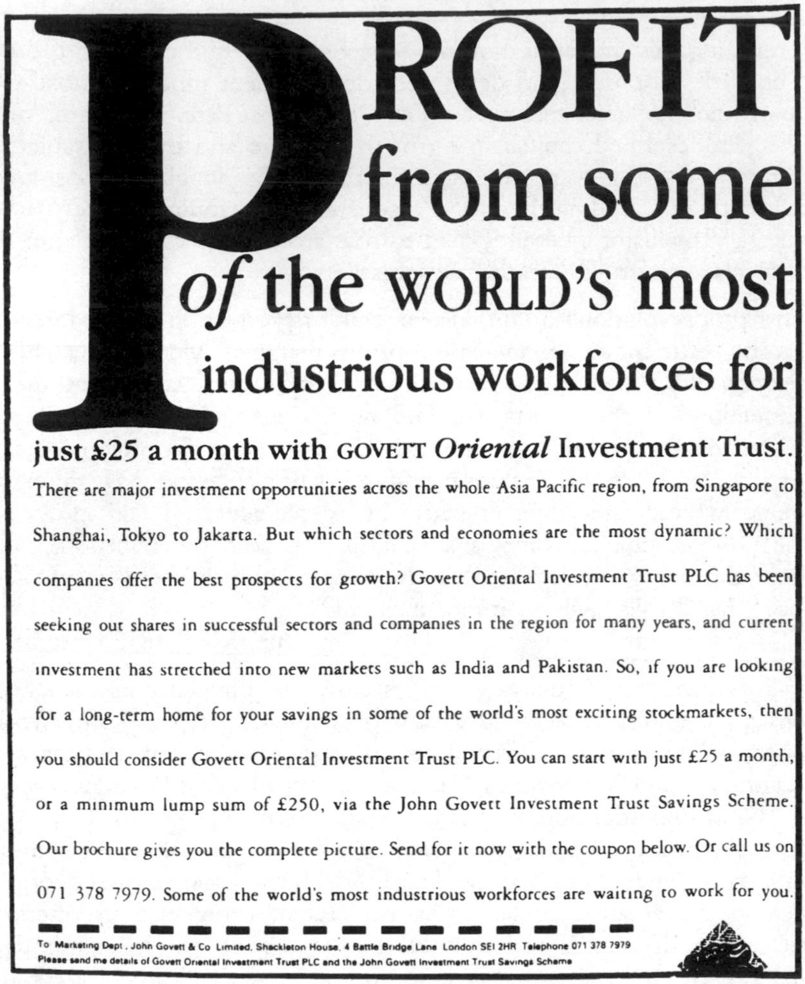

PROFIT from some of the WORLD'S most industrious workforces for

just £25 a month with GOVETT *Oriental* Investment Trust.

There are major investment opportunities across the whole Asia Pacific region, from Singapore to Shanghai, Tokyo to Jakarta. But which sectors and economies are the most dynamic? Which companies offer the best prospects for growth? Govett Oriental Investment Trust PLC has been seeking out shares in successful sectors and companies in the region for many years, and current investment has stretched into new markets such as India and Pakistan. So, if you are looking for a long-term home for your savings in some of the world's most exciting stockmarkets, then you should consider Govett Oriental Investment Trust PLC. You can start with just £25 a month, or a minimum lump sum of £250, via the John Govett Investment Trust Savings Scheme. Our brochure gives you the complete picture. Send for it now with the coupon below. Or call us on 071 378 7979. Some of the world's most industrious workforces are waiting to work for you.

To Marketing Dept , John Govett & Co Limited, Shackleton House, 4 Battle Bridge Lane London SE1 2HR Telephone 071 378 7979
Please send me details of Govett Oriental Investment Trust PLC and the John Govett Investment Trust Savings Scheme

Figure 2.5 Making money from cheap Third World labour: a 1990s financial press advertisement
London Socialist Standard, August 1993

It must be emphasised that any analysis of development and how problems of aid, trade, debt and sustainability should be approached must be related back to the ideology of the proposer, to be properly understood (Elliot 1994, 108). Implicit, too, is the proposer's basic theory of value: the subjective preference theory shows up clearly in modernisation, the Marxist theory in dependent development and the green theory in radical SD (see Corbridge 1993).

Radical environmentalism and the Third World

Poverty was linked with Third World environmental degradation well before Brundtland made it 'respectable' to do so, for instance by George (1976) and Buchanan (1982). As Sen (1981) put it, it was not that there was insufficient food, it was a loss of *entitlement* to good land to grow it on which was crucial. Dispossession of peasants from common land, replicating what happened when Europe was modernised (Goldsmith *et al.* 1992), is particularly problematic in Asia and South America. Hecht and Cockburn (1990, 239) say of deforestation that

> Any program for the Amazon basin begins with basic human rights; an end to the debt bondage, violence, enslavement and killing practised by those who would seize the lands these forest people have occupied for generations.

In Africa it is more a problem of peasant production coopted into a cash crop economy where surpluses are creamed off in taxation or their value is eroded by falling prices on the world market (Jackson 1990).

All these forces create migration into towns, where there is little work, or onto marginal land which is then overfarmed, deforested and degraded. This process is part of a larger destructive picture that includes international trade and the rise of middlemen, cash cropping for markets rather than subsistence, capital-intensive agribusiness aided by 'green revolution' technology (which is inappropriate and expensive), the attitude of men towards women (who form the majority of the agricultural workforce, yet do not own the land) and the activities of Northern governments and transnational corporations. These, says Bennett (1987) are what produces the cycle of hunger and enforced environmental degradation, not 'overpopulation', the weather, or Malthusian insufficiency – all of which are myths.

Fundamental to all this is the 'free' trade system. There is evidence that before colonialisation Third World countries, under 'primitive' communism, did not experience wide-scale famine, poverty and overpopulation (Redclift 1986, Omo-Fadaka 1990). But, says Morris (1990) echoing Marx and Engels, free trade, essential to global modernisation, makes all local allegiances expendable. Globalisation of the capitalist market economy destroys local communities, making them compete with faraway people. Unprotected from the vagaries of world market prices, local economies suffer because as more countries are encouraged to grow cash crops so the prices of these crops fall. To offset this more land is deforested to enable yet more cash crops to be grown.

Any notion that free trade could actually be used to promote environmental protection, for instance by allowing sanctions against countries with low standards (Williams 1993), is undermined by the 1994 GATT agreement. This prohibited individual countries imposing comparative competitive disadvantage on their own industries through demanding higher, more expensive, environmental standards. Only standards *universally* agreed – by over one hundred countries – could be enforced: these are inevitably low standards.

GATT liberalised trade in services, international investment flows and agriculture, encouraging each country to specialise according to 'comparative advantage'. This effectively restructured world trade around the interests of the most powerful multinationals, say Goldsmith *et al.* (1993). By removing protection for local agriculture (e.g. tariffs against cheap imports), GATT will 'pauperise millions in the South' increasing environmental degradation. Simultaneously the agreement made it easier for Northern firms to relocate in areas of cheap labour and lower environmental standards: a 1993 survey of 10,000 German firms showed that one in three intended to relocate to Eastern Europe or Asia within three years (Lang and Hines 1993). Other free trade agreements, like that for North America, have also contributed to poverty (in Mexico) and undermined environmental protection – Canada's spending on wetland protection and reforestation being disallowed because it was 'trade distorting' (Ritchie 1992).

What GATT did universally had previously been happening in individual countries under structural adjustment programmes (SAPs). These have been prescribed for nearly all one hundred developing countries in the past twenty years by the Western-financed and Western-staffed World Bank and International Monetary Fund (FoE 1993). They are programmes of economic reform demanded in return for lending money in order to pay interest on, or otherwise extend, debts originally incurred in the 1970s and 1980s (George 1989, 1992). They require recipients to earn more in foreign revenue and spend less in social welfare and environmental protection, to create more 'efficient' economies. Studies of the environmental impact of SAPs in Thailand, the Philippines, Cote d'Ivoire and Malawi challenge the notion that 'efficient' economies (in the neoclassical sense) are the most environmentally friendly ones (Devlin and Yap 1993). SAPs have, for instance, been a key factor in deforestation in the name of increased timber and cash crop exports.

Devlin and Yap conclude that radical SD based on localism and equity is not consistent with the SAP and GATT prescription of free trade, unrestrained markets and minimal state interference which characterise global modernisation. Adams (1990) agrees that 'green' modernisation 'is almost a contradiction in terms'. This includes reformist SD, 'done for' people by outside agencies such as non-government organisations, or by a reformed, green-thinking World Bank (see Pearce *et al.* 1991). But, conversely, Adams also warns against glib prescriptions of small-scale, low technology, bottom-up development models (see, for example, Ghai and Vivian 1992, Ekins 1992c). Local irrigation projects in Kenya, for instance, have been no more successful than the large-scale irrigation projects so despised by environ-mentalists. There is no magic formula, Adams concludes, none the less the key to truly sustainable development is the local model, when this implies people having power to do things for themselves.

This self-determination theme was echoed strongly in the Green Party European Manifesto (1994, 4, 5, 8), which endorsed localised development for both Europe and the Third World, 'creating and encouraging local sources of investment and finance' including independent local curren-cies, local ownership and control of businesses and community-led pro-jects which achieve diversity and self-sufficiency. The Greens are the only party to advocate reduced long-distance trade and to oppose the European single market and the trading of money as a commodity.

Their ecocentric analysis, then, does not propose eliminating capitalist development. Rather, it would 'sidestep' its worst features via anarchistic *independent development*, whereby regions would substantially go their own economic way, cutting down trade and specialisation. The important neoclassical economic principle of complementarity which underlies 'free' trade (countries specialise in what they do best and exchange their products with other countries, who have specialised in what they do best) would be substantially abandoned, so making nations far less beholden to each other and to world market prices. A 'new protectionism' would be adopted, whereby superblocs and multinational corporations had their power severely curtailed and environmental and public protection standards are raised (Lang and Hines 1993).

The response to Rio

In 1992 the United Nations Conference on Environment and Develop-ment held at Rio de Janeiro adopted Agenda 21, a work programme for the twenty-first century agreed by 179 states and based on principles defined in an accompanying Declaration on Environment and Development (see Table 2.8). Following Brundtland, UNCED affirmed the need to eradicate global poverty via ecologically benign economic development. National governments trumpeted Rio as a success, but ecocentrics derided the whole approach of Agenda 21. *The Ecologist* magazine said that for grass-roots groups around the world the question was not

> *how* the environment should be managed – they have the experience of the past as their guide – but *who* will manage it and in *whose* interest. They reject UNCED's rhetoric of a world where all humanity is united by a common interest in survival, and in which conflicts of race, class, gender and culture are characterised as of secondary importance to humanity's supposedly common goal.
>
> (Goldsmith *et al.* 1992, 122)

Table 2.8 The Rio Declaration on Environment and Development

Recognising the integral and interdependent nature of the Earth, our home, the nations meeting at the Earth Summit in Rio de Janeiro adopted a set of principles to guide future development. These principles define the rights of people to development, and their responsibilities to safeguard the common environment. They build on ideas from the Stockholm Declaration at the 1972 United Nations Conference on the Human Environment.

The Rio Declaration states that the only way to have long-term economic progress is to link it with environmental protection. This will only happen if nations establish a new and equitable global partnership involving governments, their people and key sectors of societies. They must build international agreements that protect the integrity of the global environment and the development system.

The Rio principles include the following ideas:

- People are entitled to a healthy and productive life in harmony with nature.
- Development today must not undermine the development and environment needs of present and future generations.
- Nations have the sovereign right to exploit their own resources, but without causing environmental damage beyond their borders.
- Nations shall develop international laws to provide compensation for damage that activities under their control cause to areas beyond their borders.
- Nations shall use the precautionary approach to protect the environment. Where there are threats of serious or irreversible damage, scientific uncertainty shall not be used to postpone cost-effective measures to prevent environmental degradation.
- In order to achieve sustainable development, environmental protection shall constitute an integral part of the development process, and cannot be considered in isolation from it.
- Eradicating poverty and reducing disparities in living standards in different parts of the world are essential to achieve sustainable development and meet the needs of the majority of people.
- Nations shall cooperate to conserve, protect and restore the health and integrity of the Earth's ecosystem. The developed countries acknowledge the responsibility

that they bear in the international pursuit of sustainable development in view of the pressures their societies place on the global environment and of the technologies and financial resources they command.

- Nations should reduce and eliminate unsustainable patterns of production and consumption, and promote appropriate demographic policies.
- Environmental issues are best handled with the participation of all concerned citizens. Nations shall facilitate and encourage public awareness and participation by making environmental information widely available.
- Nations shall enact effective environmental laws, and develop national law regarding liability for the victims of pollution and other environmental damage. Where they have authority, nations shall assess the environmental impact of proposed activities that are likely to have a significant adverse impact.
- Nations should cooperate to promote an open international economic system that will lead to economic growth and sustainable development in all countries. Environmental policies should not be used as an unjustifiable means of restricting international trade.
- The polluter should, in principle, bear the cost of pollution.
- Nations shall warn one another of natural disasters or activities that may have harmful transboundary impacts.
- Sustainable development requires better scientific understanding of the problems. Nations should share knowledge and innovative technologies to achieve the goal of sustainability.
- The full participation of women is essential to achieve sustainable development. The creativity, ideals and courage of youth and the knowledge of indigenous people are needed too. Nations should recognize and support the identity, culture and interests of indigenous people.
- Warfare is inherently destructive of sustainable development, and nations shall respect international laws protecting the environment in times of armed conflict, and shall cooperate in their further establishment.
- Peace, development and environmental protection are interdependent and indivisible.

Source: Keating 1993

This was a remarkable statement from a bastion of ecocentrism, which for so long had pushed the image of 'one world' as representing its own holistic, organic mission. For it suggests recognition that the 'one world' notion can actually be used to oppose this mission, by promoting universally a late-modern American development model (see Cosgrove 1994).

The Ecologist repudiates Rio's six 'mainstream' responses to environmental crisis. First, it is not 'poverty', defined as absence of the American way of life, which is the root cause of environmental degradation, it is American-style 'wealth'. Second, 'overpopulation' is *caused*, not cured, by modernisation, destroying the traditional balance between people and their environment. Third, the 'open international economic system' of the Declaration will extinguish cultural and ecological diversity. Fourth, the problem of externalisation of pollution etc. is not solvable by pricing the environment, but by reversing enclosure of the commons, so there is nowhere to 'externalise' to. Fifth, Rio's calls for more 'global management' effectively constitute Western cultural imperialism. This approach would anyway be ineffective because of the impossibility of verifying and enforcing global agreements (Greene 1993). Sixth, the attitude that transfers of Western technology to the Third World are the most urgent need smacks of traditional Western scientific imperialist arrogance – effectively presuming that ignorance and laziness characterise Third World people.

Others are more sanguine about Rio, impressed that Agenda 21 encourages localism by calling for consultation, negotiation and participation among 'stakeholders' – including women, youth, indigenous people, local authorities, workers and farmers – along with world democratisation (Roddick and Dodds 1993). In reality, though, the follow-up work by the UN Commission on SD has *not* invited local governments to participate in its future programme. Despite Rio, nation states consider themselves the only important actors (Gordon 1993). Though this attitude is mistaken it persists, because real power has passed up, not down, the scale; to the Catholic Church, the IMF and transnational corporations (whose activities UNCED merely expects to be regulated voluntarily) (Thomas 1993).

Thomas fundamentally endorses *The Ecologist*'s position when she says (p. 1) of UNCED:

At the most fundamental level, the causes of environmental degradation have not been addressed, and without this, efforts to tackle the crisis are bound to fail. The crisis is rooted in the process of globalisation under way. Powerful entrenched interests impede progress in understanding the crisis and in addressing it. They marginalise rival interpretations of its origins and thereby block the discovery of possible ways forward . . . The result is that the crisis is to be tackled by a continuation of the very policies that have largely caused it in the first place.

2.6 ECOFEMINISM

Ecofeminism is a perspective within environmentalism, influenced by the general development of feminism. One kind of feminism which does not feed into ecofeminism is the 'liberal' variety. This requires no dissolution of patriarchal society, being 'androgynous'. That is, it wants women to play equal and similar roles to those of men: this in a society still dominated by values of aggression, competition and materialism. As Plumwood (1992) reminds us, liberal feminism's view of nature is unacceptable to ecocentrics, for it follows Mary Wollstonecraft's (*Vindication of the Rights of Woman,* 1792) doctrine that humans, because of their reason, are superior to and different from the 'inferior sphere' of brute creation (which lacks reason). Hence it would put women together with men in a project of dominating nature.

By contrast, ecofeminists unite in a central belief in the essential *convergence between women and nature.* This is, first, because their biological make-up inevitably associates women, more than men, with the natural functions of reproduction and nurturing. Second, women and nature have in common that they are exploited by men, both economically and in being objectified and politically marginalised. Some consider that this common oppression developed intensively during the Enlightenment (Chapter 3.4) into a 'logic of domination' (Warren 1990), geared to hierarchical, dualistic thinking.

That 'logic' says, first, men/humans are different from women/nature, second, that they are superior to them, and therefore they are justified in dominating them. However, ecofeminism, says Warren, denies that differences imply superiority or justify domination.

During and since the 1970s, debates within ecofeminism have focused on two leading schools of thought; *cultural/radical ecofeminism* and *social ecofeminism* (Plumwood calls the latter 'socialist/anarchist' ecofeminism).

Cultural/radical ecofeminism

Cultural/radical ecofeminism is typified by Pietila (1990, 232), writing of the problems of 'Our mother, Gaia'. They could be solved by a 'women's culture', providing 'practical and philosophical guidelines to sustainable development'. This culture would draw on ancient myths combining women and nature, mother and earth, in a cooperative relationship: caring, nurturing, mutually giving and receiving. Since menstrual cycles follow phases of the moon, and fertility follows the rhythm of the seasons, then (p. 236) 'women feel themselves as part of the eternal cycle of birth, growth, maturation and death, which flows through them, not outside them'. Daly's (1987) similar ecofeminism celebrates feminine 'closeness' to nature. Collard (1988) advocates going back to the Earth goddess-worshipping, non-hierarchical matriarchies that supposedly characterised some 'traditional', 'primitive' societies.

This ecofeminism says, then, that 'female culture' is concerned with the *body*, the flesh, the material, natural processes, emotions and subjective feelings and private life. By contrast, 'male culture' emphasises the *mind*, intellect, reason, culture, objectivity, economics and public life. It constantly seeks to transcend natural constraints on what humans can do: men constantly fight to conquer, exploit and mould nature, leaving their mark behind and thus achieving a form of immortality and transcendence. Merchant (1982) describes how Francis Bacon and the Royal Society pledged to reveal 'the secrets still locked in her bosom' and to 'conquer' and 'subdue' her [the Earth] (Chapter 3.4). New female reproductive technologies, developed by men, are said to constitute a continuation of this dual domination, as is most high technology (Shiva 1992).

Cultural ecofeminism means liberating nature from the repressive male ethos so that it will be respected as a sustainer of life (see Capra's (1982) plea for a balance of 'yin' and 'yang' characteristics in people and society). This may be achieved in several ways. Women, individually and in groups, can discover their authentic natures and celebrate and affirm them. Such consciousness-raising may need to exclude men, on the grounds that they could have a negative impact on it while it is still nascent and before it is strong enough to resist male domination. Then there can also be celebration of pagan myths and rituals, and associated pastimes like tarot cards and astrology, which affirm respect for mother nature and the essential interconnectedness of humans and nature.

This latter leans towards New Age thinking, and some greens and feminists (particularly of the social variety) reject it. As Plumwood argues, the whole idea of 'connecting with nature' could be regressive and insulting, portraying women as passive reproductive animals immersed in the body and in unthinking experience of life. Biehl (1991), from a 'social ecology' perspective, attacks cultural feminism as apolitical (since it rejects conventional politics because of its hierarchical power relationships), anti-rational and home- and nature-worshipping.

Some problems

Eckersley (1992) outlines some problems of cultural ecofeminism. First, if it claims that women have a 'special relationship' with nature by virtue of their biological role (birth, nurturing) then men might stand permanently condemned because of *their* biology to an 'inferior sort of relationship with nature'. In fact, men increasingly involve themselves in nurturing the young, thus departing from the Western male cultural stereotype.

Second, if the 'special relationship' is claimed on the grounds of common oppression by men, this is also problematic, for women are not the only oppressed group in Western society. Indeed, it could be argued that *men* are oppressed in capitalism. Patriarchy may not, either, explain racism or class oppression.

Third, it is difficult to prove that patriarchy is *responsible* for exploiting both women and nature. As Levin (1994) observes, it is a 'vague and loose argument' to suggest that merely because both women and nature are dominated they are so for the same reason. As Eckersley puts it (p. 68), there may be parallels in the logic or symbolic structure of different kinds of dominance but this does not prove that both come from the one source. Indeed, many, 'traditional' societies in harmony with nature are in fact patriarchal (see Young 1990). Emancipating women may not, therefore, automatically emancipate nature, and vice versa.

Fourth, any cultural/radical feminism that wishes to elevate a female stereotype rather than a male stereotype is problematic because both stereotypes are deficient. If one is over-rational/analytical, the other is under-rational/analytical and so on.

The problem of essentialism

Essentialism is the belief that abstract entities or universals exist as well as the instances and examples we meet in space and time. Essentialism, whether expressed implicitly or explicitly, is ecofeminism's 'core problem' according to Holland-Cunz (Kuletz 1992).

Essentialism might argue that throughout different historical periods, economic modes of production and cultures, there have always been power hierarchies, patriarchy and exploitation of women and nature. And, it would assert, the reason lies in a universal, determining abstract principal, such as hierarchy – an *essential* characteristic of all societies which keeps reappearing regardless of different cultural, economic and social formations and arrangements. Alternatively the argument is often couched in terms of 'biological reductionism', that is, the essential, universal determining cause of the logic of oppression lies in biological differences between men and women – differences in their 'gendered selves'. Either way, it follows logically from such arguments that humans are powerless to rid themselves of patriarchy, however much they changed their society it would reappear.

Mellor (1992), from a Marxist feminist perspective, condemns any fatalistic talk of ahistoric universals of biological sex or essential human nature. Marxism's historical materialism, by contrast, argues that apparent constraints on human development and creativity ('human nature', 'men's nature', 'women's nature', 'nature's limits', the 'nature of the environment') are in most cases really socially, rather than biologically, constructed. Thus in different historical periods, cultures and economic modes of production imagined constraints will either take different

forms, or they may not appear or be relevant at all. And, most important, since they are socially constructed they can be socially changed – they
are not therefore 'essential'. This essentialism versus social construction
debate is another version of older arguments – those of nature versus
nurture/culture, or determinism versus free will (Chapter 2.7).

It follows that ecofeminism, if it is not to preach the impossibility of
doing anything, or simplistically that hierarchical patriarchy should be
replaced by hierarchical matriarchy, must not fall into the essentialist trap.
Using tarot cards, baying the moon, looking to pre-historical societies to see
if they were matriarchal and nature-worshipping – these are all part of this
trap, because they suggest an unchanging, inherent, feminineness in opposition to an equally unchanging masculinity.

It is sometimes difficult to avoid the trap. Warren (1990), for instance,
appears to reject essentialism by emphasising how the *context* of patriarchy
differs culturally and historically, but she then seems to fall back on it when
she says that domination of all kinds is still 'located in an oppressive
patriarchal framework', implying that a universal – patriarchy – is *the* root
problem. Similarly, King (1989) and Plumwood (1990) apparently want to
deny essentialism and the idea of a 'gendered self' by focusing on socially
produced (and therefore socially changeable) gender roles and stereotypes.
Yet neither abandon completely the idea of a specific feminine-nature link.
As Evans (1993, 184) says, 'It is hard to see how the link between women
(rather than persons) and nature could be made without invoking biological
reproduction as the key'.

Materialist social ecofeminism

Holland-Cunz considers that social (socialist-anarchist) ecofeminism picks
up many traditions in non-mainstream European socialism: utopian socialism, classical anarchism, early Marx, Engels' *Dialectics of Nature*, Morris'
News from Nowhere, and the critical theory of the Frankfurt school of neo-
Marxism. They all insist that exploitation of nature relates to exploitation in
society, emphasising social and political rather than personal aspects of the
domination of women and nature. Social ecofeminism resists essentialism
in general and biological determinism in particular. Men's and women's
'nature' is held to be a political/ideological category (see Chapter 5.6). And
women's oppression is interwoven with class, race and species oppression
(Warren 1990). But social ecofeminism also rejects the crude economic
class reductionism of some Marxism; it does not accept that women's
oppression is *merely* a special case of exploitation of the 'proletariat', or
that to establish socialism would mean automatically ending women's or
nature's oppression.

But while it rejects essentialism and biological determinism it should not,
according to Mellor, ignore what she calls 'the reality of the biological and

ecological', which includes the reality that motherhood is something that *starts* with an exclusively female role (and an exclusively male role). This fact, like the physical limits and constraints of ecological systems, cannot be entirely subsumed within the social.

Mellor wants to modify or 'reconstruct' Marxist theory, to make a socialist ecofeminism. If, she argues, the way we organise ourselves to get material subsistence (relations of production) plays a key role in shaping society (see Pepper 1993, 67–70), so also must the way we organise ourselves materially to continue as a species (i.e. the relations of *reproduction*). 'If the means of survival produced definite social relations and particular forms of consciousness, why not the means of procreation?' (p. 50). Hence, she goes on to argue, the material world of motherhood, not merely the (still largely male) material world of industrial production, should provide ideas and values for shaping an alternative (socialist) society. These ideas and values are altruistic: they include taking immediate responsibility for meeting the needs of others.

Mellor is using the notion of 'standpoint' here, which relates ideas and values to their social, material context. It concedes that there is no one 'objective, true' reality. Our approach to knowledge – to history, politics, economics, or our relationship with nature – differs from different material *standpoints*. Men's perception is coloured by their lives in capitalism, where they mainly produce goods and services to be exchanged as commodities in a market, for profit. Women's perceptions and values will be very different. Their immediate lived experience, in domesticity, values usefulness (in the broadest sense) of work and sensuous activity, not profitability (Hartstock 1987). The dominance of capitalist relations in Western society ensures that the male standpoint mediates all knowledge. Its view of nature as a commodity will prevail over the female view of unity with nature.

To modify Marx is important, says Mellor. For Marx's original theory held that labour is one of the forces of production, along with nature. That labour is predominantly male labour. But concealed within it is female labour. Many men can work the hours they do only because women are there to run the home and family, thus freeing men's time. Socialist relations of reproduction, by contrast, would equalise men and women's time in various roles. In summary, Mellor argues that 'An ecological feminism that does not embrace socialism would be as theoretically and politically limited as an ecosocialism that does not embrace feminism' (p. 59).

Idealist social ecofeminism

Other calls to construct a social ecofeminism often introduce more idealism than Mellor's strictly materialist analysis (for the distinction between idealism and materialism, see Chapter 2.7). Ruether (1975), for instance,

Table 2.9 Parallels between (cultural and some social) ecofeminist beliefs and anarchism

- Disdain for materialism, industrialism and 'civilising values'
- Balancing rational and non-rational
- Diversity and balance in all spheres of life
- Eliminating dualisms
- Natural behaviour is spontaneous, loyal, unhierarchical, egalitarian, cooperative
- Social change stems from the individual
- Disdain for conventional politics and (in modern anarchism) class struggle
- Society should be run (and changed) from the bottom up
- Reform of values, attitudes, ideas vital in social change
- Don't confront capitalism/the state head on – 'bypass' via direct action and lifestyle reform
- The individual is fulfilled in the collective
- People must take increased responsibility for their own lives
- The state should be abolished
- Decentralised society essential
- Concerns for all life on earth (chain of being)
- Holistic thinking
- Revulsion towards hierarchies
- Suspicious of/anti the nuclear family
- Valuing experiential approaches to learning
- Consciousness-raising/encounter groups as an important learning device
- Skills and knowledge sharing as another learning device
- Aversion to theory, academicism and intellectualising

considers that there can be no liberation for women in a society whose fundamental model of relationships is hierarchical, so women must unite with the environmental movement to reshape the 'underlying values of this society', that is, its prevailing *ideas*, from which hierarchical organisation and domination are held to stem.

According to Plumwood, one of the principal wrong ideas in our culture is the tendency to *dualise* in our thinking, so, for instance, imagining that there is a fundamental distinction between society and nature suggests that they are separated, making it easier for the former to exploit the latter (see Chapter 3.2). Eckersley considers that patriarchy is a subset of this 'more general problem of philosophical dualism that has pervaded Western thought' (1992, 69). Both are arguing that it is a particular approach to knowledge – a *set of ideas* rather than economic and social arrangements – that is responsible for actions detrimental to nature.

Eckersley is at pains to point out how ecofeminism strongly links with deep ecology. For both emphasise the need for understanding ourselves, and realising our connectedness with a larger whole. Both seek personal contact and familiarity with the natural world. And both have strong cross-links with anarchism.

Holland-Cunz (Kuletz 1992) describes how 1970s feminist utopian literature (e.g. Ursula LeGuin 1975, Marge Piercy 1979) came inexorably to the conclusion that the only society where there is no patriarchy must be a decentralised, ecological one. This ecofeminist utopia is also non-hierarchical and directly democratic, practising rural subsistence through small-scale technology. The comparisons with anarchism are obvious (see Table 2.9).

Women and development

Ecofeminism shares with socialism an internationalist mentality opposing women's oppression world-wide. It recognises that in the Third World women, not men, constitute the backbone of 'production' as well as 'reproduction'.

In rural societies, before full 'development' (modernisation and mass urbanisation), women bring up families, run households, grow the crops and have many more babies than they want to. Ecofeminism, applied to this situation, has meant, first, pushing for women to be involved in decision making about how land is used and who controls it. Women have resisted land appropriation by government and (Western) commercial firms. So, for instance, they are the backbone of the Chipko movement. It is, however, the men who own the land, and who have given in to the seductions behind the modernisation model.

However, many Third World women have been coopted by modernisation. They help to provide the cheap labour which builds wholly inappropriate dams and nuclear power stations. They assemble hi-fis, CDs and TVs for Western consumers, and (unwittingly and often unwillingly) cooperate in producing mass unemployment in the West because they work for such a pittance. Ecofeminism opposes the modernisation model, and resists 'Women in Development'-style movements (Simmons 1992). This last was a 1970s liberal feminism initiative wanting women to join the world market by increasing their access to paid employment. In the 1980s, however, many Third World women instead adopted the 'by-pass' strategy: setting up enterprises and movements that attempted to exclude international capital (e.g. Chipko, South African cooperatives, 'green zones' in Mozambique, cooperatives in India). They supported, therefore, a localised development model (Chapter 2.5). In India, most ecological wisdom lies in the hands and brains of women, so their struggle for development which is alternative to that of Western modernisation is also a struggle for ecologically sustainable development (Shiva 1988).

2.7 SOCIAL CHANGE

Ecologism argues for radical social change. Hence it must be concerned with a very old political issue. That is, *how* can we change a society which looks to most people in it as if it is the most 'common sense' way of organising any society – the most 'natural' way of doing things: a system of social organisation which, even if we do not like it, seems deeply entrenched and inevitable?

To approach this, we need two insights. The first is into the ways that present society functions. The second is an appreciation, through studying history, that radical change actually *does* happen constantly, and the 'eternal' nature of present society is illusory – what we take as having always been so may have only very recent historical origins. Before addressing this historical perspective in Chapter 3, we now briefly consider some fundamental questions about social mechanisms and change which radical greens must address.

Determinism and free will

Just how free are humans to control, collectively or individually, their lives, their social and economic arrangements and their relationship with nature? This is a crucial political as well as philosophical question. For to argue that there are limits on human action set by external, uncontrollable forces, for example 'laws' of economics or history, God's design, technological developments, or the physical environment, can powerfully licence the status quo. If we are *determined* by forces beyond our control, attempts to go against the grain of such forces are not only ill-advised, they are futile. Thus conservative governments often argue that features of our society which we may not like (unemployment, low pay, for instance) cannot be changed and must be accepted because they result from outside forces (economic laws of competition, world recession) beyond anyone's control, including theirs.

It is thus against the interests of any group wanting radical social change to support deterministic (therefore perhaps fatalistic) arguments, rather than the idea that humans can freely shape their own society – 'make their own history' in Marx's phrase. This is relevant to environmental determinism or biological determinism (which argues that we are shaped and constrained by genetically inherited traits of 'human nature' – see Chapter 5.6).

Environmental determinism has appeared in many guises (Glacken 1967), from the Malthusian limits to growth thesis to ideas of early geographers (Peet 1985) that human nature, physiognomy and national and social characteristics are more or less determined by climate, soil, relief and geographical position (still popular notions with many people). The view that the *built* environment controls human character and nature is also environmental determinism. However, this does not amount to fatalism, since we *can* change and control our built environment, as attempts at social engineering, from utopian socialist communities to twentieth century urban planning and architecture, testify.

113

Ecocentrics have had a neo-Malthusian tradition (e.g. Goldsmith *et al.* 1972, Meadows *et al.* 1972; 1992) of emphasising environmental (resource) limits to social and economic growth and development (Chapter 2.3). But cornucopian technocentrics and free market advocates often reject the limits to growth thesis (Simon and Kahn 1984), emphasising the 'Baconian' creed that scientific knowledge equals power over nature: a power which should be used to improve humankind's lot by extending the boundaries of nature's 'limits' (Chapter 3.2).

In a way the latter arguments are also deterministic, suggesting that *humans* now determine *nature* – its form and behaviour – through knowing and manipulating the cause-and-effect laws governing its various components and their relationships. But in another way they support the notion of freedom of human will – freedom, for instance, to control the environment.

Less technocentric Western philosophies which also emphasise human freedom of will in relation to society and nature have been developed in the last hundred or so years, as phenomenology and existentialism. The science of *phenomenology* assumes that we are not separate from the rest of the world, but neither are we predetermined by 'external' forces. Indeed it emphasises the way *we* shape the world: imposing structure, meaning and value onto it via our consciousness. This is not to deny the existence of an 'objective' nature 'out there' (though more extreme idealist philosophies like that of George Berkeley (1685–1753) and those of New Ageism did and do see matter purely as a manifestation of mental activity (Lacey 1986, 97)). But it is to suggest that this is not really relevant (Warnock 1970, 26–8). For important knowledge of the world is knowledge of how the consciousness and intentions of individuals and groups interpret, mediate and indeed structure it (this is knowledge of *hermeneutics* – understanding of the significance of human actions and thought).

Since consciousness and perception vary between individuals and groups, there can be no universal, agreed, 'objective' way of knowing about things. Phenomenology therefore emphasises *subjective* ways of knowing the world, through intuitive understanding. How different people and cultural groups know and understand their own world of immediate experiences – their lifeworld – is vital.

All this suggests a relativist view of knowledge, understanding and, indeed, ethics concerning how the world should be. That is, it implies that the knowledges of different individuals and groups can be regarded as equally important and valid – there can be no general ethical laws. This perspective is very much in tune with the condition known as 'postmodern'.

By extension, the individualist philosophy of *existentialism* says that there are no objective, external facts or laws governing our social existence, save that we are born and one day will die. We are not helpless playthings of

114

historical forces, or social laws and codes of conduct. *We* have control and choice over most facets of our existence; not being bound by economic or social conventions. This is not to deny totally that our environment, including culture, society and economics, *conditions* our situation. But 'condition' does not mean 'determine'. So we must accept that while on one hand we have been thrown into a world which is not of our making, on the other we are free to decipher the meaning of that world for ourselves, not as interpreted by others or supposedly external factors beyond our control.

Not to recognise this is to lead an alienated and 'inauthentic' existence. But if we do recognise it we open up a horizon of possibilities – of living how we desire to live. This carries implications for our relations with other people and nature. On the one hand it could be interpreted as a doctrine of selfish individualism, also giving *carte blanche* to treat nature how we wish. Alternatively, it can imply that since we have been free to make our world, if it is polluted or socially unjust *we* are responsible and we could change things (Tuan 1972).

By emphasising the individual, existentialism may appeal to the personal-is-political motif in ecologism, as it may also appeal to anarchists (Brown 1988). But many anarchists are keener to emphasise the potential of the collective in social change.

Individual lifestyle or collective politics?

To advocate radical social change involves questioning the political power of those who benefit from present socio-economic arrangements. This power is so formidable that it might only be confronted by people acting *en masse* in conventional political ways, ranging from parliamentary politics to extra-parliamentary pressure group action or even revolution through withdrawing labour and/or seizing the instruments of power. All these routes suggest *collective* approaches, by contrast with the popular green idea that significant political change starts from individuals.

According to the latter, it is no good urging the masses to take political power if you have not changed your own lifestyle. For since 'the personal is political' every individual's thoughts and actions (e.g. in choosing what food to eat) have political repercussions. In a way this could be regarded as a collectivist view, because it emphasises how individuals are part of wider society. But in practice this implication of the personal lifestyle reform approach is usually submerged in notions of the individual self having a pivotal role in social change. Such individualism mistrusts mass revolution, arguing that it usually involves violence and oppression, the very ills that revolution intended to conquer in the first place (though in the late 1980s mass revolution in Eastern Europe entailed little violence). And it mistrusts party politics, arguing that the search for political power irrevocably corrupts politicians, and that political parties always have to compromise their ideals. Individualism places faith, instead, in a continuous process of individuals changing their values and lifestyles, which should then produce a new society overall. This concept rests on an essentially liberal view of society, as the aggregate of the individuals in it.

In Britain, collective action for social change is most readily associated with the trade union and labour movement. But it could also imply the kinds of local community politics which are effective on the European mainland, and are strongly advocated by the Green Party (Wall 1990). However, collectivism is not fashionable in today's political climate. Popularly (unpopularly), it is equated with the establishment of the regulatory state in Britain in the second half of the nineteenth century. Then, *laissez-faire* was not regarded as a principle of sound legislation, and government intervention was seen as beneficial – even when it limited individual choice or liberty. But today

> under the prevailing political and economic philosophy, public and collective action is denigrated . . . The present (Conservative) government is . . . wholly committed to this disastrous pursuit of self-interest

which exacerbates environmental and social crises that can only be solved only by collective action (Griffiths 1990).

Paradoxically, approaches to creating a green society that hinge on individualism could be regarded as fundamentally at odds with the holistic philosophy of radical ecology. That philosophy, Lucardie (1993) argues, emerges in at least three ways. First, ecologism says that what we are is determined by our relation to the whole cosmos. Deep ecologists, like Callicott (1989), Arne Naess and Holmes Rolston, influenced by interpreters of Eastern philosophy such as Snyder (1969, 1977) and Watts (1968), embrace this view – and they now use the 'energy monism' which they read into modern physics to give it scientific authority (Chapter 5.2). Second, ecologism argues for holism in the psychological sense. As atomised individuals we are unfulfilled and alienated: we only realise our full 'self' interest when relating to the rest of the human and animal community (Naess 1989), perhaps putting consideration for them above our own immediate gratification. Third, ecologism argues that all organisms are interconnected through belonging to ecosystems – a pragmatic, scientific reason for thinking holistically of the collective (Sagoff 1988). Given all this, Lucardie suggests, ecological individualism is merely 'light' green: merely reflecting the individualism rife in mainstream society.

Idealism and materialism

If radical social change is possible, how will it come: by first changing people's *material circumstances*, or their *ideas*, or both simultaneously? Where should the emphasis lie in the strategies of radical groups? Does what we *think* about nature condition what we *do* to it, as White (1967) influentially proposed, or does what we do to nature condition what we think about it, as has historically been the case (Thomas 1983, 23–5)? This debate – essentially about strategy – has dogged much of red–green politics.

An idealist might claim that the world can be changed by thinking about it. If people decide, for instance, that it is a good idea to start behaving cooperatively, non-aggressively and benignly towards nature, then they can so behave. Hence if you want to change society in these directions, then you need to change social attitudes and values. Western liberal education encourages us to believe in this model of social change, often presenting history as parades of, and conflicts between, ideas – usually as articulated by 'great men'. Goldsmith (1987), typically for greens, argues that action for change follows changes in consciousness as night follows day:

> I honestly believe that if people knew the truth about the pollution caused by nuclear power stations and the dangers of pesticide residues in their food they would not tolerate either the nuclear or the chemical industries.

And notwithstanding the generally reactionary nature of institutionalised education and media, some greens persist in arguing that educational change, by 'greening the curriculum', is the panacea for environmentally malign behaviour.

A materialist would argue, along opposite lines, that it is the material, especially economic, organisation of society which leads to specific social and economic relations between the people engaged in production, and between them and nature. Social institutions like the education or legal system, which condition most people's ideas, are in turn shaped by these relations. Eventually most people come, in this way, to accept particular socio-economic arrangements as being judicious and 'natural'. So if people compete with each other (for jobs, resources, markets) and exploit nature, because such behaviour is inherent in the economic system, then these competitive, exploitative relationships, often eulogised via schools and media, will incline most people to believe that competition or nature exploitation are good ideas, or common sense or 'natural', hence unavoidable. Only under different material circumstances will ideas radically incongruous with the current material basis of society, like cooperation instead of competition, become widely accepted, as distinct from being just 'counter-cultural' minority opinion.

A compromise between idealist and materialist positions might argue that new ideas and consciousness can change the world provided that people act on them. But the extent that people will act on them will be conditioned by how much they are compatible with what people already are doing. This represents the position of many radical greens today, whose original idealism has been tempered by years of campaigning experience. This has taught them that, constrained by the economic system, people often behave in ways that they know to be unwise. During the initial publicity about rainforest destruction in the 1980s, one Brazilian landowner interviewed on television said that he knew well that burning forests caused global warming, but he intended to continue doing it.

Of course, ideas do not just reflect people's material interests. They form independently in response to our aesthetic preferences as well as in response to our material circumstances, because people search for symmetry, coherence, harmony and order in their lives. And, historically, countless people have stuck to their ideas even though to do so damaged their material interests (Atkinson 1991). But at the very least there is a 'dialectical' interplay between actions and ideas such that you cannot change people's ideas fundamentally without also helping them to change what they do at the same time.

Consensus or conflict: pluralism, élitism or Marxism?

Many greens still reject conventional politics for individual lifestyle-ism, and this often goes hand in hand with rejecting conflict models of social change. These argue that conflict is an inevitable part of any revolution. Groups who want radical change must acknowledge that there will be conflict between those who have power and do not want to give it up, and those who seek power. Hence there may be conflict between 'ruling class' and 'employee class', or between men and women, or between different race/ethnic groups, or geographical core regions and peripheries, and so on.

One important conflict model is Marxism. This argues that though modern society may be structured into classes or groups in various ways, two classes particularly are significant in any future change from capitalism to socialism. Despite the complications of the rise of the middle classes and widespread share ownership, it may be still broadly true in advanced, global capitalism that there are two main classes: those who effectively own and control the means of production (including natural resources), distribution and exchange, and those who do not, having only their labour to sell.

Marxism's conflict perspective sees social change arising from an inherent, latent struggle between these two groups. And since this struggle has been the main concern of socialists and the labour movement, it would follow to Marxists that new energies for social change – coming through green concerns, for instance – should be directed through these traditional channels. Therefore, new radicals should demonstrate consciousness of how their new concerns relate to the class struggle: the 'old' politics of poverty and wealth, left and right.

Some regard this approach as simplistic and/or denying the idea that we live in a democratic, *pluralist* society. This society, they think, comprises a plurality of groups, all related in a system, and when one group is particularly alienated or disadvantaged the system will adjust: not through revolutionary conflict but through appeal to the law, or through government responding to pressure group protest, or firms responding to consumer pressure and so on – to lessen that group's grievances. Thus a new consensus is reached and the system remains stable, though changing and evolving. It will be the task of any new interest group, articulating new concerns, to enter and use this process by pressure group politics, applying rational argument and 'reasonableness' in its lobbying. This will challenge and change the old consensus, yet the broad features and structures of social policy and decision making will remain in place. Like any natural system, the social system in this model remains robust and stable by accommodation and adjustment, not by being forcibly impelled, via positive feedbacks, over new thresholds. Hence the judicial, parliamentary and bureaucratic system in a Western 'democracy' like Britain is ostensibly based on the idea that when two sides dispute something a satisfactory resolution will not necessarily involve 'natural justice' but an outcome where each side gets *something* of what it wants.

Pluralism, then, believes that 'democracies' are indeed democratic. In them all citizens have the right to seek, and the opportunity of seeking, access to the political process in pursuing their own preferences. Disputes (say, about environmental matters) are settled within a planning and parliamentary framework which has majority support. So, by implication, do the decisions which planners and politicians reach. Additionally, the concept of pluralism is

> frequently used to denote any situation in which no particular political, ideological, cultural or ethnic group is dominant. Such a situation normally involves competition between rival elites or interest groups, and the plural society in which it arises is often contrasted with a society dominated by a single elite where such competition is not free to develop.
>
> (Bullock and Stallybrass (1988, 656)

In the latter, élitist view, society is indeed composed of competing interest groups, but the process of accommodation to group interests is biased – towards particular groups who have 'unfair' (from a pluralist perspective) advantages. Hence, the resources of money, articulacy, education and time which rural preservationists possess may give them overwhelming advantage over the interests of those in lower socio-economic groups when it comes to decisions about where some environmentally damaging project may be placed (see, for example, case studies in Kimber and Richardson 1974).

A *Marxist* view takes this élitist analysis one stage further, accommodating it to a conflict model based on material economic interests. Hence the fact that a particular group is an élite, able to manipulate the system to its own advantage, is thought to be due to that group's *economic* power. For Marxist analysis considers that the division of labour under capitalism inhibits upward mobility between economic classes, maintaining the unequal society where the capital-owners are more enabled than others to realise their own interests. So structural constraints operate in favour of the ruling class and against oppressed pressure groups – the techniques of planning, for example, reflect and reinforce the social order and world view of capitalism.

More generally, the interests of the state and the ruling economic élite closely coincide: indeed the personnel of big business and government often interchange. Thus it is naive for environmental protest groups to appeal to supposedly neutral state authorities, ostensibly established to balance and reconcile conflicting interests. For these authorities – planners, bureaucrats or Members of Parliament – *cannot* act as environmental managers in a way which is free from the constraints of a social-economic structure that is designed to further the interests of capital.

Pluralism implies that capitalism is broadly democratic, élitism that it is not but might be, and Marxism that it is not and cannot be. There is much evidence against the pluralist perspective on environmental issues. Hamer (1987), for instance, shows how the road transport lobby in Britain controls parliamentary decision-making on transport (a leading figure in that lobby has actually been Treasurer of the Conservative Party). Blowers and Lowry (1987) demonstrate that the Anglo-American nuclear industry shapes scientific research and central government decisions on nuclear power and waste. Blowers (1984), in analysing the history of decision making about the siting and scale of Bedfordshire's brickworks, finds that elements of pluralist, élitist and Marxist models all apply to that case study at different times.

Structuralism

The above leads into a related issue: should we see society from a structuralist or non-structuralist perspective? That is, is what we see in the way of social events and individual and group behaviour to be interpreted in terms of deeper and less visible underlying structures in the human mind and/or in society? Are we to regard surface structures as conditioned by underlying ones? Or is what we see simply all that there is, as 'post-structuralists' would have it?

Our answer to this question will obviously determine whether we work on what we see, or on what we think underlies what we see. And if we do adopt a structuralist perspective, then we have to decide what the significant underlying structures are – are they cultural, economic, or consisting of the basic characteristics of the human mind, for instance?

Structuralism is preoccupied not simply with structures, but such structures as can be held to underlie and generate the phenomena that we see (Bullock and Stallybrass 1988, 821). This is a danger, for structuralism can lead us to reduce the world to a matter of what we perceive to be its deep underlying principles, denying any independent value or meaning which the surface level might have. The opposite danger can lie in postmodernist perspectives (see Harvey 1990). These may deny that observed phenomena reflect any deeper underlying principle at all (see the passage from Edward Abbey introducing the Preface of this book), implying that there are no principles, no grand explanations of how society works. By extension it is senseless to stand up for universal moral principles, because they do not exist.

In a broad sense, any theory is 'structuralist' if it posits that deep, unobservable, only subconsciously-apprehended realities give rise to observable realities. This describes Marxist theory, which tries to discover the causes behind social events, as well as social relationships hidden in apparent 'objects' such as commodities. These causes and relations are seen, in Marxism, largely in materialist terms. The *material*, economic structure is thought to influence strongly the roles which groups and institutions play. And different economic modes of production (feudalism, capitalism, socialism, for instance) each have different social goals that relate strongly to the economic (hence political) aspirations of their dominant classes.

Some structural Marxists (e.g. Althusser) make this model quite economistic, that is, reducible entirely to economic factors. It therefore conceives of a 'superstructure' of social beliefs, ideas, relations, institutions, practices and rituals which is quite rigidly *determined* by the underlying economic 'base'. For other Marxists (Peet 1991, 176) the relationship is more subtle and dialectical. We are not all drones, behaving strictly according to our economic class imperatives, and individual volition and consciousness is allowed in social theory. But structural Marxism never allows the substantial role of material, economic forces and structures to be forgotten, either.

Atkinson (1991), from an ecological perspective, is more interested in the structuralism pioneered by Levi-Strauss and inspired by Saussure's theories of linguistics. Levi-Strauss was concerned with relating behaviour and institutions to basic (i.e. universal) characteristics of the human mind, reflecting how it imposes structures onto reality:

the unconscious activity of the mind consists in imposing forms on content, and if these forms are fundamentally the same for all minds ancient and modern, primitive and civilised (as the study of the symbolic function, expressed in language, so strikingly indicates) – it is necessary and sufficient to grasp the unconscious structure underlying each institution and each custom, in order to obtain a principle of interpretation valid for other institutions and other customs.

(Levi-Strauss, *Structural Anthropology*, cited in Atkinson, p. 73)

Thus, Levi-Strauss argued that the apparent meaning and order of the natural world is not innate. It is humans who impose that order, through a mental capacity to classify. And, since all humans (he argues) have the same kind of brain, including those who live in 'advanced', 'civilised' societies and those who do not, then the mental imposition of structure is universal: everywhere there is a tendency, for instance, to create binary opposites and see one side (hot, clean) in terms of the other (cold, dirty). Thus the exact symbolic meanings in human social behaviour may be specific to given societies – but they are also reiterated through all other cultural subsystems.

All this illustrates a principal division between different structuralist approaches. On the one hand are versions that emphasise structures whose limiting logic resides in allegedly 'universal' (i.e. similar throughout history) characteristics of the human mind. On the other hand are those, like structural Marxists, who locate their structures within particular economic, political and social arrangements (all subsumed under the term 'modes of production') which clearly vary through space and human history, so that they cannot be described as 'universal'.

Green explanations of society have tended towards the former approach (though less so nowadays). Consequently they have been accused, especially by Marxists, of either embracing shallow, non-structural analyses, or structural analyses which rely on a universal human nature: ignoring how humans, individually and socially, constantly change their own nature (as well as external nature) through history. The next two chapters aim to reinforce a historical perspective by examining the history of some of the ideas and actions which have fed into perspectives on the society–nature relationship.

Chapter 3

PRE-MODERN AND MODERN IDEAS ABOUT NATURE AND SCIENCE

The roots of technocentrism

PRE-MODERN AND MODERN IDEAS ABOUT NATURE AND SCIENCE

The roots of technocentrism

3.1 MEDIEVAL AND RENAISSANCE NATURE: A LIVING BEING

Pre- and postmodern parallels

Technocentrism has constituted the *official*, dominant set of attitudes towards nature and environmental issues in modern Western society. It has not only coloured the outlook of the most powerful social groups, but has also underlain what seems to most of us to be 'common sense'. Basic to its view that environmental problems must be approached and managed scientifically, objectively and rationally is a conception of nature as machine-like and fundamentally separate from humans, and open to control and manipulation once it is understood. The roots of this perspective are surprisingly recent, and spatially restricted to the West. They lie in the *scientific revolution* of the sixteenth to eighteenth centuries, which was concurrent with the beginnings of industrial capitalism. This period, from Renaissance (fourteenth to sixteenth century) to eighteenth-century Enlightenment, laid the grounds for the 'modern' period, from the mid-eighteenth to twentieth century (such dates may be rather arbitrary, and not undisputed, so they should be regarded as indicative only). This latter was dominated by faith in linear reasoning and progress in science and technology to achieve material progress, and in the sorts of values associated with liberalism and the French Revolution. Modernism, said Whitehead critically (1926, 5) is a mentality which searches after 'irreducible and stubborn facts' and abstract principles; it has a 'distinctive faith that there is an order of nature which can be traced in every detailed occurrence'.

We are calling the science associated with the scientific revolution 'classical', to distinguish it from the newer scientific perspectives that have arisen from the work of twentieth-century physicists. According to ecocentrics the latter encompasses relativism, subjectivity and even non-rational knowledge (Chapter 5). It is therefore appropriate for the late twentieth century, which many see as a 'postmodern' period in which there is a reaction against many of the beliefs and presuppositions of modernism.

Indeed, Cosgrove (1990) speculates that modernist perspectives may be historically and culturally a sort of temporary aberrance. Our experiences over most of time have usually led us to see ourselves and nature as a unified whole. The modern period was but an interruption of a view which held sway in pre-modern times and is now regaining some currency.

Cosgrove points up the similarities between pre- and postmodern conceptions of nature. He says that Renaissance natural philosophy went back to people such as Nicholas of Padua, in the fifteenth century. Nicholas accepted the medieval cosmography and cosmology derived from Aristotle and Ptolemy (see p. 126), but together with other 'neoplatonists' (reviving Plato's philosophies) he also argued for a pattern of correspondences and sympathies that united the whole cosmos through the power of divine love. This very 'New Age' sentiment saw the cosmos as a living soul, imbued with spirit, where humans were a microcosm of the larger order ('all things are in all', was Agrippa's aphorism – compare the holograph view of the universe today discussed in Chapter 5.3).

The job of alchemy and natural magic was to know and manipulate, via images, the occult (hidden) forces which inspired the cosmos, so as to bring together and dissolve, synthesise and unify, mind and matter and all other opposites. Thus Renaissance science was monistic, rather than the dualisms of classical science and modernism (see p. 141). Its approach to knowing nature refused, as in postmodernism, to distinguish between 'signifier and signified'. This means that images and metaphors were not considered to be surface ways of representing a deeper underlying reality, they *were* that reality. As Mills (1982) has stressed, it was not a matter of *likening* nature to, for instance, a book. To the pre-modern mind this metaphor meant that nature *was* a book. Nature and the cosmos – the macrocosm – itself was made up of a system of signs, which needed to be read accurately in order to guide how humans – the microcosm – would live. It followed, too, that language, ritual, spectacle, image and metaphor became active ways by which people could transform nature.

Again, this resonates strongly with today's 'New Age' perspectives, which are (sometimes conscious) resurrections of an older, pre-modern view (see Button and Bloom 1992). This pre-modern view was suppressed during the modern period. But although astrology, magic and the occult went from respectable mainstream ways of understanding nature in the early 1600s to being 'cranky' by 1700, it would be wrong to see them as having totally vanished afterwards. They kept resurfacing in various forms, in, for instance, aspects of the Romantic movement, in the nineteenth-century vitalist and monist movements and in Steiner's bio-dynamic agriculture of the 1920s and 1930s (see Bramwell 1989).

But in stressing the continuity of such ideas there is, too, a danger that one might under-represent the depth of the changes in world view which also occurred between medieval and Renaissance perspectives on the one hand and that of classical science on the other. Hence the ease with which any change *back* might now occur, to a more unified conception of the society–nature relationship could also be underestimated. This is why we now need to consider those differences between pre-modern and modern perspectives in more detail.

Medieval and Renaissance cosmologies

The medieval (fifth to fifteenth century) view of the universe – what educated people thought about how it functioned and their position in it – was governed in its physical aspects by the ideas of Aristotle. These were integrated with evolving Judaeo-Christian ideas over a long period. The integration was a very close one, and was accomplished by the twelfth and thirteenth centuries. There was an almost perfect mapping of Aristotle's physical picture onto Christian theology – onto the Christian moral universe.

The cosmography was geocentric. The earth was at the centre of the universe, and, as all evidence suggested, it was solid, stationary, finite and spherical. The stars rotated around and were equidistant from Earth (see Figure 3.1). They were attached to the inner surface of a rotating sphere which looked like a dome from Earth, and marked the edge of the universe. Outside this sphere – beyond the universe's edge – was nothing, or *non ens* (non-being). This did not mean just empty space, it meant literally that nothing could exist. If one could travel to this boundary and stick one's hand beyond it, the hand would become non-existent. To put it another way, one could not ask questions about this region.

Within the sphere of the fixed stars the universe was divided into two zones, celestial and terrestrial, with the moon's orbit forming the boundary between them. The behaviour of the celestial objects was very predictable; they moved in circular orbits around the earth at constant speeds. But in the terrestrial region things moved randomly or in straight lines. Terrestrial things were born, died and decayed – they changed. This did not happen to celestial objects. Non-changingness suggested no need for change because of already-achieved perfection. Thus change meant imperfection. Circular motion suggested perfection – the perfect geometrical figure being a sphere – randomness suggested imperfection. So the celestial zone was one of perfection, the terrestrial zone was one of imperfection. This was an observational *and* a value difference. Observational (empirical) evidence and values were combined.

Between the moon and the sphere of stars were, in order, Mercury, Venus, the Sun, Mars, Jupiter and Saturn. Each of their orbits was part

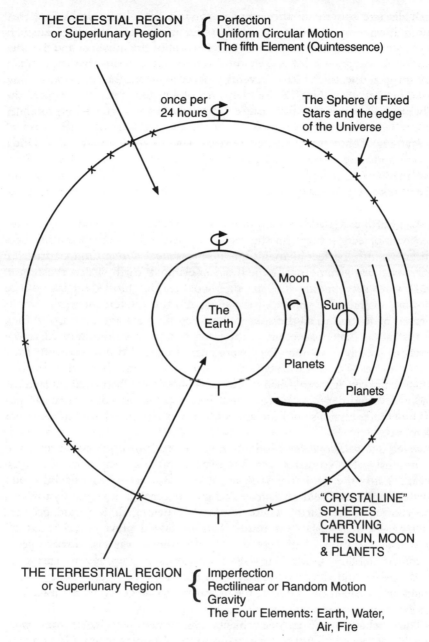

THE CELESTIAL REGION
or Superlunary Region
{ Perfection
Uniform Circular Motion
The fifth Element (Quintessence)

once per
24 hours ♀

The Sphere of Fixed
Stars and the edge
of the Universe

The
Earth

Moon

Sun

Planets

Planets

"CRYSTALLINE"
SPHERES
CARRYING
THE SUN, MOON
& PLANETS

THE TERRESTRIAL REGION
or Superlunary Region
{ Imperfection
Rectilinear or Random Motion
Gravity
The Four Elements: Earth, Water,
Air, Fire

Figure 3.1 The medieval cosmology

of a discrete sphere, so the arrangement was of spheres within spheres. When it came to explaining observed phenomena such as gravity, answers lay in a combination of this observed structure of the universe and the idea that there was *purpose and design* in it. This *teleological* view – that there was a distant goal towards which all events worked – meant that there was a *final cause* behind everything. Since the cosmology was a Christian one, the Christian God was the final cause. The universe was ruled by principles which helped to achieve God's purposes and he was the final cause of everything. These principles, or physical laws, were a function of God's design and were to be explained through understanding that design. Such explanations of the physical world, which reside in God, are known by historians as *physico-theological*. Thus the explanation of gravity was a physico-theological one, as follows.

The Earth was made up of four elements, earth, air, fire and water, while heavenly objects – being in the region of perfection – must consist of a different, fifth, perfect element: the 'quintessence'. When God created the universe it was symmetrical – the heaviest element (earth) was at the centre, with water surrounding it, then air, then fire. But God then introduced motion, spinning the sphere of the stars, and this movement was communicated by friction to all the spheres between the stars and moon, and then to the moon. The movement of the last was in turn communicated to the series of earthly elements. They were stirred up and dislodged from their *natural* places (i.e. those designed for them by God). Now the Earth – which had been smooth – had mountains where earth had been thrown up, and valleys and depressions whence it had been dislodged – water had got into these depressions, and fire and air had become mixed in with the other elements.

Being rational, however, God intended that the universe should return to its original state. No restoration was needed in the already perfect celestial region, but in the terrestrial zone all the objects were trying to fulfil God's desire. Thus if one held up a stone and then released it, it would fly towards the centre of the Earth, where it naturally belonged. It would go in a straight line – the shortest route. This explained gravity, and the well-known acceleration of objects as they descended was also explicable in terms of fulfilling God's purpose. Since the stone wanted to carry out God's desire, and since (as we all know) the nearer one gets to the desired object the stronger one's desire, the nearer the stone got to its natural place, the faster it went.

Thus behaviour of natural objects conformed with what they were mainly composed of and to their position in relation to where God wanted them to be. Subterranean air, water and fire travelled upwards with force (in volcanoes and springs and geysers) to get to their appointed positions, as did a candle flame point upwards. Water in the region which was appointed for air fell down. The explanations formed part of a remarkably logically consistent, coherent and complete system of theory. Several significant points may be noted about it.

First, the explanations which scientists gave for physical phenomena followed from, and were compatible with, theology. The 'paradigm' (Chapter 5.4) for science was set by religion. This was inherent in the academic and power structure of the medieval universities. The Faculty of Theology was the senior one, dominating the others in the university hierarchy. And though there was some latitude for varying interpretations, Aristotelian science had in general to conform to Christian theology. So to challenge Aristotle's science would be very difficult: a challenge to this meant also a challenge to the whole theology which encompassed it. The problems of throwing out the cosmology were therefore enormous. Second, despite the importance of God in the scheme, the medieval cosmology was *anthropocentric*. It was replete with human views imposed onto nature. Categories used to explain human experience were transferred to explain physical ones; hence the universe was said to be purposeful, in which objects desired to fit in with God's plan – purpose and desire being human attributes. Space had human values attributed to it. The celestial zone was most valued, but within the terrestrial zone the nearer one went to the Earth's centre the greater was the imperfection – with Hell at the centre. And on the Earth's surface there were areas more valued than others (sacred spaces, like Jerusalem, or the precincts of a cathedral). Scientific knowledge of these spaces, and of behaviour within them, could not be divorced from human values – the values of the society in which the science was done.

Third, and arising from this principle, the world could be seen very much in terms of analogy with human experience. Metaphors could be used to interpret it: it *was* a series of metaphors. Nature as a book was one of the principal medieval metaphors. It meant that since nature was a result of God's design and desires, one read two books, the Bible and nature, to discover God's purposes and act accordingly. So nature was not there just for people's sensuous enjoyment; it carried in it instructions and hidden meanings. The industry of the ant or the bee would be an example to us. The fly was a reminder of the shortness of life, and the glow-worm of the light of the Holy Spirit (Thomas 1983, 64, who reminds us that this metaphor was carried forward until well into the eighteenth century).

Thus 'There is nothing in visible and corporeal things that does not signify something incorporated and invisible' (the Irish ninth-century philosopher Erigena, cited in Mills 1982). In nature one read not only how to be saved but also, because God had created the Earth for humans, other knowledge of benefit to humans. So where a plant, stone or animal resembled in shape or colour or behaviour a human organ or disease, it could be used in healing. This was the doctrine of *signatures*, examples being that milky plants helped new mothers to produce milk, bony plants were good for the bones, summer plants helped to cure summer complaints (Mills 1982), spotted herbs cured spots, yellow ones healed jaundice and adder's tongue was good for snake bites (Thomas 1983).

If the book metaphor derived from human experience, so, too, did the *organic metaphor*. This was rife up to the eighteenth century, and it has reappeared today as a cornerstone of deep ecology. The medieval version saw the world as a huge animal with feelings, in which men and women, like intestinal parasites, lived:

> the Earth was perceived as a living body: the circulation of water through rivers and seas was comparable to the circulation of blood; the circulation of air through wind was the breath of the planet; volcanoes and geysers were seen as corresponding to the Earth's digestive system – eruptions were like belches and farts issuing out of a central stomach.
>
> (Gold 1984, 13)

Earth, which was female, also gave birth to stones and metals from her womb, having been fertilised by the (male) sky/heavens. Base metals then grew into silver and gold.

There was a spectrum of Renaissance organic philosophies. Common to them all were the notions that all parts of the living cosmos were connected into a unity, in mutual interdependence, and that they were all alive. This *vitalism* saw everything permeated by life – by the 'vital principle' – so that it was impossible to distinguish between living and non-living beings. The Earth was another living being among humans. Sometimes she nourished and nursed us, and for that she deserved respect and reverence. But like most females she also could be wild, passionate and frenzied. Men then argued that she needed to be tamed, says Merchant (1982), who draws parallels between the growth of attempts to control women through witch trials and the growth of attempts to control nature through classical science.

Merchant distinguishes three variants of the Renaissance organic view of nature, which also had implications for the view of human society (because nature was a macrocosm of the human microcosm). The first was that of a designed hierarchy – a Chain of Being – in nature (Figure 3.2), which also did and should exist in society. This clearly lent itself to conservatism, that is, maintenance of the existing social order. The second saw nature as an active unity of opposites in dialectical tension: a view which emphasised constant change and lent itself to revolutionary ideas (e.g. in Renaissance utopias or the utopian religious sects of the Civil War). The third considered nature as benevolent, peaceful and rustic; an 'arcadian' view (Chapter 4) prompting both romantic escapism and the notion of 'female' passivity – yielding itself up to manipulative ploughing, fertilisation and cultivation to provide a garden for the comfort and nurture of men exhausted by urban commerce.

The organic view stemmed from the pervasive cosmology which paralleled that of the physical world, whereby the medieval and Renaissance biological world was organised. That was the cosmology of the Great Chain of Being. In it the parallel was drawn between all elements in the universe, animate and inanimate, spiritual and material, and the links in a chain. They were joined together in a fixed hierarchy, and were interdependent. Everything in the Chain was alive, including stones, earth and 'slime'. For it followed that if everything had a sense of desire to fulfil God's purpose, then they had *living* attributes – desire and purpose are what living things show. They were given life by the overflowing of the Soul from the perfect God at the head of the hierarchy.

The idea of the Great Chain of Being originated with the Greeks, and was transmitted down to medieval writers, who adapted it to their cosmology. According to Lovejoy (1974), despite waning faith in metaphysics and increasing faith in Baconian empiricism (Chapter 3.2), the Chain of Being cosmology was widely diffused and accepted as late as the eighteenth century. John Locke, for example, restated it in his (1690) *Essay Concerning Human Understanding*, and late seventeenth-century scientists spent much time discussing which monarchical animals – lion, eagle, or whale – were at the top of the hierarchies into which they were assumed to be organised (Thomas 1983). A version by the Genevan naturalist Charles Bonnet (1720–93) illustrates that the concept continued as a way of conceptualising nature (Figure 3.2) and the supernatural:

> Above the Earth in Bonnet's chain were conceived higher worlds and even higher universes. In these higher realms also the thread of the chain continued, through angels, archangels, seraphim, cherubim, choirs of angels, virtues, principalities, dominations and powers. At the very summit of the hierarchy was 'the Eternal – that which *is*, possessing alone [seul] the plenitude of being'. The parallel with the ancient Neo-Platonic hierarchy of Plotinus is quite remarkable.
>
> (Oldroyd 1980, 11)

As Lovejoy describes it, for most men of science 'the theorems implicit in the conception of the Chain of Being continued to constitute essential presuppositions in the framing of scientific hypotheses' into the eighteenth century. And Thomas reports that as late as 1834 a zoologist named William Swainson thought that zoology's aim was to discover animals' stations in the scale of creation.

This Chain of Being metaphor and its related ideas tended to place people and nature into an intimate relationship, as may be true of medieval views in general, since each link of the Chain was vital for the whole chain's existence. Interconnections were important:

131

The eternal – that which *is*, possessing alone (seul) the plenitude of its being.

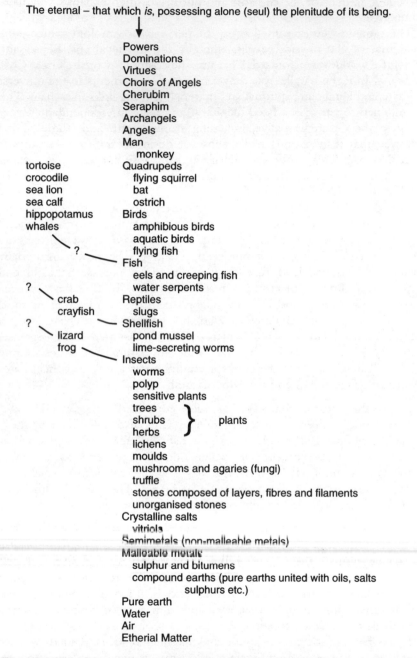

Powers
Dominations
Virtues
Choirs of Angels
Cherubim
Seraphim
Archangels
Angels
Man
  monkey
tortoise   Quadrupeds
crocodile   flying squirrel
sea lion   bat
sea calf   ostrich
hippopotamus Birds
whales   amphibious birds
            aquatic birds
   ?   flying fish
        Fish
         eels and creeping fish
?   water serpents
   crab   Reptiles
   crayfish slugs
?   Shellfish
   lizard   pond mussel
   frog   lime-secreting worms
        Insects
        worms
        polyp
        sensitive plants
        trees
        shrubs   }   plants
        herbs
        lichens
        moulds
        mushrooms and agaries (fungi)
        truffle
        stones composed of layers, fibres and filaments
        unorganised stones
Crystalline salts
  vitriols
Semimetals (non-malleable metals)
Malleable metals
  sulphur and bitumens
  compound earths (pure earths united with oils, salts
           sulphurs etc.)
Pure earth
Water
Air
Etherial Matter

Figure 3.2 The Great Chain of Being: according to Charles Bonnett, naturalist,
1720–93
Source: Oldroyd 1980

Since, from the Supreme God Mind arises, and from Mind, soul, and since this in turn creates all subsequent things and fills them all with life, and since this single radiance illumines all and is reflected in each, as a single face might be reflected in many mirrors placed in a series; and since all things follow in continuous succession, degenerating in a sequence to the very bottom of the series, the attentive observer will discover a connection of parts, from the Supreme Good down to the last dregs of things, mutually linked together and without a break. And this is Homer's golden chain, which God, he says, bade hang down from heaven to earth.

(Macrobius, cited by Lovejoy, p. 63)

Thus to eliminate one link – one creature or one part of inanimate matter – would fully dissolve the 'cosmical order', and, ceasing to be full, the world would be incoherent.

These ideas, of continuity and gradation, are fused with that of 'plenitude', or fullness. This held that the universe is filled by diverse living things, such that all species that *could* theoretically exist do in fact exist. Not only is there diversity, but there is also an abundance of creatures, stemming from their theoretically infinite fecundity (which was curbed only by competition and nature's limits). Plenitude, too, held that the world was better the more it contained.

The link between plenitude and the Great Chain of Being was that at the top of the hierarchy was a transcendent *hypostasis* (personality or underlying substance) – a First Principle: the *One* or *Good* in the Greek version; which became God in the Christianised version. Emanating from this was a Universal Soul ruling the universe at higher (spiritual) and lower (material) levels. Being perfect, then, the transcendent First Principle had not remained in itself, but spilled over, generating in plenty the things below it in the chain – the soul, the archangels and angels, humans, animals, plants, metals, stones, mud and slime. This was why all things were linked and interdependent as an organic whole.

The Chain of Being gave people several good reasons to feel humble in relation to the rest of nature. First, removing *any* link from it would destroy the chain, hence all links were equally important for the Chain's completeness. Second, humankind, in medieval and Renaissance versions, was situated only in the middle of the chain, at the transition between being purely animal-like creatures of instinct and physical senses, and thinking beings with a soul who transcended the physical. Superior to humans were the intelligences of the angels. Third, humankind's distinctiveness from the creatures below was hardly chasm-like because of the continuity principle, which said that there was an almost imperceptible transition between each element of the chain.

Lovejoy (p. 200) concludes that despite other strains in the eighteenth century which worked against humility,

> and prepared the way for those disastrous illusions of man about himself which were to be so characteristic of the century that followed . . . the immense influence of the complex of ideas which was summed up in the cosmological conception of the Great Chain of Being tended . . . to make man not unbecomingly sensible of his littleness in the scheme of things.

Related to the element of humility towards and respect for nature, which is common to both Chain of Being and contemporary ecocentrism's bioethic, is that of *animism:* attributing souls to animals, plants and inanimate objects. Mills (1982) says that the organic view permeating medieval and Renaissance cosmologies is essentially animistic. If the cosmos is an organism stemming from the immanence of the One – the Absolute Being which is in everything – then it is but a small step to have nature and natural objects endowed with the attributes of organisms, especially humans. This was done not merely by giving, for example, 'brows', 'shoulders' and 'feet' to mountains, or 'heads', 'gorges' and 'mouths' to rivers, or a circulatory system to the whole earth, it also went on to give each part of nature some soul – some of the universal spirit which had flowed down into them (see also *pantheism*, Chapter 1.2).

These ideas were carried forward into seventeenth-century practices of worshipping plants, talking to them and offering them libations, according to Thomas. They also look back to the pre-medieval and medieval 'pagan animism' described by White (1967), where every tree, spring, hill and stream had its own guardian spirit which was accessible to but unlike them – centaurs, fauns and mermaids for example. 'Before one cut a tree, mined a mountain, or dammed a brook,' says White, 'it was important to placate the spirit in charge of that particular situation'. Animism and pantheism resurfaced in romanticism, as they resurface today. Many people talk to their plants; New Agers negotiate with the specific 'deva' of each type of plant they cultivate (see Thompkins and Bird 1972), while other ecocentrics often point to the animism of 'traditional' cultures such as the Navajo Indians as examples for Western cultures to emulate. The theme is strong in Callenbach's *Ecotopia* (1978), where trees are given a heart-to-heart talk before felling to explain and apologise for the regrettable circumstances in which the atrocity is to be perpetrated.

The ideas of the cosmos as a purposeful, divine organism, and of the Earth as a living being were seen strongly in that system of natural knowledge known as *natural magic*. This exerted a powerful hold on European thought, in the form of alchemy and astrology, during the Renaissance and the beginning of the scientific revolution, although it was increasingly opposed by Aristotelian/Christian cosmology and by that of classical science.

The basic tenet of natural magic was, again, the universe as an organism, fully alive and active. It was permeated with influences, forces and correspondences that linked everything in nature, people included, to everything else, forming a multidimensional network that was not only material but also mystical and spiritual. Thus, everything that was found in the 'small universe' that was a human being (the microcosm) was linked with, and corresponded to, the various parts of the larger universe of nature (the macrocosm). For example, the human heart corresponded to the sun; the sun was at the heart of the universe – in a sense the heart and the sun were the same. The microcosm and macrocosm were images and symbols of each other: two levels of the same reality (compare Bohm's ideas of 'implicate order' in Chapter 5.3). Hence one could study and understand humanity by studying and understanding the universe. Furthermore, violating nature (say, through mining) constituted the violation of a human, female, body.

Natural magic opposed any suggestion of distinguishing between humans and nature. On the contrary, natural magicians had to recognise that they were inextricably part of the nature they were studying (again, compare the perspective of twentieth-century physics). Their knowledge was therefore not (and they would not have wished to claim that it was) objective and impersonal: it was subjective and personal. Furthermore, reason had to be inadequate to the task of understanding a universe permeated with occult forces and symbolic relationships; only the non-logical faculties of intuition and empathy were capable of doing so. These faculties were brought into play when natural magicians immersed themselves in nature and opened themselves to the interplay of her forces and influences. They did this by manipulating these forces, that is, by experimentation. Natural magic, alchemy in particular, was profoundly experimental, but in all other aspects it was deeply opposed to the classical science emerging alongside it in the seventeenth century.

3.2 THE SCIENTIFIC REVOLUTION AND NATURE AS A MACHINE

The period of the 'scientific revolution' covered about 150 years, from the time of Copernicus, who published *On the Revolutions of the Heavenly Orbs* in 1543, to the end of the seventeenth century, with Isaac Newton's *Mathematical Principles of Natural Philosophy* (1687) and his *Opticks* (1703). During this time, the beginning of the 'modern' period, the principles of 'classical' science were established in contradistinction to the pre-modern cosmologies sketched above. These principles are sometimes known as the 'Newtonian paradigm', that is, the model derived from Newton's work, of what questions it was legitimate for science to ask of nature, how they should be asked, and what constituted valid answers.

Despite Einstein, quantum theory and all the other advances associated with twentieth-century physics, the Newtonian paradigm still constitutes conventional wisdom about science, and it is this conception which is therefore the basis of technocentrism. Indeed, the idea of society in technocentric, liberal political philosophy is that it is composed of discrete individuals interacting with other discrete individuals. The analogy with Newtonian physics' view of the physical world is clear, and not fortuitous: both pictures grew up together (Zohar and Marshall 1993).

In some ways the scientific revolution drew on preceding thought. Whitehead (1926, 15) considered that one great contribution which medi-evalism made to the scientific movement was 'the inexpugnable belief that every detailed occurrence can be correlated with its antecedents in a perfectly definite manner, exemplifying general principles'. Coupled with this belief in cause and effect – right back to the 'final cause' of all things, which was God's will – was the medieval insistence on God's rationality, (compared with the Asian view of God as arbitrary and inscrutable). All this made the growing, naive, faith in the possibilities of rational science an unconscious derivative from faith in medieval theology, says Whitehead.

Despite this, the scientific revolution did fundamentally challenge the medieval cosmology, based as the latter was on divine purpose. Copernicus suggested a simple revision to the cosmography, swapping the positions of the earth and sun. But the implications of this were so enormous that it needed 150 years to construct the new cosmology which was required. For the new ideas were not only an intellectual challenge to the established science. They also ate away at the theology from which it stemmed – which in turn supported a particular social structure.

Indeed, were it not for social and economic challenges to that structure which were simultaneously occurring, the intellectual ideas represented by the Newtonian paradigm might not have triumphed in the eighteenth and nineteenth centuries as they did. We should also bear in mind that although through 300 years of socialisation the ideas of the Newtonian paradigm constitute 'common sense' to us, they are not that 'sensible'. They do not accord with what the most immediate evidence of our senses tells us about the world. When we look around us it does not *appear* that we or the clouds or the other objects we see are speeding along at about 1000 mph. And it is equally obvious that a 'solid' object like a stone *is* solid, and not composed partly of empty space. Classical Newtonian science tells us that we are wrong, and that what we think we see is unreal.

Copernicus' heliocentric solar system opened up, then, an enormous can of worms, leaving science to explain difficult problems that had previously been easy to account for. If, for instance, earth moved through the universe, space could not easily be divided up into perfect and imperfect regions. Removing this distinction also meant revising theories of matter, which had spatially separated the four elements from a quintessence. The new theories, it seemed, would have to be based on *spatially universal* principles.

Also, night and day could not be explained in terms of the movement of the stars and sun around the earth, therefore the latter had to be conceived of as rotating. But this then raised questions of why we do not encounter a 1000 mph wind as we rush around on our axis, and why objects which we throw vertically upwards also land vertically instead of some distance away. Neither could this vertical fall be explained, as before, as a function of a body's position in space, and its desire to reach its 'natural' position.

Johannnes Kepler (1571–1630) sought universal principles in trying to explain planetary motion. He argued that the sun was the cause; both it and the planets functioned as magnets, and in rotating the sun pushed the planets around with it. So the sun was the driving force to a kind of machine:

> I am much occupied with the investigation of the physical causes. My aim in this is to show that the celestial machine is to be likened not to a divine organism but rather to a clockwork . . . insofar as nearly all the manifold movements are carried out by means of a single, quite simple magnetic force, as in the case of a clockwork all motions [are caused] by a simple weight. Moreover I show how this physical conception is to be presented through calculation and geometry.
>
> (Kepler, letter to Herwart von Hohenburg, 1605, cited by Holton (1956))

Thus Kepler was using a new metaphor for nature: the organism was replaced by a clock. This *mechanistic* conception of nature is a principal component of the classical scientific (technocentric) view. Kepler was religious, and he believed he could understand nature and, through it, God's intention, via mathematics and geometry:

> Why waste words? Geometry existed before the Creation, is co-eternal with the mind of God, *is God himself* (what exists in God that is not God himself?); geometry provided God with a model for the creation and was implanted into man, together with God's own likeness – and not merely conveyed to his mind through the eyes.
>
> (Kepler, *Harmonici Mundi*, 1619, cited by Koestler (1964))

This was, therefore, a conception of God as a Creator and nothing more. He was not ever-present and did not intervene in the workings of his creation. Rather, he was an engineer, using geometry to make a plan from which he had constructed a machine. He set the machine going and then left it.

This was an essentially *deterministic* view of nature. For the way a machine works is that its structure and past configuration determine its present behaviour. A clock is a series of cause-effect (deterministic) mechanisms (cogs and springs) linked together. And, most important, if you know enough about how the clock works, you can exactly predict its future behaviour. In this cosmology, the past now determines present and future, rather than, as in the medieval view of a final cause, a future goal (fulfilling God's will) determining present behaviour.

Kepler's view was developed by Galileo Galilei (1564–1642). He believed that the 'book of nature' was written in the language of mathematics and therefore had to be read and understood via mathematics. The physical problems of motion were to be treated as geometrical problems, and objects like falling stones as geometrical entities. This followed from the fact that God had structured the universe according to mathematical principles. So the way to understand it was through deduction from these principles in conjunction with observation and experimentation. The logical outcome of this mechanisation and mathematisation of nature was that when we look at nature we do not see what it really is (a machine). The true reality of nature is mathematical and regular. From this followed what has become a basic tenet of classical scientific philosophy: *what is truly real is mathematical and measurable, but what cannot be measured cannot have true existence.* In essence it was Galileo who first distinguished between what was measurable and therefore 'objectively' ascertainable – being the same for everyone – and what was not measurable and therefore varied from person (subject) to person (subject) – that is, was 'subjective':

> Now I say that whenever I conceive any material or corporeal sub-stance, I immediately feel the need to think of it as bounded, and as having this or that shape; as being large or small in relation to other things, and in some specific place any given time; as being in motion or at rest; as touching or not touching some other body; and as being one in number or few, or many. From these conditions I cannot separate such a substance by any stretch of my imagination. But that it must be white or red, bitter or sweet, noisy or silent, and of sweet or foul odour, my mind does not feel compelled to bring in as necessary accompani-ments. Without the senses as our guides, reason or imagination unaided would probably never arrive at qualities like these. Hence I think that tastes, odours, colours, and so on are no more than mere names so far as the object in which we place them is concerned, and that they reside only in the consciousness. Hence if the living creatures were removed, all these qualities would be wiped away and annihilated.
>
> (Galileo, *The Assayer*, 1632)

Galileo here argued that whatever else he was unsure of he could not doubt that objects had shape, size, motion and quantity: he could not, in other words, imagine objects without these (measurable) properties. But he could conceive of bodies with or without one or other sort of smell or taste or colour. So these sorts of properties were variable and existed only in human consciousness. They were not totally necessary or inherent qualities of the objects, as witness the fact that what they were (and whether they existed at all) was a matter for different individuals to judge differently. Thus if humans were removed from the scene, objects would still have shape, size position, motion and quantity – these were what really existed 'out there', and they were *primary qualities*. But the rest of the qualities would not continue to exist without the existence of humans. They were not really 'out there' and were therefore *secondary qualities*, 'in here'. Hence, to understand what really existed one would have to try to cut out the influence of the secondary qualities. One had to cover up one's personality and the influence of the (subjectively ascertainable) secondary qualities and restrict one's attention to the (objectively calculable) primary qualities.

As Whitehead ironically described it, science presupposed the

> ultimate fact of an irreducible brute matter, or material, spread throughout space . . . In itself, such a material is senseless, valueless, purposeless. It just does what it does do, following a fixed routine imposed by external relations which do not spring from the nature of its being . . . but . . . the bodies are perceived as with qualities which in reality do not belong to them, qualities which in fact are purely the offspring of the mind. Thus nature gets credit which should in truth be reserved for ourselves: the rose for its scent; the nightingale for his song; and the sun for his radiance. The poets are entirely mistaken. They should address their lyrics to themselves and should turn them into odes of self congratulation on the excellency of the human mind. Nature is a dull affair; soundless, scentless, colourless; merely the hurrying of material, endlessly, meaninglessly.
>
> (1926, 26, 69)

Like the deep ecologists who followed him, Whitehead here traces and denounces in classical science what he considers to be the foundations of an anthropocentric view of nature as lacking intrinsic value and worth (Chapter 2.1), and being essentially dead. As Merchant (1982) put it, this scientific view spelled 'the death of nature: no longer did *all* its parts have spirits or the vital life principle in them'.

Classical science, then, tells us that primary qualities can be measured; secondary qualities cannot. Primary qualities have objective existence; secondary qualities are subjective. Primary qualities are not challengeable; secondary qualities, by contrast, are not 'true' or 'real'; they are 'just' a matter of subjective impressions, feelings and opinions. Therefore the truth of scientific theory has less to do with how soundly or cleverly that theory is argued than with how *objective* is the knowledge from which it stems: *objective knowledge is 'true', and correct, while subjective knowledge is not*. It is in this distinction that classical science has its powerful appeal as being in some way independent. For it can give 'objective' knowledge of nature. Being objective rather than subjective, this knowledge must be free from the sectional interests of any particular person or group – it can therefore be *trusted* (Chapter 5).

With this distinction between what is objectively 'out there' and what is subjectively in human perception arises also a fundamental distinction between society and nature. An enormous gulf, in fact, opens up between the two. The gulf was most starkly revealed by Descartes, the most radical of the thinkers of the scientific revolution and one of the creators of modern philosophy.

Developing on Kepler and Galileo, he argued that matter was nothing more than extension in space. It was geometry. It did not consist of sensible qualities like hardness, weight or colour, but was extension in breadth, length and depth. What was mathematical was real and what was real was mathematical. The implications of this were that the universe must be infinite (because the space of a geometer was infinite) and that since matter was extension in space the universe must be full: there could not be empty space.

The only way to have a full universe which also contained motion was to have matter divided into particles. For Descartes, matter was infinitely divisible, and the universe contained nothing except particles with size, shape, weight, motion and position. Explanations had to be in terms of matter – particles – in motion. No occult action, in terms of sympathy or desire, was possible.

Descartes extended the concepts of nature as a machine and the unreality of that which was not measurable. He viewed animals, and the human body too, as machines. They were automata, and their workings could be fully known by reducing them to matters of physics and chemistry, which in turn would be understood in terms of mathematics. This is a *reductionist* view, whereby, through analysis (breaking down of things into component parts), everything can be eventually reduced to the same basic quantities and qualities, all of which are measurable. It holds also that there is no vital principle necessary to life: life differs from non-life only in its degree of complexity.

Reductionism and universalism abound in the classical scientific view, for example in the Royal Society standards for 1976. These chose seven physical qualities as 'dimensionally independent base qualities' (length, mass, time, etc.). 'All other physical qualities are regarded as derived from these basic qualities' – cited by Hales (1982, 124–5) who goes on to say 'These plus the ten base digits 0–9 . . . are the sum total of the legitimate language of observation in hard physical science, and the stuff of all explanation.'

Modern ecocentrics criticise classical science for its mechanistic and reductionist tendencies. These tendencies made it possible for scientists to argue that consciousness could be explained mechanically and that a person's entire psychic life 'was the product of his physical organisation. What Descartes said of animals would one day be said of man' (Thomas 1983, 33).

Reductionism also poses the question: 'If everything consists of the same basic stuff, with the same basic form (atoms), in what way, if any, are humans distinguishable from the rest of nature?'. Descartes solved this problem in the course of his methodological procedure of systematic doubt:

> But then, immediately as I strove to think of everything as false, I realised that, in the very act of thinking everything false, I was aware of myself as something real; and observing that the truth: *I think therefore I am*, was so firm and so assured that the most extravagant arguments of the sceptics were incapable of shaking it, I concluded that I might have no scruple in taking it as that first principle of philosophy for which I was looking.
>
> My next step was to examine attentively what I was, and here I saw that, although I could pretend that I had no body, and that there was neither world nor place in which I existed, I could by no means pretend that I myself was non-existent; on the contrary, from the mere fact that I could think of doubting the truth of other things, it followed quite clearly and evidently that I existed; whereas I should have had no reason to believe in my existence, had I but ceased to think for a moment . . . I concluded that I was a substance whose whole essence or nature consists in thinking, and whose existence depends neither on its location in space nor on any material thing. Thus the self, or rather the soul, by which I am what I am, is entirely distinct from the body, is indeed easier to know than the body, and would not cease to be what it is, even if there were no body.
>
> (Descartes, *Discourse on Method*, 1637)

So Descartes reasoned that the very *act* of doubting was a process of thinking, that the one thing that could not be doubted was that *he* was thinking and that this thinking was the act which established the fact of human existence. Humans were therefore definable as no more nor less than thinking beings. What separated them from the rest of nature, and from their own bodies, was this thought process. Whereas the body could be analysed into its component parts, the mind could not. Descartes thereby had introduced a most fundamental dualism in modern thought: that between mind and matter. Whereas matter was composed of primary, objectively knowable, qualities, the mind was subjective and it attributed secondary qualities to nature.

Thus this *Cartesian dualism* involved mind and matter, subject and object, and it had a profound implication for the society–nature relationship because nature became composed of *objects metaphysically separated from*

141

humans. These objects had primary qualities and no others. They were reducible to atoms, whose unthinking, machine-like behaviour was universally the same and explicable in terms of mathematical laws. Humans, by contrast, were defined as rational thinking beings – subjects who observed objects, including nature, and could impart secondary qualities to them.

Ecocentrics (Capra 1982, Skolimowski 1992) now consider that this dualism paved the way for a society–nature separation in which the former was conceived of as *superior* to the latter. 'Cartesian' thinking is regarded as a prime culprit in creating a scientific world view where humans are separate from and above nature, a mere machine. Descartes is even cited as a villain of the piece by animal rights protagonists (e.g. Singer 1983, 218–20). Barbara Adam (1993, 409–10), however, points out that the Western conception of the environment as external and 'other' has a number of very deep-rooted sources in pre-scientific praxis.

> Central among these are the linear-perspective vision and the creation of a clock-time. Both are powerful externalisers. Both separate subject from object. Both are devices that distance us from experience. Both facilitate mathematical description, quantification, and standardisation.

Linear perspective was developed from the fifteenth century to represent three-dimensional space on to a two-dimensional plane. But it was more than just an artistic technique; it was a way of conceiving the world. Before it the observer was always part of the observed, but 'With the linear perspective the body was moved from the centre to the outside . . . This entailed a transformation of participants into spectators'. Similarly, the development of clocks created a linear time, independent and disembedded from other conceptions of time (for instance as rhythms and cycles of nature), and capable of being measured objectively and used as a social tool. Time became distance travelled through history and space (compatible with the later notion of history as linear progress). Prescientific as these concepts were, however, Adam stresses that

> The linear-perspective and clock-time vision find their coherent and full expression in the scientific world view. As conceptual tools . . . they encompass a number of principles that can be observed in scientific theories and designs: emphasis on abstraction, separation and otherness, elimination of surrounding context, and the allied pursuit of permanence and timeless truths.

But today, thinks Adam, 'Engagement with environmental matters requires a different tool kit'.

Descartes' method involved accepting as true only what could be clearly seen as beyond doubt; dividing every problem into as many parts as was necessary to solve it; beginning with the simple and then increasing the levels of complexity and making a rigorous review, omitting nothing

(Merchant 1982). This scientific method was, unlike the 'science' (i.e. 'knowledge' in strict definition) which preceded it, analytical, experimental and reductionist, seeking understanding by taking the machine of nature to pieces to see how it worked. Mathematics was to be its language.

It set out to explain the phenomena of nature in terms of nothing more than matter, composed of indivisible particles, called atoms, in motion in infinite geometrical space under the influence of forces that were measurable. Unlike medieval Aristotelian science with its universe divided into two distinct regions, these explanatory principles were to apply throughout the universe. Thus, when in the story Newton was hit by the falling apple, he did not conclude that he had learnt merely about the behaviour of apples in orchards, but that he had also understood how the moon moved, because his science was universal. The same principles, explanations and causes were seen to operate on earth, on the moon, everywhere: Newton formulated not a law of gravity, but *the* law of universal gravitation.

The scientific revolution might be seen as a three-cornered conflict between classical Newtonian science, medieval Aristotelian science, and natural magic. Though the first one 'won' this does not mean that the other two approaches to knowing nature – especially natural magic – ceased altogether to exist, particularly in popular thought. For the foundations of modern technocentric attitudes about the *purposes* of that science and its methods, and the social role of the scientists, we need to turn next to Francis Bacon (1561–1626).

The Baconian creed and its high priests

Bacon was the first figure in the scientific revolution to draw out the full implications for the society–nature relationship of the emerging principles of the new science. As we have noted, pre-modern cosmology asserted the unity of society and nature; Cartesian dualism effectively separated the two. But it was Bacon who asserted the creed that *scientific knowledge equals power over nature.*

Bacon grappled with the same issue as Descartes: the separateness of humans from their object of study, nature. Human subjectivity stood in the way of this separation. Men or women, the subjects, were forced to interpret nature through human feelings and experiences: they had selfish interests, assumptions and presuppositions. All these militated against being 'objective' – seeing nature without prejudice, as a separate object. Because of this, and also because written and spoken words were imprecise, the unfettered exercise of the human mind in reasoning constituted an inadequate scientific method. The flights of the mind had to be controlled by appealing only to evidence – the evidence of the senses.

Descartes started with reason, by formulating hypotheses based on assumptions about the laws governing relationships between objects, and he deduced from them the expected behaviour of the objects. But then there needed to be observation of evidence. If observed matched expected behaviour, the prior assumptions were deemed correct. His method was thus *deductive*.

Bacon's method was the reverse; it was *inductive*. It held that the scientist should first make many observations of nature, and from them it would then be possible to draw out, or induce, laws governing their relationships. Having been formulated via observation, the initial hypotheses could then be tested and verified collecting more data, and once verified these hypotheses would gain the status of laws of nature. Further observation and experiment would lead to the formulation of laws that were ever wider and more general in scope, with the distant goal of a single law that would embrace all phenomena in the universe.

Bacon's scientific method can be visualised as pyramidal – founded on a broad base of *empirical* (based on observation and experiment) knowledge, and building on to it laws of nature of increasing generality; the whole pyramid crowned with the universal law that will explain everything. In practice, classical science has followed Newton, by mixing and combining both deductive and inductive approaches. But the importance of Bacon's induction lay in its implications that 'truth is the daughter of time' and that science is progressive.

For collecting empirical knowledge took a long time, hence inductive science was based on the steady accumulation of data from experimentation and from direct ('field') observations of nature. New knowledge was built upon old knowledge. Thus total scientific knowledge increased with time; it could not be encompassed by any one individual or age. It involved communal activity, where the community of scholars worked towards a future goal – establishing truth (although the ultimate truth could never be known, for this was the exclusive province of God).

Such activity could not be accomplished within the existing scientific institutions, since the medieval tradition was too strong in them whereby argument rather than objectively assessed facts constituted the hallmark of valid knowledge. Hence Bacon advocated a new start for science, in separate research institutions that would be state supported.

Bacon further argued that science was central to human endeavour, because of its goals and the motives associated with doing science. First and foremost, the true understanding of nature that only inductive method could achieve was the means of glorifying God, who had created nature.

But the case for state support required more justification, and Bacon embodied this in his definition of the purpose of scientific activity. For its second aim was to relieve 'man's estate'. It was a *philanthropic* activity where the scientist should assume the moral duty of improving society's material lot. This could come about by understanding how the machine of nature

144

worked. This was the first stage in using the laws of nature for society's benefit. So science's purpose was to 'command nature in action':

> The end of our foundation is the knowledge of causes and secret motions of things and the enlarging of the bounds of the human empire to the effecting of all things possible.
>
> (Bacon, *New Atlantis*)

And there was no compunction about doing this:

> For the whole world works together in the service of man; and there is nothing from which he does not derive use and fruit . . . insomuch that all things seem to be going about man's business and not their own.
>
> (Bacon, *De Sapientiae Veterum*)

Science was therefore progressive in two senses. First it built upon the secure basis of facts, advancing from them towards greater and greater truth. Second, this steady march to truth would be the way to obtain progress in humanity's material circumstances. So science was *equatable with human progress*, a powerful argument for supporting a community of scientific scholars – of professional scientists.

With the professionalisation of science (which did not happen much before the nineteenth century) its role became increasingly identified with this utilitarian humanist objective, and less tied to the medieval one of knowledge of God. And the scientist/expert has become someone usually looked up to. 'Democratic', 'compassionate', 'humble', 'radical', 'socially aware', 'philanthropic', 'honest', 'unselfish', 'serene', 'noble', 'dedicated', 'priestly', 'good', 'cosmopolitan', 'apolitical': all these epithets are suggested in Bacon's descriptions of scientists (Prior 1954).

In his essay *New Atlantis* (1624) he portrayed a mythical island where a community of scientists was given high social status. Their leader bore the stamp of dignity – he (and it was a male group) was peaceful and serene, pitying and democratic. The community was dedicated to gaining knowledge not for profit, fame or power, but for the benefit of all. The members accepted no authority except that of the scientific method. They bore no allegiance to any particular social group, even keeping their knowledge secret if they felt it undesirable for the state to have it. So objective knowledge was produced and protected by a group dedicated to humanity as a whole. Therefore anything leading to the development of science was, by implication, good. And also by implication it was scientists who should have the leading decision-making roles.

This view of the professional scientist – objective, undogmatic, internationalist and committed to improving society's lot – can easily be equated with that of a humanist 'priest', replacing the established religious priesthood. He was working, it seemed, for a universal good, for the interests of science are universal. This was science's self-justification, and it was a powerful one, helping classical science to become the dominant ideology over the past 250 years.

Its success has been phenomenal. Its perspective has come to be equated with 'natural', 'normal' vision, and it has become the pursuit of most European intellectuals, in place of natural magic. Science has also become the arbiter of most environmental and many social issues: it is appealed to as a source of objective truth on which to base decisions.

A typical modern technocentric view of the reasons for this is that

it works – one only has to look at the things science has made possible . . . satellite communication, organ transplants, bombs that can destroy entire cities. These things suggest that, for good or ill, scientists get their sums right. Science has much in common with common sense.

(*The Economist*, 1981)

As has been shown, this last sentence is quite inaccurate, and we should be sceptical about the rest of the statement as well. For if Hiroshima resulted from scientists getting their sums right, then they were doing the wrong sums. This suggestion, that science can somehow be viewed apart from its social context, as merely a function of its ability to achieve technically workable and sophisticated designs, is quite unacceptable for twenty-first century thinking. Science never has, in fact, been independent of its social context (Chapters 3.4, 5).

But if social factors always influenced scientific development, the reverse was also the case. In fact science figured centrally in the emergence of a new secular view of society in the eighteenth century. Fundamental to that view was this concept of *progress*.

Eighteenth-century writers from Voltaire (1694–1778) to Condorcet (1743–94) argued that since the end of the Middle Ages society had been changing at an ever-faster rate and that this change was for the better. They held that this had not been true for the Middle Ages because the superstitions of the Bible and the false science of Aristotle had kept humankind in darkness and ignorance, oppressed and enslaved. But now the new science of Copernicus, Galileo, Descartes, and above all Newton, had banished these errors and superstitions, discovered the truth and brought humankind out of the Dark Ages. The success of Newton's science had shown that we must no longer seek knowledge in the old authorities – Aristotle and the scriptures – but must seek it in nature herself, by using reason. People must not accept just what they were told, but must free themselves by the independent use of their reason from the dead hand of the past.

Newton had set the paradigm for modern thought, in showing that the enormous complexity of the universe was intelligible and that its true reality lay not in the chaos of its surface appearances but in the harmony and simplicity of its inner nature, which could be discovered by reason and experiment and expressed in mathematical laws. When the eighteenth-century Enlightenment *philosophes* looked at their own society they found injustice and inhumanity, oppression and slavery, but they had learned from science that the appearance of things is not the way they really are. Reason and experiment would enable them to penetrate beneath this surface to discover the natural laws of a harmonious social life. This was to be the task of a new *social science*, modelled on Newtonian science. Once social laws had been discovered it would then be possible for rational people to act to modify society in such a way as to bring present social arrangements into closer conformity with them. And this would be an improvement – progress towards a more just and humane society.

Reason and science were thus the instruments by which error and superstition, tyranny and oppression were to be attacked and removed, and the means by which society was to control its own destiny and set out to create a better future. The Enlightenment thereby extended Bacon's argument: science was to be not just the means of improving society's material circumstances, but also the means of commanding *human* nature in action so as to improve social and moral conditions. Just as, in consequence, Bacon had assigned to the scientist a superior social role and position, so the Enlightenment gave an élite composed of rationalist intellectuals (*philosophes*) and social scientists (though the term was not yet invented) the prime role in the research, education and action that would bring about progress.

The improvement of society's moral condition, the perfectibility of humanity itself, became possible through adapting John Locke's theory of knowledge, which owed much to the philosophy of Francis Bacon and seventeenth-century science. In his *Essay Concerning Human Understanding* (1690) Locke argued that at birth the human mind was a clean slate, and that knowledge consisted of the accumulation on it of impressions transmitted from the environment by the senses. It followed that all people were in this respect equal at birth and developed entirely by interaction with their environment. People *were* their experiences. Changing those experiences, by instituting the rational natural laws of social organisation and by rational scientific education, would change people for the better. The potential for progress was infinite. As Condorcet, mathematician and secretary of the Paris Academy of Sciences, said in his *Sketch for a Historical Picture of the Progress of the Human Mind* (1794):

Nature has set no term to the perfection of human faculties, the perfectibility of man is truly infinite; and . . . the progress of this perfectibility, from now on independent of any power that might wish to halt it, has no other limit than the duration of the globe upon which nature has cast us.

Condorcet's *Sketch* is the most powerful statement of the new eighteenth-century view of people and society. It is one where all have equal claims on natural justice, rights and the law. The scientific view of this society was expressed by Thomas Hobbes in the seventeenth century. He was a materialist monist: to him soul and body, mind and matter, were all reducible to material – to the same matter in motion. He regarded society, therefore, as atomistic and mechanical in its nature and operation. The 'atoms' in the social machine were individual people: this was the beginning of the rise of liberal political philosophy, which gave primacy to the role and status of the individual.

However, like atoms, individuals were not totally isolated, although they were discrete. They were bound together, in this *Gesellschaft* society, into a body politic. The cement consisted of social contracts which were to the mutual benefit of those involved. Hobbes described why such contracts were necessary in *Leviathan* (1651). This painted a dismal picture of human nature; one which reflected the market economy that was then beginning to develop. Chaos and fear abounded because of each individual's self-interested appetite to compete with others for resources, for domination and for glory. The only answer to this disordered nastiness was an ordered social machine, where people were to be controlled and their appetites and behaviour curbed by contract laws overseen by the sovereign. The latter was outside and above the machine – a technician operating it according to the social laws that governed it: laws, dictated by logic and rationality, that were analogous to natural laws (Merchant 1982).

The rise of classical science thus appears particularly associated with the rise of secular values: notions of progress and liberalism which were increasingly to regard nature as something to be controlled and manipulated for utilitarian, material purposes. But there is also a view which emphasises the importance of religion as part of modern attitudes: this blames Judaeo-Christianity, allied with science, for the ecological 'crisis' of the late modern period.

3.3 CHRISTIANITY AND NATURE: DESPOTISM OR STEWARDSHIP?

Christianity as arrogant and despotic

In 1967 Lynn White Jnr published an article blaming the West's ecological 'crisis' on Judaeo-Christianity (that branch of Christianity, as opposed to Graeco-Christianity, which permeated the West). White's views influenced ecocentrism's development in the 1970s. Christianity's influence on Western attitudes was much debated in ecocentric literature. But another of White's assertions, and one assumption, was not much discussed. The assumption was that there *is* a crisis, and that it is a crisis in humankind's relationship with 'nature' (as other-than-what-is-human). The assertion was that 'What we do about ecology depends on our ideas of the man–nature relationship'. That is, the view of historical change was that ideas determine actions and not the other way around (Chapter 2.7). But the importance of actions – of what was *happening* in society during the development of attitudes connected with the scientific world view – is inescapable, and will be further discussed in Chapter 3.4 (for a longer discussion of this and White's thesis, see Ferkiss 1993).

White argued, first, that Judaeo-Christianity preaches that humans are separate from and superior to the rest of nature, which is there to be used and dominated by humankind. Second, he proposed that such attitudes were translated into harmful actions towards nature with the aid of technological developments made well before the scientific and industrial revolutions, but which coincided with the rise of Christianity in medieval Europe.

However, matters reached 'crisis' proportions, he thought, with the 'marriage of science and technology' around the 1850s. This marked the end of a period of slow technological evolution, based mainly on trial and error, and the beginning of huge leaps in technological power based on the ability to predict and calculate very accurately what would be the effect of building machines in one way rather than another. (A simple way of looking at this might be to contrast how trial-and-error technology could have got to the moon; by firing off a sequence of rockets and missing the moon at first, but continually correcting the trajectory until the target was found. This would compare with how the Apollo missions actually *did* it; essentially playing out, in minute detail, everything which had already been planned and foreseen on Earth, so that the first landing was within just a few metres of where it had been intended.)

White maintained that the Western scientific and industrial revolutions were not sudden events. Technologies for manipulating and exploiting nature stretched back to the seventh century, when a heavy iron plough had already replaced an ineffective scratch plough, enabling farmers to attack the land with unprecedented violence.

All this happened within the context of developing Christian axioms which have underpinned Western thought since the Middle Ages. Early technology and nature exploitation were in harmony with these 'larger intellectual patterns'. They regarded time as linear, history as progressive and nature as created for human benefit – God *wanted* us to exploit nature for our use.

The linchpin of such ideas was the Book of Genesis, with its command to humans to be fruitful, multiply and subdue the Earth. And while Adam was placed in the Garden of Eden as its caretaker; to 'till it and keep it' (Genesis 2:15), after the flood 'Every moving thing that lives shall be food for you; and as I gave you the green plants, I give you everything' (Genesis

9:3). The victory of this version of Christianity (by contrast with Graeco-Christianity, which was more passive and contemplative towards nature) over paganism destroyed the animistic beliefs which had made people think twice before plundering and destroying nature. As White put it:

> Man named the animals, thus establishing his dominion over them. God planned all of this explicitly for man's benefit and rule: no item in the physical creation had any purpose save to serve man's purposes.

With the loss of paganism, nature was shorn of spirits and could be treated as an object. Ette and Waller (1978) point out that religion and ritual was thus transferred from the physical world and fixated onto a metaphysical world, and this meant that to save our soul we had to follow God's will, not nature's. Nature merely became a reflection of the will, intentions and actions of God: God looked like humans, so religion was essentially anthropomorphic.

Humans became regarded as superior to nature, while God was located outside and above nature. Hence to worship nature, as pagans and witches did (and do today) became a Christian heresy. Witchcraft was equated inevitably (therefore often wrongly) with Satanism.

Merchant (1982) stresses how Christianity took on from Aristotle a notion of the superiority of *form* (which was associated with mind, maleness, activeness and power) over *matter* (material, female, passiveness and inertness, as in the atoms which comprised the universe). She says that in the sixteenth century biological generation in nature was described as the impregnation of the female Earth by the higher, celestial, masculine heavens: Copernicus said in 1543 that the 'earth conceives by the sun and becomes pregnant with offspring'. The superiority of mind, intellect and the soul over matter is an important part of today's Western world view: feminists complain that it is part of patriarchal and environmentally malign ideology. By contrast, earlier 'Gnostic' Christianity – based on a *unity* of opposites and an equality of male and female principles, with God as both father and mother – was, says Merchant, condemned and marginalised. Most Renaissance thinkers who saw the human position in the universe as less than central were persecuted, and it was the Church that persecuted them.

White considered that when such Christian attitudes were combined with the explosion of technological power, as applied science in the nineteenth century, the way was open for apocalyptic ecological despoliation:

> the population explosion, the carcinoma of planless urbanism, the now geological deposits of sewage and garbage, surely no creature other than man has ever managed to foul its nest in such short order.

Even though the West is now ostensibly atheistic, White considered that its world view is still essentially this 'Christian one'. Indeed, Soviet 'communism' was also underpinned by Christian beliefs in progress via subjugating nature.

Others agree that Christianity's guilt continues. Ette and Waller (1978) think that the Church, through property owning and dealing, has removed sacredness from land helping to commoditise it. And, as a landowner, the Church also condones the factory farming, monoculture and deforestation which its tenants practise.

Singer (1983) is particularly concerned about Christian 'speciesism'. Both Old and New Testaments, he thinks, lack proper consideration of animal interests, telling us that killing animals is permissible because humans have a special place in the creation whereby other animals will fear and dread us. He condemns contemporary Catholicism for discouraging concern for the welfare of non-humans. Even Christians who opposed animal cruelty, like Thomas Aquinas, did so from an anthropocentric viewpoint, arguing speciesistically that animal cruelty is bad because it leads to cruelty against humans. Francis of Assisi's much-vaunted love for nature was also suspect because it still affirmed that nature was made *for* humankind, and it valued equally animate and inanimate nature. In such criticisms, Singer shows affinity with deep ecologists. They often argue (e.g. Devall and Sessions 1985) that even if it could be shown, as some assert, that Christianity encourages us to be stewards of nature rather than despots over it this is still not good enough because it is not *eco*centric. To cast humans as stewards, charged with careful management of nature as God's 'agents', merely values other creatures for their value to (the anthropocentric) God, not for their intrinsic value.

Against White's thesis

Most of the many repudiations of White's thesis focus on the possibility of interpreting the Judaeo-Christian message in exactly the reverse way to White. Glacken (1967, 152), for instance, considers that the Book of Job and Paul's Epistles to the Romans set

> man apart from all other forms of life and matter because God has willed this role for him; he is . . . set here as a steward, responsible to his Creator for all he does with the world over which he is given dominion.

Doughty (1981), too, believes there is a theological perspective in Christian thought that is sympathetic to the environment. He regards White's argument as 'partisan and overgeneralised', and cites scholars like Santmire and Gowan, who maintain that Genesis characterises humans as life tenants of nature, not freeholders.

Attfield (1983) also maintains that 'there is much more evidence than is usually acknowledged for other, more beneficent Christian attitudes to the environment and to nonhuman nature'. Christian teachings about nature have been diverse and contradictory, but they have, he says, not typically been exploitative. He considers that Psalms 104 and 148 cast doubt on Passmore's (1980) claim that the Old Testament suggests that everything exists to serve humanity. For these Psalms admire God's handiwork, express his care for various creatures, and suggest that human domination of nature means ruling it in a way consistent with being responsible to God for his realm. Similarly, the New Testament, in the gospels of Matthew and Luke, also witnesses God's care for animals and plants.

As to White's thesis that Western science picked up exploitative attitudes to nature, Attfield points out that the earliest modern scientists, such as John Evelyn, were far from believing that they could treat nature as they pleased, while even Descartes, who regarded animals as machine-like, did not suppose that they had no feelings. Attfield concludes that adverse interpretations of Christian attitudes to nature come from exaggeration and selective use of evidence. In fact all interpretations of the Bible are open to question because of translation difficulties, especially the Old Testament written in ancient Hebrew (Kay 1988).

Palmer (1990) offers an alternative interpretation of Judaism. The Books of the Torah (Moses, Genesis, Exodus, Leviticus, Numbers and Deuteronomy) and the commentaries and stories of the Talmud all make it clear, he says, that we rule animals only as God rules us; with love and compassion. And the fact that time is linear and each life is lived once means that we must get our lives right, and we have strong responsibilities to future generations. Christianity shares such assumptions, while the very notion of resurrection signifies that there will be a future here on earth when all nature is in harmony. This will redeem the present sin of disharmony.

Moore (1990) considers that Bacon's domineering sentiment, 'I am come in very truth leading you to nature with all her children to bind her to your service and make her your slave' definitely echoed Genesis 1:28. But, *today's* Christianity is rediscovering the Holy Spirit, as the vehicle by which God makes a cosmic community of all created beings, and through which he enters creation. Because God permeates all creation the relation between humans and the creation must be one of reciprocity and respect, not domination and exploitation. This, says Moore, is a 'more mature understanding of the meaning of creation' showing that humans are to be stewards, not despots. For Genesis also says (2:15) that 'man' was in the Garden of Eden to dress and keep it, that is, to be a creative interventionist. The garden, with the sabbath – a day set aside for non-intervention and enjoying nature – are neglected symbols, but they, together with St Francis' view of humans as 'first among equals', and with Old Testament symbols of the lion lying down with the lamb; all these must now be rediscovered and reaffirmed.

Indeed, Christian theologians are now at pains to emphasise that 'God is green' (Bradley 1990, Cooper 1990), and that there is an activist tradition of respect for nature in Christianity. McDonagh (1988, 129–37), traces how various leading figures in shaping Christian consciousness lived out themes of caring for the Earth, fellowship with the rest of creation and joy and wonder at the material Earth. The Benedictines, for instance, organised their monasteries on the basis of self-sufficiency and stewardship; improving soil fertility and developing good husbandry so as to ensure continuing fruitfulness. For St Francis (1182–1226) every creature mirrored God's presence, which fact meant fellowship and kinship with animals, and with the sun, moon and Earth. He also preached anti-consumerism and a wilderness ethic. All this is why St Francis should become the 'patron saint of ecology'.

Hildegarde of Bingen (1098–1178) added a unique dimension to Christianity, McDonagh continues. For she brought into it pagan and feminist dimensions, celebrating Earth's fertility and sensuality and therefore not only the spiritual but also the material and earthly.

If these were minority and persecuted perspectives, that of the Catholic priest Teilhard de Chardin (1965a and b) seems, by contrast, to have the potential to enter the mainstream of contemporary Christian thought. De Chardin told the story of creation in such a way as to dispense with dualisms between mind and matter and the spiritual and material, so that the inner spiritual nature of things evolved intimately with their material aspects. In his monistic Christianity the psychic dimension evolved with and is part of the material dimensions of atoms and molecules. Every atom has its own 'value', regardless of its worth to humans. And from de Chardin's perspective the cosmos cannot be properly understood simply through analysis, reductionism and rationality. It has to be known holistically through intuitional insight and understanding symbolic meanings. In this idealistic conception of the universe, where the Earth is surrounded by a 'noosphere' or realm of thought, 'The transformative power of love is now pushing human consciousness to reach up and achieve a new level of union at Omega point'.

Such sentiments, which are very pre-modern, are key parts, too, of the New Age litany (see Chapters 1.2 and 6.1). Some Christians also welcome and develop them as part of their interpretation of relevant Christianity – although other Christians specifically resist New Ageism because it does not believe in original sin, and because it believes in all gods, not just one God.

Yet other Christians take on materialist 'social ecology' dimensions of ecologism. Moore cites in this respect the Roman Catholic Bishops of the Philippines, whose letter in 1988 not only held up the Filipinos as exemplars of respect for the living world because of their animist traditions, but also advocated *political* education of their flock and action to create environmental and social justice. This resonates with liberation theology, which emerged in Latin America in the 1960s and regards freedom from oppression as a pivotal Christian message. It uses materialist Marxist analysis to understand how to achieve liberation and establish the kingdom of God on Earth. Because it is centrally involved with the struggle of peasants against multinationals and the capitalist market economy, attempting to reappropriate former peasant land, many ecocentrics regard liberation theology as crucial in establishing an ecologically sound society.

Repudiations of White may also highlight various flaws in the nature of his thesis. Other, non-Christian, cultures also do and have abused nature, although their religions actually preach respect for nature. Yi-Fu Tuan (1968, 1970) points out that the high ideals towards nature in Eastern cultural tradition were frequently abused, and much pollution and ecological degradation resulted from what people actually did in Asian countries, never mind what they professed. Indeed these societies had as much impact on nature as the West did until the early modern period. Thomas (1983) adds the point that ancient Romans exploited nature more effectively than did their Christian medieval successors, while the worship of nature in modern Japan has not prevented industrial pollution.

Thomas makes a further observation; one that Passmore also raises. This is that other, non-Christian religions have also given to humans authority from their gods to destroy nature. He discusses the American 'Indian', much loved as a 'noble savage' by ecocentrics (Chapter 1.2). A 1632 report about the Indian, says Thomas (p. 24), told how, in reality,

> they have it amongst them by tradition that God made one man and one woman and bade them live together and get children, kill deer, beasts, birds, fish and fowl and what they would at their pleasure.

White also possibly exaggerated the extent to which *ordinary* people before the Reformation were influenced by the official religion of Christianity. In fact most could not read or understand Latin, so their knowledge of Christianity must have been limited. Their lives received more meaning from the magical, astrological and spiritual traditions of paganism. These remained strong into the nineteenth century being eventually overcome by urbanisation rather than theology (Atkinson 1991). White tends to forget this continuing importance of other influences, like the Great Chain of Being concept and the Renaissance philosophy of the nurturing Earth, all of which made European culture more complex and varied than White's judgement allowed.

And sociological research indicates that White overestimated how much religious values impact on *contemporary* environmental values and actions towards the environment. Shaiko (1987), for instance, shows that both domination and stewardship attitudes can be found among professing Christians. However, it is possible that the more literally one takes the

Bible, or the more rigid one's religious beliefs, the less one is concerned about the environment anyway (Greeley 1993, Eckberg and Blocker 1989).

Hornsby-Smith and Proctor (1993) apply path analysis to the 1990 European Values Surveys, to test the links between religiosity and other variables, like political interest, approval of the green movement, willingness to pay to safeguard the environment and anxiety about environmental issues. They conclude (pp. 39–40) that

> there is no direct effect of religiosity on . . . the potential for action, for example in the ecology movement . . . the indirect effects of religiosity are also relatively small . . . Our findings give no support to those who espouse the position that Christianity has successfully promoted a stewardship orientation to the world. In the same way there is very little evidence to support the alternative theoretical position of those who claim religion favours a dominating and exploitative orientation to economic activity in the world

All this supports Tuan's (1970) conclusion:

> A culture's published ethos about its environment seldom covers more than a fraction of the total range of its attributes and practices pertaining to that environment. In the play of forces that govern the world, aesthetic and religious ideas rarely have a major role.

3.4 THE IMPORTANCE OF MATERIAL CHANGES

Keith Thomas (1983) rightly refers to White's article as 'almost a sacred text for modern ecologists'. An aspect of White's analysis rarely challenged in green literature is the idealistic approach to social change referred to above. 'What we do about ecology depends on our ideas about the man–nature relationship' is now an ecocentric orthodoxy, underlying the strategy of working particularly to change people's ideas, attitudes and values.

But Thomas considers that whether Christianity is intrinsically anthropocentric is irrelevant: 'The point is that in the early modern period its leading exponents, the preachers and commentators, undoubtedly were [anthropocentric].' Thomas says (pp. 22–3) that White

> almost certainly overrated the extent to which human actions have been determined by official religion alone. In the 1680s the English secretary Thomas Tryon had also contrasted the moderate demands made on nature by the North American Indians with the ruthlessly manipulative approach of the European invaders. But he recognised that it was new commercial incentives that had made the difference: it was less the replacement of pagan animism by Christianity, than the pressures of the international fur trade which led to overhunting and the unprecedented onslaught on Canadian wildlife. As Karl Marx would note, it was not

their religion, but the coming of private property and a money economy which led Christians to exploit the natural world in a way the Jews had never done; it was what he called the 'great civilising influence of capital' which finally ended the 'deification of nature'.

Thomas's (p. 34) materialist approach to the growth of domineering attitudes asserts that the most powerful argument for the Cartesian position that animals were machine-like was that this was the best rationalisation of how people were *actually treating* animals. Hence what they were doing conditioned what they thought: the latter legitimating what they were *doing*: 'denying the immortality of beasts . . . removed any lingering doubts about the human right to exploit the brute creation'.

As Heilbroner (1980, 87) put it (emphasis added):

the Cartesian-Newtonian world view of the seventeenth and eighteenth centuries *sprang from* and supported a society of prosperous commerce and banking.

This is not to say that there was a simple or direct relationship between emerging science and industrial capitalism, especially in its earlier development. Landes (1969) has shown that the links are diffuse. Barnes (1985, 15) believes that it is no coincidence that the rise of science was paralleled by the rise of industrialism, but it is difficult to pin down the connections – at least until the nineteenth century, when science's professionalisation was more clearly linked to capitalist growth. Even then 'It will not do to claim that professional science thrived and prospered purely as a response to a demand for useful knowledge'. As Atkinson (1991, 133) puts it: 'Marx's suggestion that men's social existence determines their ideology remains a useful explanatory insight', but this must not be reduced to too simple a functionalist argument. Barnes concludes that science's link to industrialisation was more as a world view appropriate to and legitimating the new class of industrial entrepreneurs: the foundations of an alternative culture (Chapter 5.6).

However, this does not alter the fact that big changes in what people were doing to nature were gaining momentum during the scientific revolution. *Why* they were doing it also changed, as production moved from a largely subsistence-needs basis in the medieval manorial system to that of capitalism, where production (changing nature) was primarily of commodities – of goods produced to realise profit.

Some accounts of the changes in attitudes to nature during the scientific revolution neglect or underemphasise this importance of material, economic events (e.g. Capra 1982, Whitehead 1926). To underline the need not to do this, the rest of this chapter first reminds us of some of the changes which were occurring. It then highlights analyses, particularly that of Carolyn Merchant, of the growth of scientific attitudes which underline the possible significance of these material changes.

Social, economic and environmental changes from pre-modern to modern periods

If the way we regard society's relationship to nature relates to what we are *doing* to nature at any particular time, then we can equate economic modes and social relations of production with different conceptions of nature. Johnston (1989, 43–5) describes how in 'reciprocal societies' like those of primitive communism there was collective ownership of land and resources, products were allocated more or less equitably and there was limited division of labour. Such societies were (and are still in some vestigial 'traditional' societies) in relative equilibrium with their environment. But this equilibrium was always haunted by the Malthusian spectre of population outstripping resources, since the technology to mitigate environmental vicissitudes like drought or flood was lacking.

When such societies gave way to what Johnston (pp. 47–8) terms 'rank redistribution' economic and social arrangements, power became unequal and resources were distributed according to social rank. The goal of such hierarchical societies was not maximising the benefits and survival chances of all, but reproducing an élite via the work of others. In medieval Europe a manorial system evolved into full-blown feudalism, to which land ownership by individuals was basic, as was serfdom. The monarch delegated land ownership, and therefore power, down to an aristocracy, which also had a right to govern the serfs. The latter were tied workers – they were not free to move about from 'employer' to employer. The manor and village constituted the unit of production: the serfs had to work so many days on the lord's land, but could work the rest on their own land (farmed collectively as strips for efficiency), and on the common land of woods, river meadows, heath and extensive pasture.

The determinant of class and power was birth, rather than accumulation of capital. Indeed, usury conflicted with the feudal economy and was declared a sin by the Church, together with avarice. The theory of the just price meant that prices were not left merely to the seller – no one should sell produce or lend money to receive any interest over the cost of production. Even as late as the 1760s, popular dissent and disrespect for authority was based partly on a feeling that custom and face-to-face bargaining, rather than a market, should regulate prices and wages (Thompson 1968).

The manorial system initially aimed for general stability and subsistence and production's aim was to fulfil social need: though within this there was unequal distribution of surplus. Much produce was not bought and sold. Indeed money was limited in its range, currencies being local.

157

As the system evolved serfs increasingly worked their own land as 'free' peasants, that is, tenants whose surpluses were creamed off via taxes to landlords. The village's two large open fields often became three, to increase productivity by more sophisticated crop rotations and other methods. As populations grew up to the fourteenth century, common land was cleared and taken into individual ownership.

Initially, this manorial system was in relative equilibrium with its environment, compared to modern agriculture. High crop productivity and soil fertility were maintained, while family and communal holding of farmland and sharing of the commons all militated for self-regulation of the degree of exploitation of nature. However, the hierarchical power structure eventually became a source of environmental pressure. In stable conditions landlords did not try to maximise their extraction of labour, taxes, rent and services, from the peasantry. But for various reasons this stability did not last (political infighting and aggrandisement amongst landlords, and foreign campaigns by the monarch, required an army and navy, for instance; the exertion of landlord's rights was in tension with the ethos of community control by the peasants – especially over technology – leading to struggle and repression; and there was general population growth except during the famines and plagues of 1350–1450). Hence there was increasing 'interference' with the ecosystem. As Johnston stresses, this in itself might not be problematic, but it was badly managed interference.

Developing two- into three-field systems secured only short-term productivity gains, at the expense of longer term substantial soil nutrient losses, soil erosion consequent on structure and texture degradation, decline in populations of nitrogen-fixing bacteria and spread in soil pathogens from monoculture. Cooter (1978, 469) says that 'At best, open-field husbandry offered a means for sustaining a mediocre level of productivity at the price of judiciously slow depletion of the nutrient reserves of the arable's hinterland.' That is, this (strikingly modern-sounding) litany of soil problems could be offset, but only (except during fourteenth-century population decline) at the expense of greater exploitation of the surrounding commons.

That assault on the commons is well-documented in the classic works of Darby (1956, 1973) and Hoskins (1955). It was uneven in Europe. In Poland, Hungary and parts of Germany and Austria, for instance, a collective peasant agriculture strongly resisted the pretensions of landlords for hundreds of years. But in Britain, Holland and other parts of Western Europe there was the spread of a capitalist market economy from the city states of the Renaissance (Dobb 1946). By the end of the fifteenth century, classical feudalism was fast disappearing, and a landless peasantry was being created. The drive to remove peasants from the land by enclosing it was triggered off with the rise of the Flemish woollen industry, making sheep-raising and wool export potentially a greater source of money and power than peasant-farmed arable.

The Flemish wool trade was capitalist: it depended on companies which were no longer local but international, and a local decentralised economy did not suit it. Feudal landlords in Britain, Spain, Portugal, Italy, Holland and France joined the ranks of capitalist entrepreneurs. The new class made gains in political power, wealth and status in the seventeenth and eighteenth centuries which meant that it could legalise and accelerate enclosure.

Trade quickened during this 'mercantile' phase of capitalism, stimulated by the age of discovery, as did mining, iron-making and craft production. The spread and standardisation of money facilitated not only exchange but also its open-ended accumulation for re-investment, and the commodification which is the hallmark of capitalism. Labour on the land and in craft production, which had previously been tied to landlord or guild master, now became 'free' to be bought and sold as a commodity on the open market. Land and other natural resources, too, became commodities, that is, valued mainly, though not exclusively, by their marketability. Today it is not only land, but also landscape which, with history, is packaged and commoditised in tourism and the heritage 'industry'.

Of course, raw price is not alone the measure of a commodity's worth in the eyes of someone whose job it is to accumulate capital. The gap between price and the cost of production (i.e. the profit) is what counts. So that capital accumulation can be maximised profits must be maximised – hence all means of increasing the productivity of labour and land must be relentlessly pursued (Johnston 1989). Thus with the transformation from feudalism to capitalism previous pressures to get more out of the land were much intensified, as notions of productivity became more explicit and imperative. The ideology of scientific agricultural improvement gained sway.

Fen drainage accelerated from the seventeenth century. Merchant (1982) describes the resultant destruction of wetland ecosystems, lifestyles and occupations as 'a striking example of the effects of early capitalist agriculture on ecology and the poor'. As with soil 'improvement', fen drainage benefited mainly landed and monied classes.

Again, Merchant considers that the pressures on woodland were much multiplied by capitalism's growth. Mercantile capitalism drained timber resources for shipbuilding (trade, and trade-inspired wars with other nations) and as a source of fuel for smelting and processing metals (there was a big growth in mining in the sixteenth century). Shipbuilding devastated Mediterranean oak woodlands. With growing use of coal, pollution was emerging as a problem of capitalist growth by the seventeenth century. Capitalism's rise in Europe and America, says Merchant, directly depended on exploiting natural resources.

Whereas in the twelfth century half the land in Britain belonged to villagers, was unfenced and was worked and managed with varying degrees of collectivisation, by 1876 0.6 per cent of the population owned 98.5 per

159

cent of the land in England and Wales (Goldsmith *et al.*. 1992, 132). In the English revolution of 1649–60, landowners who had benefited from enclosure gained more power, and the previously piecemeal process recommenced with renewed vigour. From the early 1700s to the General Enclosure Act of 1845, Parliament passed 4000 Acts of Enclosure, covering seven million acres of open field and wood, while a similar amount of land was enclosed without application to Parliament. Thus, in the eighteenth and nineteenth centuries large amounts of heath and forest were destroyed, and a dispossessed urban proletariat was created, ready to serve the emerging factory system necessary to industrial capitalism.

Goldsmith *et al.*. consider that this process continues today across the whole world in response to the expansionary drive inherent in capitalism, with similar consequences for environmental stability and social justice. As E. P. Thompson (1968, 14) put it:

> the greater part of the world today is still undergoing problems of industrialisation . . . analogous in many ways to our own experiences during the industrial revolution. Causes which were lost in England might, in Asia or Africa, yet be won.

The scientific world view as a reflection of material changes

Atkinson (1991, 135) identifies some link between elements of the emerging scientific world view and that cluster of practical and ideological interests which were part of Protestantism and nascent capitalism (analysed by Max Weber, 1976):

> Science was from the beginning a tool with which to advance the interests of certain individuals, classes and nations – and in the end a particular culture – against all others.

Atkinson cites C. S. Lewis's *The Abolition of Man* (1947): 'What we call Man's power over nature turns out to be a power exercised by some men over other men with nature as its instrument'. This means that the *ideology* of power over nature (the Baconian creed) may not reflect the interests of all, as Bacon envisaged, but instead may serve the material vested interests of élite minorities.

The connection between science and ideology is discussed in Chapter 5: suffice to stress here that the rise of the scientific world view to become eventually Western conventional wisdom cannot be divorced from these material changes already described. Johnston (1989, 83) describes how

> the agricultural and industrial revolutions of the sixteenth century onwards, that most associate with the transition to capitalism, saw the technocratic interpretation of the environment come into its own.

160

Merchant stresses how this interpretation was associated with the change in how most people *experienced* nature – as altered by machine technology – with the rise of capitalism.

Being ousted from the land alienated people from a direct organic relationship with it, and this was accompanied by changes in popular and, more particularly, official ideas about humankind's place in nature. For most sixteenth-century Europeans the root metaphor for self, society and cosmos was still the organism, emphasising interdependence, subordination of the individual to the common purpose in the (extended) family, the community and the state. These ideas were all very compatible with feudal socio-economic organisation. This metaphor and its associated image of Earth as female (usually nurturing, but sometimes wild and uncontrollable) changed to the metaphor of nature as a machine, and, says Merchant, this change in controlling imagery directly related to behavioural changes. The image of Earth as a mother had constrained its exploitation. Mining was digging into 'her' entrails. If minerals and metals ripened in the uterus of the Earth, mining constituted abortion. Miners therefore often offered propitiation to the soil and subterranean deities, performed ceremonial sacrifices, fasted, were clean and sexually abstinent before sinking a mine. For many centuries this organic framework was compatible with a relatively low level of exploitation, commercial development and technological innovation.

But when these levels rose in the sixteenth and seventeenth centuries, a new set of images, based on a new metaphor, was more suitable as a cultural sanction for denuding nature:

> Society needed these new images as it continued the processes of commercialisation and industrialisation which depended on directly altering the Earth . . . as the economy became modernised and the scientific revolution proceeded the dominion metaphor spread beyond the religious sphere and assumed ascendency into the social and political spheres as well.
>
> (Merchant, pp. 2–3)

However, Merchant's work confirms that the change was not a simple one, as discussed above. There had been several possible social and political-economic implications of the organic metaphor.

One emphasised, in natural magic, the idea of a hierarchical cosmos, properly mirrored by a hierarchical society. Generally, society was likened to a body: the peasantry were the feet, while princes and clergy were part of the soul. Taking the bees and ants as the natural analogy, each must work for the common good while the queen superintends. The lower's purpose was to serve the higher, while the latter's duty was to guide the former towards the common moral good. This interpretation of the organic analogy was rife by the sixteenth century, and it clearly supported a conservative, feudal society. Indeed, elements of this model, with its ideas of *noblesse oblige* from the higher to the lower, can still be discerned among traditional conservatives today.

Other aspects of the organic metaphor were also carried forward, for example the assumption in natural magic that nature could be altered by manipulating 'her' (alchemy), and that matter is passive (the feminine earth). Though declared heretical by the Catholic church this notion was ultimately assimilated into the scientific world view, with its emphasis on controlling nature. Indeed Bacon's first formulations of science were made to assist his attempts to be a successful alchemist.

A second interpretation of organicism echoed forward to the under-pinning themes of Marxism and the socialist left. This challenged the hierarchical view of the cosmos by emphasising its unity, acknowledging that the whole was greater than the sum of the parts, and regarding the parts as of equal value. It saw nature as constantly growing and changing, reflecting the break up of old social classes in parts of Europe (e.g. Spain and Italy). And change came out of unification and opposition of the contraries of matter and spirit. Here are pre-empted notions of dialectics, conflict as a motor of history and egalitarianism: all cornerstones of socialism.

A third interpretation of organicism, says Merchant, saw matter and spirit united in a single active, vital substance (vitalism and monism). It became associated with radical libertarian ideas: suggesting as it did the anti-establishment view that individuals could understand and manipulate nature (again in magic) for themselves. Priests or magicians were not the only ones who could do it: this was a folk tradition. Such ideas were revived in the mid-seventeenth century in the radical doctrines of the dissenting sects. The Diggers seized common lands and established egalitarian communities. The Ranters practised pantheistic religion – every creature was God, and to know nature was to know God's work. Such mysticism, vitalism and anarchism were also carried forward but only as a minority strain in Western culture: at least until their reappearance in ecocentrism today.

The first and second interpretations of organicism, described above, although criticised by mechanists, were eventually adopted into the mainstream of the scientific world view. Why were they 'selected' while the third, libertarian-related, interpretation was not? Merchant (p. xviii) says that this is because of wider social and economic changes:

> An array of ideas exists, available to a given age; some of these for inarticulated or unconscious reasons seem plausible to individuals or social groups; others do not . . . the direction and cumulation of *social* changes begin to differentiate among the spectrum of possibilities so

that some ideas assume a more central role in the array, while others move to the periphery.

The Baconian creed held that civilisation was to be created by harnessing nature, the machine. The social changes which this was consonant with include the exploitation and 'taming' of woodland, moor and fen referred to above, the growing phase of discovery and colonialism which was associated with mercantilism, and the mining which was also essential to it. Merchant says that the new economic and scientific order had as its ideological core the idea that nature and women (spheres of production and reproduction) must be passive and controlled in order to be exploited. Hence notions of wilderness and wildness (non-civilisation) began to be attributed to both when not controlled. What were really commonly owned lands among aboriginal people, for instance, were dubbed 'wilderness' by European colonists (Young 1990), making it legitimate to appropriate them. Women who were anti-hierarchical, nature worshipping animist healers were branded as wild and satanic, making it legitimate to persecute them as witches and keep them under social control.

Merchant (p. 143) points out that 'As the unifying bonds of the old hierarchical cosmos were severed, European culture increasingly set itself above and apart from all that was symbolised by nature'. Baconian science's often sexual language, style, nuance and metaphor reduced female nature to a resource for economic production: the secrets of nature's womb were wrested, and nature was raped, by dint of 'hard' facts, 'penetrating' minds and 'thrusting' arguments. Merchant (pp. 179–90) is in no doubt about Bacon's patriarchal and class-ridden perspectives, which grafted themselves on to his seminal vision of a scientific utopia in *New Atlantis*. Here Bacon placed progress in the hands of scientific and technical *males*. Descartes, meanwhile, wrote of rendering 'ourselves the masters and possessors of nature'. Such attitudes

> reinforced the tendencies towards growth and progress inherent in early capitalism ... The constraints against penetration associated with the Earth mother image were transformed into sanctions for denudation.
>
> (Merchant pp. 185, 190)

In France the rise of mechanical philosophy also coincided with central-isation of political power in the name of 'rational management' of nature. As we have noted, there were substantial problems of pollution and resource depletion by the seventeenth century. In Britain, in response, John Evelyn's *Sylva* (1664) called for conservational management of the forest.

Merchant describes how the environmental management concept grew as an alliance of older organicism (notions of what we would now call holism, systems and interdependence) with the rationalising tendencies in mechanism. Environmental management and the 'systems' approach apply human values (of efficiency and production) to nature but reject any 'irrational' ideas of intrinsic value in it. This is the utilitarian ethic that underlay the nineteenth-century conservation movements (where Gifford Pinchot's pragmatism rather than John Muir's romanticism was applied to American wilderness), and which underlies twentieth-century scientific ecology (Chapter 4.6).

Long-term planning for sustainable economic gain aims to maximise energy production, economic yields and environmental quality through ecosystems modelling, prediction and manipulation. Such manipulation, with its associated ecosystems view of the society–nature relationship, sits nicely with the aspirations of a more 'developed' capitalism: one which has appreciated that continued unregulated mining of resources produces the contradiction of economics which destroy their own resource base. The current softening of the Christian message of domination into one of stewardship, to 'garden' the world and improve it for God, also lends itself well to this phase of capitalism.

So the lesson of history is that radical changes in ideas, attitudes and values generally correspond with radical changes in social and economic arrangements. This argument does not have to be crudely economistic, where all change is simply determined by economics. However, the economic element in historical and cultural change can never be left out. As Merchant (1987, 265–6) puts it in her study of the ecological changes in New England consequent on the seventeenth-century collapse of indigenous Indian ecologies and on eighteenth-century changes in European capitalist farming approaches:

> Ecological revolutions are major transformations in human relations with non-human nature, and arise from changes, tensions and contradictions that develop between a society's mode of production and its ecology and between its modes of production and reproduction. These dynamics, in turn, support the acceptance of new forms of consciousness, ideas, images and world views.

Hence, 'Social values and realities subtly guided the choices and paths to truths and certainty taken by European philosophers' (Merchant 1982, 227).

Furthermore, and perhaps most fascinatingly:

> Between 1500 and 1700 an incredible transformation took place. A 'natural' point of view about the world in which bodies did not move unless activated either by an inherent organic mover or a 'contrary to nature' superimposed 'force', was replaced by a non-natural non-experiential 'law' that bodies move uniformly unless hindered. The

'natural' perception of a geocentric earth in a finite cosmos was superseded by the 'non-natural' commonsense 'fact' of a heliocentric infinite universe. A subsistence economy in which resources, goods, money or labour were exchanged for commodities was replaced in many areas by the open-ended accumulation of profits in an international market. Living inanimate nature died while dead inanimate money was endowed with life. Increasingly capital and the world market would assume the organic attributes of growth, strength, activity, pregnancy, weakness, decay and collapse, obscuring and mystifying the view underlying social relations of production and reproduction that make economic growth and progress possible . . . Perhaps the ultimate irony in these transformations was the new name given them: rationality.

(Merchant 1982, 288)

But as we have said, not all pre-modern views of the society–nature relationship disappeared in the modern period. We now go on to consider how some of those ideas were kept alive and developed as the modern roots of ecocentrism.

Chapter 4

MODERN ROOTS OF ECOCENTRISM

MODERN ROOTS OF ECOCENTRISM

4.1 ECOLOGY'S EARLY DEVELOPMENT

Reacting to capitalism's development

Chapter 3 emphasised the need to set changing attitudes to nature within the context of what people were actually doing to it. It described how capitalism's exploitation of nature supported, and was supported by, exploitative attitudes inherent in the scientific world view. But it also noted how, by the seventeenth century, any ideas of unbridled exploitation of nature were already being tempered. Forest and fen destruction were well advanced and the need to manage what was left in the interests of sustainable, rational exploitation, was becoming accepted. Out of this need developed an important strand in what was to become the modern science, and moral philosophy, of ecology – a strand emphasising conservation as a management strategy.

In Richard Grove's (1990) view it was developments abroad rather than in Europe which gave the main impetus to what we now call ecologism. European attitudes to nature were modified by experiences of it in the expanding colonies.

On the one hand there was a revival of the myth of the garden of Eden: the accounts of travellers and commercial exploiters suggested that Eden might be found in the colonies of India, Africa or America, where there were 'wild' landscapes apparently little altered by humans. This encouraged nature mystification, making tropical forests and islands symbolic locations 'for the idealised landscapes and aspirations of the Western imagination' (p. 11), for instance in *The Tempest*.

On the other hand, the damaging effects of deforestation and other practices of commercial exploitation were apparent. They had been noted from as early as 1300. But it was not until the mid-seventeenth century that

> awareness of the ecological price of capitalism started to grow into a fully fledged theory about the limits of the natural resources of the earth and the need for conservation . . . Ironically, this new sensitivity

developed as a product of the specific, and ecologically destructive, conditions of the commercial expansion of the Dutch and English East India Companies.

(Grove, p. 12)

This expansion had not been a free-for-all, being controlled by 'the absolutist nature of colonial rule' exercised by the state. That control was exerted through conservation strategies – applied to Mauritius' forest between 1768 and 1810, for instance. They were informed by the ideas and observations of a growing group of experts on tropical 'ecology'. (The term 'ecology' was not actually coined until 1866.) These scientists and policymakers – 'pioneers of modern environmentalism' – were motivated not just by pragmatism and the fear of extinctions of useful species; they were also 'romantic' followers of Jean Jacques Rousseau and 'adherents of the kind of rigorous empiricism associated with mid-eighteenth century French Enlightenment botany'. They 'held that responsible stewardship of the environment was a moral and aesthetic priority, as well as a matter of economic necessity'. From 1820 their views were reinforced by the writings of Alexander von Humboldt, himself influenced by orientalist thinkers:

> Humboldt strove, in successive books, to promulgate a new ecological concept of the relations between man and the natural world which were drawn almost entirely from the holist and unitary thinking of Hindu philosophers. His subordination of man to other forces in the cosmos formed the basis for a wide-ranging and scientifically reasoned interpretation of the ecological threat posed by the unrestrained activities of man.

(Grove, p. 12)

Influenced by such 'ecologists', the directors of the East India Company were expressing concern, by 1847, about the dangers of artificially induced climatic change. By 1852 the British Association was reporting on economic and physical effects of tropical deforestation, and by 1858 it was publishing papers on *global* climatic desiccation and changes in atmospheric composition. Grove (p. 14) concludes that the modern day

> awareness of a global environmental threat has, to date, consisted almost entirely in a reiteration of a set of ideas that had reached full maturity over a century ago.

Ecology's eclecticism

There were various influences on the development of thought in ecology and its associated political philosophy of ecologism. Some could be regarded as 'scientific': inherited particularly from empiricism and rational thinking in biology and ecology, and encouraged by the growth of global trade and travel. Others were 'non-scientific', partly derived from nineteenth-century romantic thought.

Such a dichotomy instantly poses problems: some romantics, like Thoreau or Ruskin, also expressed themselves 'scientifically' (e.g. in rigorous detailed empirical accounts of nature), while some scientists, such as Malthus or Haeckel, did not always conform to the objectivity of the Baconian stereotype. Nonetheless to think for a moment of ecocentrism as having both scientific and non-scientific roots can be useful in reminding us of how this mode of thought is an eclectic mix.

Ecocentrism's Janus-like tendencies are emphasised in Donald Worster's (1985; see also Worster 1988) fascinating account of ecology's development in the eighteenth century. Two principle figures symbolise a contradictory mix of what Worster calls 'arcadian' and 'imperialistic' attitudes to nature. The first, which we might regard as ecocentric, searched for intrinsic value, order and purpose in nature and life. The second was technocentric, emphasising management of nature for human instrumental purposes, by desacralising it and viewing it as a kind of machine.

White and Linnaeus

Gilbert White (1720–93), the naturalist and parson of Selborne in southern England's chalk country, was more of an arcadian. His *Natural History of Selborne* (1789) both described local plants and animals and 'grasped the complex unity in diversity that made of the Selborne environs an ecological whole'. He noted how cattle's excrement furnished food for insects and therefore eventually fish, and that 'nature is a great economist for she converts the recreation of one animal to the support of another' (Worster 1985, 7). And he saw the importance of each apparently humble creature to the 'chain of nature'.

In expressing the rudiments of such ecological concepts as interdependence and food chains, White was aware of natural science's potential usefulness to humans. But, much more, he desired to rediscover the spirit of harmony with nature which he had found in classical literature such as Virgil's *Eclogues* and *Georgics*. Like the poets Cowper, Gray and Thompson, he saw the English countryside as counterpart to Virgil's arcadian ideal: 'a compelling idyll of contentment and peace'.

H. J. Massingham, one of White's interpreters, emphasises how he portrayed Selborne as a community of people as well as animals and plants. Contemporary developments beyond Selborne – the French and American revolutions, factory-based industrialism, scientific agriculture – were of no consequence to White. But such developments forged the climate in which White's writing eventually became popularly regarded as an antidote to the evils of modernism.

The whole genre of scientific-literary, synthesising naturalist writing which White ultimately spawned lasted into the twentieth century (through, for instance, the geological/geographical surveys of Jukes-Browne 1895, Osbourne White 1909, or Wooldridge and Ewing 1935). It shows a 'feel' for the countryside which is uncharacteristic of more clinical, dispassionate, analytical science (compare modern geological monographs). It could, Worster says, be seen as a route back to nature from industrialism, but via science rather than romanticism.

If White's ecology was thus in an 'arcadian pastoral' tradition, Linnaeus represented a purer 'reasoned' and instrumental approach to its object of study. Though he too was a mix, holding his local Scandinavian countryside in reverential awe, his main contribution was to classify nature into 'neat rows of shelves and boxes' (Worster, p. 32). His *Oeconomy of Nature* (1749) was 'the single most important summary of an ecological point of view still in its infancy'. It describes the cyclical nature of geological-biological interactions and how this pattern was uniform through nature. Nature was a rational order and harmony of God's design: God had benignly assigned to each creature a specific place (niche) so that competition for resources could be minimal in a community of peaceful coexistence and abundance. 'There are no scarcities in nature' was his anti-Hobbesian message.

But humans had a special place in all this, by which they could and should exploit fellow species to their own advantage: 'All things are made for the sake of man'. Hence the world was an *economy* to be managed rationally for maximum output. The political-economic subtext here is technocentric, contrasting with White's ecocentric ecology.

Worster (1985) describes a Linnaean school that looked to reason and nature to express its Christianity (epitomised in the title of William Derham's book, *Physico-theology*, 1713, or William Paley's *Natural Theology* 1802). One of its axioms was that God had created in nature an integrated order functioning like a well-oiled machine.

Some went further, bringing in the idea of the vital spirit – that which distinguished organic from inorganic things, and could not be analysed, reduced or otherwise split into component parts. It was a mysterious force, for which only a God could be responsible. John Ray, for instance, in *Wisdom of God Manifested in the Works of Creation* (1691) introduced the concept of a 'plastical power', so that living creatures could not be seen as mere automatons. Worster says that this life force, which came from the creator's goodness overspilling from the top of the Chain of Being (see Chapter 3.1) was a compromise between the mechanical model of nature and pagan animism.

Such vitalism continued from pre-modern tradition, and can be traced forward as a minority current throughout most of the modern period. Merchant (1982) considers Anne Conway to be one of the prime linking philosophical figures in this movement. Her *Principles of the Most Ancient and*

171

Modern Philosophy (1690) was fundamentally anti-Cartesian. She was a monist, who saw matter and spirit, body and soul, as different aspects of the same substance; a substance infused with the vital spirit. Much of her thought converged with that of the philosopher Gottfried Leibniz (1646–1716), who considered that reality consisted of independent monads (indivisible units) that mirrored but did not influence each other. While contributing to the growing rationalist and mechanist trend in classical science, Leibniz was also informed by vitalism. He considered that life and perception permeated all things, and that monads moved and acted according to their own internal will. Merchant sees him, therefore, as part of this lineage of 'dynamic vitalism' that effectively opposed the metaphor of the dead mechanical universe in classical science. Leibnitz's principle of *self-contained internal development* in nature, which counterpoises the thermo-dynamic view that nature moves towards increasing disorder, is also strongly reiterated in contemporary mainstream green literature and in the eco-anarchism of Murray Bookchin (1990) (Chapters 5.3, 6.1).

Another Linnaean axiom was God's beneficence in having ensured plenitude by creating niches – fixed places in nature's chain – for each creature. Conservative notions of hierarchy, anti-democracy and sacrifice of the individual to the *Gemeinschaft* came in on the coat-tails of this (again, pre-modern) principle.

But a third axiom was fully progressive and modern: the idea noted above that humankind has God's permission to manage nature's economy – its ecology – for its own profit. This was born of a technocentric dislike of primitivism and a desire for material progress, following the developing mainstream utilitarian culture of the time. It echoed, Worster points out, the values of industrialists, and of the Linnaean disciple Adam Smith, who conceived of nature as a storehouse of raw materials for society's ingenuity to unlock.

4.2 MALTHUS AND THE ROOTS OF THE ECOCENTRIC–TECHNOCENTRIC DEBATE

Ecocentrics and technocentrics may be poles apart in some respects, but they do, paradoxically, both start from Malthusian premises. The neoclassical economics underlying technocentric managerialism assumes that economics is fundamentally about allocating inherently scarce resources (Chapter 2.3). Ecocentrism, too, is also 'founded on a fundamental commitment to the principle of scarcity as an unsurmountable fact of life, and the consequent limits to growth imposed by a finite system' (Dobson 1990, 80). Furthermore it believes that 'Environmental impact is a function of not just technology and affluence but *absolute* human numbers, which is why we cannot afford simply to wait for the "demographic transition"' (Eckersley 1992, 159).

For the roots of the modern limits to growth argument (Chapter 2.3) we need to turn to Thomas Malthus (1766–1834). We could go further back in tracing concern about a possible conflict between population growth and subsistence. The Linnaeans proposed that there were limits to the carrying capacity of an area. But, as Hollingsworth (1973) says: 'It was the attempt to be specific and mathematical about the dilemma of an expanding population that is Malthus' special contribution to thought.' As Worster (1985, 152) has it: 'what were unprecedented in Malthus' argument were the ironclad ratios and his warnings of impending national apocalypse'.

Malthus' *Essay on the Principle of Population* can be seen, then, as an attempt to make a *scientific* contribution to the debate. It was also the first one which attempted to take a *global* perspective: one that is now very much the hallmark of ecocentrism.

Malthus wrote at the end of the eighteenth century, when industrial capitalism, bolstered by science and technology, was beginning to pre dominate. A debate about how far it could take society towards progress was already under way. This debate pre-empted the modern one between ecocentrics and technocentrics, and it contained elements of the fundamental philosophical question concerning free will and determinism (Chapter 2.7).

In the debate many see Malthus as a pessimist among a growing band of technological optimists. He proposed that the earth sets limits to growth in human populations and in material well-being. According to Glacken (1967) Malthus' pessimism stems from contemporary observations by travellers and explorers of just how teeming the Earth was with a rich variety of life. This abundance and fecundity of living organisms – Earth's plenitude – was so great as to suggest that each species could multiply indefinitely were it not for the existence of checks to multiplication. It led people towards what Glacken calls 'absurd extravagances', such as Count Bouffon's observations in 1749 that a single species could cover the entire earth, were no obstacles placed in its way, or Darwin's view that the elephant (the slowest breeder) could stock the whole earth in a few thousand years – or Malthus' own statement in his *Principles of Political Economy* that humans could theoretically fill not only earth but all the planets in the solar system. It was clear, however, that such extravagances had not happened, so therefore populations were being held in check by wars, famines and other necessary but 'unhealthful events', and by competition within and between species.

Indeed, some felt that such events had been only too effective, so that the number of people on Earth had actually declined since classical Greek and Roman times. This thesis, of natural *senescence* is the opposite of the idea of progress. It involves the organic metaphor, arguing that earth is decaying, just as all organisms do from the day of their birth:

173

You are surprised that the world is losing its grip? That the world is grown old? Think of a man: he is born, he grows up, he becomes old. Old age has its many complaints: coughing, shaking, failing eyesight, anxiety, terrible tiredness. A man grows old: he is full of complaints. The world is old; it is full of pressing tribulations.

(St Augustine, AD 410, cited by Mills 1982, 243)

In 1755 Robert Wallace postulated that the ancients had been more numerous because of their moral superiority over the moderns. His thesis was that the environment limits the numbers and wellbeing of humankind, and Malthus developed this into the view that there were insurmountable natural barriers blocking the realisation of utopian dreams of unrelenting social progress via reforming human institutions.

Malthus' position particularly opposed William Godwin and Condorcet (Chapter 3.2). Both felt that 'overpopulation' in relation to subsistence needs would never seriously threaten progress, for if such a contingency arose in the distant future it could be overcome with technological skills, or by the wisdom of future societies who would see the danger and refrain from overbreeding. As has been suggested, such optimistic assertions corresponded to the spirit of the times, and they were associated with the reality of rising bourgeois revolutionary fervour in Europe.

It was to counter such revolution that Malthus preached in the first edition (1798) of his *Essay* what was essentially a polemic against Godwin, Condorcet and revolutionaries like Tom Paine (*The Rights of Man*, 1791). It consisted of what became Books III and IV of later editions of the work, and it did not have the 'scientific data' which appeared in the 1803 (second) and subsequent editions as Books I and II. In eventually producing such data, Malthus showed himself also to be at least partly a man of his time. For, as Petersen (1979) observes, the age was one where moral philosophy (the direct application of Christian values to current issues) was giving way to political economy: the defence of 'natural' rights set in the context of a 'natural' order of things with less and less reference to a deity. In other words, it was a period where the empirically derived laws of nature (and the classical economic laws being formulated by Adam Smith) were attaining ascendency over the supernatural in a clear break from religious to secular views.

Establishing such laws required data collection. Because he talked of natural rather than Christian laws, Malthus was seen by some as an atheist – especially since he seemed to be warning against the injunction in Genesis to 'be fruitful and multiply'. However, we need to distinguish here between primary and secondary causes. Malthus held the laws of nature as secondary causes or determinants on society. The primary cause, or ultimate determinant, was still God, working through natural laws.

174

Malthus' thesis

Malthus said that human population

> When unchecked, goes on doubling itself every twenty-five years, or increases in a geometrical ratio . . . the food to support the increase . . . will by no means be obtained with the same facility. Man is necessarily confined in room when acre has been added to acre till all the fertile land is occupied, the yearly increase of food must depend on the melioration of the land already in possession. This is a fund, which, from the nature of all soils, instead of increasing must be gradually diminishing. But population, could it be supplied with food, would go on with unexhausted vigour; and the increase of one period would furnish the power of a greater increase of the next, and this without any limit.
>
> (Seventh edition of the *Essay*, 1872, Book 1, Chapter 1)

This tendency for population to push against the bounds of food supply results from the 'passion between the sexes', and is inescapable, although it can be contained. Agricultural production, at best, can increase only in arithmetic proportions. Assuming the best

> the human species would increase as the numbers 1, 2, 4, 8, 16, 32, 64, 128, 256, and subsistence as 1, 2, 3, 4, 5, 6, 7, 8, 9. In two centuries the population would be to the means of subsistence as 256 to 9; in three centuries as 4096 to 13, and in two thousand years the difference would be incalculable.

Malthus was sceptical that agricultural production could be indefinitely increased, even arithmetically, because of the increasing need to use marginal land. Obviously, therefore, this 'principle' or law of population represents an impediment to 'the progress of mankind towards happiness'.

However, while in plants and irrational animals potential population increases are often reached, to be reversed afterwards by lack of room or nourishment, humans, by process of reason, react in a more complicated way to 'the constant tendency in all animated life to increase beyond the nourishment prepared for it'. Impelled by instinct to increase, humans ask if inability to provide for their children will not bring them down economically and make them obliged to seek charity. This reasoning prevents early marriage, and such 'moral restraint' forms one of a series of *'preventive checks'* to unbounded population increase. Malthus approved of such chaste restraint, but some other preventive checks constituted 'vice' in his view, for example contraception, abortion, promiscuity with prostitutes, homosexuality.

Malthus thought he observed that preventive checks were applied unevenly: more so in Europe than in 'primitive' societies. Control over food supply in the latter was slight and therefore population tended to oscillate around its means of subsistence (as with animals).

Malthus also deemed that preventive checks were practiced less by poor and uneducated 'classes' in Europe, whereas wealthier and better educated people married later and had smaller families. Their wealth was unharmful – indeed a good thing – but for the poor any increase in income would lead to increased family size because they were not likely to practise preventive checks and because other restraints on procreation, such as starvation, disease and illness, would have been lifted. These kinds of restraints, together with wars, Malthus called *positive checks* on population growth, or 'misery'. Where preventive checks were not applied, then positive checks would naturally and inevitably come more into play – unless poor relief interfered temporarily with their operation.

Nowadays, technocentric objectors to Malthusianism argue that while human populations may *tend* to increase geometrically, human ingenuity and the technical ability to get round the problem can grow infinitely. Demographers, meanwhile, often criticise the economic aspect of Malthus' thesis. This argued that rates of population growth for some groups increased steadily as their income rose. Widespread economic growth, particularly if it benefited the poor, would therefore lead automatically to increased breeding, which would defeat the ends of economic growth because the total wealth would have to be shared among so many more people. Population increase therefore increases 'misery'. It lowers wages, for when there are more people the labour market is depressed. Lower wages induce cultivators and landowners to employ more labour, thus agricultural productivity rises, increasing carrying capacity and encouraging a further population surge. This will go on until ultimate limits are reached. During depression poverty leads to postponement of procreation (of 'marriage'), but with the return of prosperity more people will begin families, building up population pressure on subsistence. Since this positive correlation between agricultural production and the number of marriages and children, thought Malthus, applied mainly to the lower classes, it was hidden from demographers and historians, because most recounted history tended to be the history of the upper classes.

History seems to have disproved Malthus' thesis, although ecocentrics often hold on to it. For with economic growth in the West has come increased contraception and population stabilisation and/or *decrease* among all socio-economic groups. Modern demographers use the term 'demographic transition' to describe the observed fact that as a nation's average family income and living standards rise and death rates fall, birth rates also fall.

Political critics of Malthus from both left and right (Chase 1980, Simon 1981) seize on his very clear ideological message: that because of the *scientific* law of population (and not because of uncaring or incompetent governments) attempts to raise the economic status of the poor as a whole were doomed to be self-defeating, as were other liberal and socialistic ideas of social improvement. Malthus' use of his science as ideology is discussed in Chapter 5.5.

Malthus as a scientist

Although Malthus was apparently motivated by politics, his *Essay* has considerable claim to be regarded as a scientific work, in the Baconian sense. For his 'principle' or scientific law was ostensibly

> a general statement of fact, methodologically established by induction on the basis of observation and experiment and usually . . . expressed in a mathematical form . . . Ideally, scientific laws are strictly universal or deterministic in form, asserting something about all members of a certain class of things.
>
> (Bullock and Stallybrass 1988, 761)

Modern biographers and critics do not seem to doubt that Malthus can indeed be viewed as a scientist. Hollingsworth (1973), a demographer, describes him as a moralist and politician, but also as a social scientist and a demographer 'tracing the population trends in various countries as accurately as he could, and arguing about the reasons for their differences'. Schnaiberg (1980) called Malthus an 'early and distinguished' social scientist, and Malthus seems to have achieved some distinction among contemporary scientists, being an FRS, a Fellow of the Royal Statistical Society and a member of the Political Economy Club.

His *Essay* appears to have replaced deduction by induction as it developed. The first, polemical, edition started with two postulates of 'fixed laws of nature': that 'food is necessary to the existence of man' and that the 'passion between the sexes is necessary and will remain nearly in its present state'. From these it deduced that the power of the population to increase was greater than that of Earth to provide food, introducing geometrical and arithmetic ratios as additional postulates.

The approach, however, changed for the second and subsequent editions, where Malthus laid out empirically derived evidence from which he induced the principle of population. No less an economist that Alfred Marshall rated it as 'One of the most crushing answers that patient and hardworking science has ever given to the reckless assertions of its adversaries' (cited in Petersen 1979, 53). Hollingsworth describes Books I and II as devoted to a remarkably complete demographic survey for the time, and Petersen (p. 19) remarks that 'Malthus spent a lifetime labouring to improve the statement of his theory and gathering facts on which to base these emendations', though he admits that some thought the demographic data 'of dubious quality'.

Indeed the survey seems very complete, so that any law which derives from it can be thought of as universal in application. Book I was about the checks to population in 'the less civilized parts of the world' and in past times. It covered the American Indians, South Sea Islands, ancient Northern Europe, modern 'pastoral nations' (Asia), parts of Africa, Siberia, the Turkish dominions and Persia, Indostan and Tibet, China and Japan, and the ancient Greeks and Romans. Book II, on the checks to population in modern Europe, covered Scandinavia, Russia, Middle Europe, Switzerland, France and the UK, and it ended with a chapter of general conclusions.

The sources of this evidence, when population data were generally sparse and inaccessible, were various, and mainly secondary. There were reports from missionaries, voyagers and explorers. There were geographies and histories and books of observations on customs and habits of far-off lands. There were also population registers and indices of prices. James (1979) says that 102 authorities were consulted. In addition, anecdotal material was collected assiduously from travellers: especially from a lifelong friend, Edward Clarke, who visited Russia and many other places, seeking all the time the answers to Malthus' written questions on checks to population. And there were Malthus' own travels to France and Switzerland, and to Norway in 1799. Petersen (p. 50) paints a picture of him as an enthusiastic scientist, 'earnest without being pedantic; interested in just about everything', and making jottings on finance, economic trends, profits and wages and population – including the institutional arrangements to prevent early marriage.

Petersen admits, however, that his Scandinavian trip was 'a rather casual affair', that he knew no Norwegian and that he visited only portions of that country. Certainly we might today think his evidence too unsystematic and anecdotal. James (p. 97) records how his chapter on Switzerland used the

> Philosophic discourse of the driver of our char on the overpopulation of his country, he [the driver] complained much of the extreme early marriages, which he said was the 'vice du pays'.
>
> (extract from Harriet Eckersall's diary)

And Chapter 4, Book I, on the checks to population among the American Indians is typical of the long rambling accounts with their tenuous sweeping generalisations which make it difficult for modern readers to see his strengths as a scientist:

> It was *generally remarked* that American women were far from being prolific. This unfruitfulness *has been attributed by some* to a want of ardour in the men towards their women . . . it . . . *probably* exists in a great degree among all barbarous nations whose food is poor and insufficient, and who live in constant apprehension of being pressed by famine or by an enemy. Bruce *frequently takes notice of it,* particularly in

reference to the Galla and Shangalla, savage nations on the borders of Abyssinia, and Valliant *mentions* the phlegmatic temperament of the Hottentots as the chief reason of their thin population. It *seems to be* generated by the hardships and dangers of savage life.

(emphases added)

So it goes on, and on – with stories of habits and customs in marriage, of the kinds of common diseases in different countries, of cannibalism (positive or preventive check?), of propensity to wars, of agricultures and soils and climates. And, despite this attempt at empiricism, we cannot but agree with Hollingsworth when he says: 'The real weakness of Malthus as a demographer is, to a modern mind, that he is too fond of general theories and not interested enough in empirical results'. This verdict is supported by a letter, cited in James, from Malthus to the Director of the Statistical Bureau in Vienna:

It is quite true as you observe that the vast provinces belonging to the Austrian Empire have come very little under my notice. One reason was . . . that the statistical information relating to them was not very easily accessible; and another was, *that the information which I could collect without much difficulty appeared to me to embrace a sufficient number of instances to establish the principle which I had in view.*

(emphases added)

This bears out Chase's (1980) contention that Malthus 'was not a man to be intimidated by mere facts. To his dying day he was to cling to his central dogma.' This was that the ratio between levels of human food and baby production was fixed by natural laws.

Certainly Malthus, in his world survey, clearly maintained that he sought to establish a natural law: one universally applicable principle operating uniformly in time and space. As Glacken remarks, the whole earth was considered the proper unit of study, and the principle of population was thought constant in its operation everywhere. Hollingsworth attributes this desire to reduce 'human demographic and social development to two mathematical laws, of geometric and arithmetic progression' to Malthus' training in mathematics.

There remain questions, however, of how valid Malthus' postulated relationships were, whether they were precisely mathematically structured or not. Although Malthus may have seen a positive correlation between economic growth and the number of marriages among the poor, to claim that the correlation evidences causal relationship is quite another matter. Indeed, from the data available Malthus could perhaps have reached different interpretations and conclusions. It was true that at the time there was little empirical justification for associating economic growth with greater health and civilisation and declining population growth. Indeed from the visual evidence it appeared that only dirty and overcrowded towns were associated with economic growth. But on the other hand Malthus did have data showing that healthier (and wealthier) communities with lower death rates also had lower birth rates. He could have induced a causal relationship between the two and gone on to make appropriate social recommendations involving substantial wealth redistribution. That he did not appear to advocate much change from the social and political status quo is perhaps attributable to a certain lack of value freedom. Whether such a lack disqualified him from being regarded as a scientist, however, is doubtful, and is a theme returned to in Chapter 5.

4.3 DARWIN AND HAECKEL: THE WEB OF LIFE AND ORGANICISM

Darwin

Along with Malthus, Charles Darwin (1809–82) can be regarded as a towering nineteenth-century figure whose ideas inform, among other things, the scientific strand in modern ecocentrism. Though the history of the ecosystems idea and systems theory is complex and shows many influences, one can recognise their fundamental outline in Chapter 3 of *The Origin of Species* (1859). One can also find

> two contradictory moral implications in Darwinism: the mainstream Victorian ethic of domination over nature and an emerging biocentric attitude that was rooted in arcadian and romantic values.
>
> (Worster 1985, 114)

The *Origin*, together with Thomas Huxley's *Evidence as to Man's Place in Nature* (1863) was born of an era of massive empirical observation of the natural world. 'I worked on Baconian principles and without any theory collected facts on a wholesale scale', wrote Darwin (cited in Ferris 1990, 233). Darwinian evolutionary theory was therefore cast in the mould of classical science – but its results and implications ran counter to that science's Cartesian view of humans as separate from nature. For it placed humans squarely *within* the rest of nature.

Darwin and Wallace, and particularly Huxley, drew a close analogy between humans and animals, emphasising, for instance, the structurally similar features between *Homo sapiens* and apes, and their common ancestry. Furthermore, humans constituted just one of many species on Earth – no more, no less. Most important, *all* species were linked intimately by a 'web of life': the modern, more sophisticated (because not simply linear) equivalent of the pre-modern Chain of Being.

In Darwin's scheme, which was but one of many evolutionary theories so popular in the nineteenth century (see Oldroyd 1980), variations occurred between individuals in a species substantially by chance. They were passed on (selected) via those who succeeded in that struggle for survival between individuals which resulted from the Malthusian problem of there not being inherently enough resources to go round: a 'problem' which Darwin accepted as read from Malthus.

Hence, individuals who were 'fittest', that is, had features which were best adapted to an environment, were more likely to survive than those who were poorly adapted. It would be natural for the latter to die out. In this aspect of the theory, Darwin and Huxley clearly and openly took the ideas of an inherent struggle for resources, competition, and survival of the fittest from Malthus. Hence ideas which came from observations of *human society* were applied to *nature* by Darwin, who asserted that it was 'natural' for animals to behave in these competitive, Hobbesian ways. In fact, the phrase 'survival of the fittest' was taken from Herbert Spencer, a sociologist constructing a theory of evolution of *society*. Darwin had metaphorically transferred the characteristics of social practice to biology. The ideological implications of doing this are discussed further in Chapter 5.

Darwin himself did not deny his social influences, but at the same time could lay some claim to resting his theory on prodigious empirical evidence gained from painstaking scientific investigation. His inductive approach was founded on the age of discovery, starting with the sixteenth century voyages of exploration and classification of nature in 'far-off' parts and in Europe. Zoologists, botanists, geologists – all were engaged on a grand empirical exercise. Bouffon, Linneus and von Humboldt were some of the leading figures. This, again, cannot be divorced from a social context – discovery was related to the age of colonialism and imperialism, for the expansionary dynamic in Western industrialism made it necessary to find new resources, sources of cheap labour, and eventually new markets.

Darwin's work heavily emphasised observation in foreign lands (e.g. the voyage of the Beagle which helped him to develop his theory of coral reefs). He says, on page 1 of his *Origin*:

> on my return home, it occurred to me in 1837 that something might be made out of the question by patiently accumulating and reflecting on all sorts of facts . . . after 5 years' work I allowed myself to speculate on the subject.

This describes, par excellence, the process of induction.

The idea of systems figures strongly in his work – it carries a message about the relationship between all the species within nature, including humans. Chapter 3 of the *Origin* is entitled 'The complex relations of all animals and plants to each other in the struggle for existence'. In it he describes how, when heathland in Staffordshire is converted to coniferous woodland, changes in bird and insect life follow. In Farnham, he notes how cattle determine the existence of the Scots Fir – when cattle are fenced out of a heathland area the firs grow. In Paraguay, he observes how the number of insectivorous birds controls the number of parasitic insects, which controls the number of flies on which they feed, which controls the number of eggs in the navels of newborn cattle, horses and dogs. If the number of flies' eggs decreased this would cause the habits of the livestock to change (they would become feral – in a wild state, roaming from captivity). This would alter the vegetation of different areas: 'this in turn would largely affect the insects and this the insectivorous birds, and so onwards in ever increasing circles of complexity'. Thus he describes a Paraguayan 'ecosystem', stressing the interdependence of all the components of that system. He also says:

> In the long run the forces [of competition and struggle] are so nicely balanced that the face of nature remains for a long time uniform, though assuredly the merest trifle would give the victory to one organic being over another.

Here is the idea of systems in dynamic equilibrium. And he stresses the intricacy and complexity of relationships in nature when describing the relationships between cats and flowers. Cats influence the number of field mice. The mice influence the number of bees, because they destroy combs and nests; and bees, through pollination, control the frequency of hearts-ease and red clover. Thus

> The dependency of an organic being on another, as of a parasite on its prey, lies generally between beings remote in the scale of nature . . . the structure of every organic being is related, in the most essential yet often hidden manner, to that of all the other organic beings with which it comes into competition for food or residence, or from which it has to escape, or on which it preys.

And we are greatly ignorant about the 'mutual relations of all organic beings', partly because of the complexity of these relationships, which are harder to establish than are those governing non-organic objects:

> Throw up a handful of feathers, and all fall to the ground according to infinite laws; but how simple is the problem where each shall fall compared to that of action and reaction of the innumerable plants and animals which have determined, in the course of centuries, the proportional numbers and kinds of trees now growing on the old Indian ruins.

Here Darwin also describes the relative indeterminacy of *biological* laws by comparison with physical and chemical laws, so that with the first the power of prediction is weaker because of the difficulty of isolating out simple cause–effect relationships.

Given Darwin's significance it is especially important to ask what his

182

influence was on ecology, and therefore on ecocentrism. The answer, in Worster's account, is ambiguous. Like the eighteenth-century naturalists, Darwin's ecology contains several sometimes conflicting views of nature and society's relation to it.

On the one hand, Darwin's early experiences inclined him to an 'anti-arcadian' hopeless view of nature. In the Galapagos Islands he saw a bleak, depraved and hostile landscape. In South America he saw fierce competition for space, the decimation of indigenous species by European invaders and a fossil record that attested to much extinction. A stay in London made him recoil from what he saw as the results of human nature and a process of economic competition and struggle. Hence by the time he read Malthus' *Essay* in 1838 he was already 'well-prepared to appreciate the struggle for existence which everywhere goes on' (cited in Worster 1985, 149) – to regard such processes as universal principles.

Malthus was the most important of several influences on Darwin. Naturalist Charles Lyell was another. In his *Principles of Geology* (1830–3), volume 2, Lyell discussed the processes by which animals and plants became dispersed, and how such dispersal everywhere put individuals and species into fierce competition for food and space. The result was violence – a universal natural principle. Along the lines of 'what is, is right', Lyell proposed that this violence was justified, and this included human violence towards nature.

Darwin's theory followed and ultimately contributed to this continuation of the 'imperial' tradition in ecology. In so doing, it reflected not merely influences from individuals, 'great thinkers' though they may have been. It also reflected the dominant intellectual milieu of the time, which in turn was inseparable from developing economic activity, that is, from what people were doing to nature and each other via industrial capitalism.

The ethic of dominating and controlling 'wild nature' identified by Merchant (see Chapter 3) was popularised in much later nineteenth-century painting, poetry and music. What Worster calls 'a superabundance of terror' was commonly found in portrayals of nature, 'red in tooth and claw'. Tennyson's aphorism here was reflected in the landscapes of natural carnage painted by the celebrated artist Edwin Landseer. As Gold (1984, 26) says of his images of thuggish, sadistic, savage animals: 'it is difficult to look through the body of his work without feeling that these images of animal violence are really statements about society'. Darwin gave such interpretations of nature – and, through social Darwinism, society (Chapter 5.6) – the support of ostensibly impersonal and objective (i.e. empirically-derived) 'facts'.

But, as suggested above, other themes in Darwinism reflect the benign ecological tradition which ecocentrics today prefer. Most of all, the web of life implied for Darwin that nature was 'one grand scheme' of cooperative integration, in which the most insignificant beings are important. Darwin's 'biocentric view' saw a brotherhood of creatures in a 'community of descent' from ultimately common origins.

Darwin also laid the foundations of the modern ecocentric regard for diversity. To avoid competition for a niche, he argued, species could carve out new niches (the term here refers not just to a particular space but also to a function or role within the ecosystem). This (underdeveloped) aspect of his theory could spell not only peace between creatures but also (as Herbert Spencer had argued for human societies) an increasing number and complexity of functions as a mark of the evolved, sophisticated society.

Worster detects such an 'arcadian' strand of Darwin's thought as a result of the influence of the naturalist von Humboldt. Darwin enthusiastically read his *Personal Narrative* (1799–1804) of his travels in Latin America. This presented a holistic view of nature; for instance in the volume on plant geography Humboldt saw plants not as individual species but as ensembles, grouped in relation to different (climatic) environments. In this, Humboldt himself was influenced by the romantic philosopher Goethe. Indeed, Worster calls Humboldt's work, with its message of ecological harmony rather than conflict, a meeting of science and romanticism.

Its influence on Darwin, however, seems to have been outweighed by the theme of conflict. For even when Darwin wrote of diversifying niches as ways of competition avoidance he did not regard this solution as permanent: sooner or later the old occupants of a niche would be thrown out by newcomers. Indeed since this was part of the evolutionary 'mechanism' it was part of the process by which all life progressed towards greater efficiency and perfection. Such a mechanism could hardly be more contrary to the gentle, arcadian relationships which ecocentrics want to see as part of their ideal society, but it is part of the legacy inherited by the modern ecological model, which ecocentrism is generally so enamoured of.

Haeckel

The science of 'ecology' was further developed by Ernst Haeckel (1834–1919), who was the first to use the term, in 1866 in his *Generelle Morphologie*. He derived the word 'oecologie' from the Greek 'oikos', meaning 'household', which is also the root of the word 'economics', a point stressed by modern environmentalists. Haeckel defined ecology as the study of all the environmental conditions of existence, or the 'science of the relations of living organisms to the external world, their habitat, customs, energies, parasites etc.' (Worster 1985, 192). Ecology was about the 'economy of nature' (p. 298).

Anna Bramwell (1989) says that Haeckel's influence on modern ecologism was considerable. He helped to shift biology away from affinities with classical science philosophy, towards a holistic view, carrying a message for humans about the society–nature relationship. Haeckel was overtly political.

He founded the 'Monist League', was a republican atheist, and practised nature worship. He influenced D. H. Lawrence, and through Lawrence, the founders of the Soil Association (now a major organisation representing organic farmers and gardeners). Bramwell considers that he created the scientific root of ecologism, giving credence and scientific legitimacy to romantics who were alienated by the effects and values of industrial capitalism.

She says (p. 43) that Haeckel was an ecologist in the political sense in three ways:

> Firstly he saw the universe as a unified and balanced organism . . . Hence, his monism, whether defined as all matter or all spirit. He also believed that man and animals had the same moral and natural status, so he was not man-centred. Thirdly, he preached the doctrine that nature was the source of truth and wise guidance about man's life. Human society should be organised along the lines suggested by scientific observation of the natural world.

Here, we recognise characteristics of today's ecocentrism which appear particularly strongly in deep ecology. They are holism (monism), the bioethic and the 'nature knows best' principle.

Haeckel very much opposed the *dualism* of the mind–body split in classical science, and by implication the human–nature, emotional–rational dualisms. As Bramwell puts it, 'Monism's opposition to dualism was that mind and matter were one, because the universe existed at one level'. You could see the universe as all mind (as in phenomenology and idealism) or all matter (as in scientific materialism), it did not matter which – it was all one. Haeckel drew on the ideas of Goethe, Lucretius, Bruno, and Spinoza, saying they had showed 'the oneness of the cosmos, the indissoluble connection between energy and matter'. There was no such thing as empty space, it was filled by 'ether' and atoms.

As an ardent Darwinist, Haeckel also argued for equality between animals and humans, who shared a common origin in the Tertiary era (indeed, he was one of those who argued that there must be a 'missing link' between humans and apes, see Coleman 1971). He denied the assumption of humanism, that humans were unique and special, believing instead that the higher vertebrates also displayed 'the first beginnings of reason' and 'the first traces of religious and ethical conduct . . . the social virtues . . . consciousness, sense of duty and conscience'.

> Monism teaches us that we are . . . children of the earth who, for one or two or three generations, have the good fortune to enjoy [its] treasures.
> (cited in Bramwell, p. 49)

Haeckel wrote this a hundred years ago, and ecocentrics are repeating it word for word today. Later ecocentrics also share Haeckel's interest in

185

Buddhism, 'a religion that gives equal status to all'. Haeckel even pre-echoed Lynn White's attack on Christianity for elevating humans above animals and separating the former from nature. He argued for a nature *impregnated with god* – not a God separate from and above nature. By extension, since we are nature too, we are god – a belief which permeates New Ageism today to the chagrin of orthodox Christians.

There are many more points of contact between Haeckel and modern ecocentrism, especially deep ecology. For instance, as a prophet of the Volkish movement (see p. 188) he proposed that an individual could belong to something greater than him- or herself. That is, people had a mystical unity with others and with the whole of the cosmos. And the process of evolution was not seen as a mechanical one, as Darwin saw it, but as a cosmic force, a manifestation of the creative energy of nature (Gasman 1971). Haeckel's evolutionary monism is recapitulated in contemporary ecocentric ideas, calling not only on romanticism but also on twentieth-century science, about humankind's relationship to nature and about social change and evolution (Chapters 5.3, 6.1).

Evolution, says Gasman, together with industrialism and the discoveries of the scientific revolution, formed the basis of the new pantheistic religion which Haeckel wanted, to be led by a biological élite forming the nucleus of strong state power which would bring the German people in harmony with the laws of nature.

In all this, Haeckel brought apparent scientific authority to 'Volkism's essentially irrational and mystical ideas' (Gasman, p. xxiv). His 'biogenetic' law of 'recapitulation' stated that the biological development of the organic individual must in abbreviated form recapitulate the development – the evolution – of its ancestors. Thus each individual re-lived and re-experienced the evolutionary process, sharing in the eternal cycles and oneness of nature. This theory, now discredited, was influential for half a century. In developing it, and other parts of his science, Haeckel 'habitually perverted scientific truth to make certain of his doctrines more easily assimilable'. These doctrines were 'a system of the crudest philosophy . . . a mass of contradictions . . . the habiliments of a religion, of which he was at once the high priest and the congregation' (Singer 1962, 487–8). Thus Haeckel's science appears to have been driven by ideological princi-ples: another trait which, perhaps, he shares with today's ecocentrics.

On one issue in particular Haeckel did not see eye to eye with some other leading Darwinists of the age. Huxley and others were mechanists, but Haeckel was a vitalist. He thought that nature, including humans, had a life force which was not understandable by reductionism and analysis. One of Haeckel's students, Bramwell tells us, was Hans Driesch, who went on to make the case for vitalism in the 1900s and 1910s. He proposed that there was a *teleology* (an inherent purpose) behind everything material, whether we recognised it as living or non-living. Indeed the only distinction between the last two states was that in living things the life force operated at a more intense level. Life was governed by a higher will and purpose – this will and purpose was the life force.

In such ideas, Haeckel had infused modern science with pre-modern views about nature, and his organicism was part of modernity's counter-cultural tradition (Table 1.6, p. 36). This tradition was carried on by the young Karl Marx (the society–nature dialectic), Henri Bergson (1859–1941, who postulated an *élan vital*), Driesch, Wilhelm Reich (1897–1957, who developed 'bio-energetics' based on a life force, as in sexual energy), Jan Christian Smuts (1926), Alfred North Whitehead (Chapter 5.2) and the 'ecology' movement of the 1920s and 1930s. It is part of the Gaianist wing of ecocentrism. Bramwell says that vitalism disappeared from scientific view in the early twentieth century and 'retreated into philosophy. It remained a vigorous subculture, finding expression in existentialism as well as popular science after the Second World War'.

Human ecology and geography

Darwin and Huxley put humans squarely within nature and therefore as part of 'ecosystems'. From 1910 the term 'human ecology' was used for the study of humans and their environment together (Stoddart 1966). This is a way of describing the science of geography, and Barrows (1923) gave an influential presidential address to the Association of American Geographers, where he described a future paradigm for geography and geographers. 'Geography,' he said,

> will aim to make clear the relationships existing between natural environments and the distribution and activities of man . . . from the standpoint of man's adjustment to environment.

The organising concept for this deterministic human ecology paradigm was to be the region. In regional geography humans and environment would be studied in a *synthetic*, holistic, way. Physical and human systems would be described as well as their interrelationship – to produce the image of a well-defined spatial identity with unique character – a living ecosystem. This type of regional geography followed the tradition of the French geographers, Vidal de la Blache and Jean Brunhes, who, at the turn of the century, described France in terms of distinctive cultural regions.

The whole region was regarded as a sort of living organism, as was the state. For if the earth is a living organism, so, too, will its natural regions be living. And if the state is based on natural regions, then it also must be an organism. This notion came to have sinister political implications in theories of the state as a Darwinian organism – struggling with others for '*lebensraum*' – espoused by Nazism.

German social Darwinism grew in the nineteenth century and was associated with the rise of the German state. According to Biehl (1993), one of its primary spokesman was Haeckel, 'a mystical racist and nationalist', and hence German social Darwinism was almost immediately married to ecology. Haeckel, says Biehl, gave scientific legitimation to volkism's irrational mysticism.

Volkisch subculture saw German national rebirth as the true flowering of the German folk. It wanted to restore an idealised pagan and medieval golden age, in a reunified Germany. It was a 'palingenic myth', says Griffin (1991, 88) – one which promulgated a new creation (cf. New Ageism).

> The volkists dreamed of binding the individual German to his natural and topographical surroundings, in short, to his regional landscape.
>
> (Gasman 1971, xxiv)

Haeckel's desire to put humans back into nature, then, translated into support for volkism's myth that, despite history and modernism, people are still basically rooted in their home locality, region and nation. This translated into the 'blood and soil' ideology, by which such imagined close bonds between 'traditional folk' and their land was used in the twentieth century to ferment that extreme nationalism that characterises fascism: recognisably 'ecocentric' ideas being enthusiastically espoused by leading Nazis (Bramwell 1989 and 1985, see Chapter 4.6). Partly because of this legacy, but more because 'natural' regions were becoming increasingly difficult to identify and sustain, geographers abandoned this kind of regionalism in their discipline.

But it is coming back into fashion with the rise of ecocentrism; particularly through the idea of the *bioregion*. This is the spatial unit which some ecocentrics (e.g. John Papworth and Kirkpatrick Sale) argue should be the basic *political* unit, to replace the nation state. It would be self-sustaining, and its people and features would be part of a close organic harmony. Its boundaries would be formed of well-defined geographical features, like mountains and watersheds. Like the volkists, bioregionalists dream today of binding us all to our regional landscapes. Their arguments also recall the ideas proposed in the 1950s by Leopold Kohr for true federations of natural regions to replace the nation state (see Chapter 6.3).

4.4 ROMANTICISM, NATURE AND ECOLOGY

Romanticism in essence

Romanticism nourishes modern ecocentrism in two senses. First, it is a particular mental disposition, which today implies a liberated imagination, emotions, passions, irrationalism and subjectivism (Williams 1983, 274–6). As such it may idealise and mythologise whatever it contemplates. The romantic idealisation of nature, countryside and folk societies constitutes a long tradition in art and popular culture, stretching from classical times to today. We have already encountered this in the 'arcadian' strand which Worster detects in ecology's history (Chapter 4.1) and modern ecocentrics tend to romanticise nature, as do Anglo-Americans generally.

Second, the late eighteenth- and nineteenth-century Romantic movement, which championed and developed romantic attitudes, has strong and direct historical links into modern ecocentrism, often through the influence of particular individuals. For instance, William Blake (1757–1827) is described by Wall (1994, 67, 90–1) as a 'poet, holistic philosopher and inspiration of much twentieth-century green thought'. His bioethical sentiments derived partly from a Western holistic tradition and partly from Eastern philosophy. Wall writes:

> The influence of William Blake in synthesising and transmitting such ancient knowledge cannot be understated. Inspired by Blake, Ginsberg helped create the Beat movement of the 1950s; the Beats in turn provided the basis of the 1960s counter-culture, out of which grew the modern Green movement.

The Romantic movement was an artistic and intellectual one, commonly expressed in literature, music, painting and drama. But it was not simply a set of ideas, unrelated to what was happening in the material world. For it was clearly a reaction *against* material changes in society, which accompanied emerging and expanding industrial capitalism in the eighteenth century. In this transition production became centralised in the city, which no longer functioned merely to gather in and consume the countryside's produce. The factory movement and mass production were founded on processes that unleashed and controlled violent natural forces. These processes, allied to the profit motive, 'degraded and despoiled', as some saw it, the environment. Cities grew unprecedentedly, into centres of squalor and deprivation. They began to symbolise the failure of *laissez-faire* liberalism's philosophy that a perfect society could be attained by essentially permitting people to follow their self-interest. Population movement from the land, allied to the rational search for economically efficient production methods (involving division of labour, timekeeping and mechanisation) led to spiritual alienation of the mass of people from the land and from each other. As Marx and Engels perceived, they became units of production – cogs in an impersonal productive machine. People and nature were objectivised, and reduced to commodity status.

Society's upper echelons were also in flux. The new bourgeoisie was displacing the old landed aristocratic order, to which Byron, Shelley and some other Romantics belonged. Partly because of this affinity, Romantics had no empathy with industrialisation and its materialistic culture. As Russell (1946, 653) put it: 'The Romantic movement is characterised, as a whole, by the substitution of aesthetic for utilitarian standards'. Romantics

189

hated how industrialisation made previously beautiful places ugly, and they rejected the vulgarity of those who made money in trade. They separated themselves from both this vulgar bourgeoisie and the working-class proletariat (but not the 'traditional' agricultural peasantry). They promoted the idea that their own labour – unlike that in capitalism – was not reducible to commodity values. *Their* labour was intellectual, and special. It was *artistic* whereas that of skilled workers was *artisan*. Artisans had plain, utilitarian skills whereas artists had aesthetic skills, and were distinguished by the sensitivity with which they applied them. Mechanical production and the associated materialist philosophy were emphatically rejected:

> The truth is, men have lost their belief in the Invisible, and believe and hope, and work only in the Visible . . . Only the material, the immediate practical, not the divine and spiritual, is important to us. The infinite, absolute character of Virtue has passed into a finite, conditional one; it is no longer a worship of the Beautiful and Good; but a calculation of the profitable.
>
> (Thomas Carlyle, Signs of the Times, 1829)

Romantics also championed *freedom of the individual*. This freedom should be expressed by articulating how different one person was from the next. The resultant emphasis on individual spirit, manifested in feelings and passion (which were different for each person), ran counter to the intellectual current of scientific rationalism, where the attempt was always to find laws, that is, *generalisations*, about nature and people, through which their behaviour could be forecast. As a unique, passionate, feeling individual your future behaviour can never be predicted: as one of a group under scientific scrutiny, it can. Hence romanticism is attacked by champions of scientific rationalism:

> They admire strong passions, of no matter what kind, and whatever may be their social consequences. Romantic love is strong enough to win their approval . . . but most of the strong passions are destructive . . . Hence the type of man encouraged by romanticism . . . is violent and anti-social, an anarchic rebel or a conquering tyrant.
>
> (Russell 1946, 656)

Many of the Romantics, notably Thoreau and Ruskin, delighted in the scientific study of nature; they were field naturalists in the Gilbert White tradition. Yet Romanticism was and is the antithesis of many things associated with *classical* science; for instance logical behaviour, order, central control, and the subject–object/human–nature separation. As Whitehead (1926, 106–8) explained, Romantics intuitively refused to accept the way that science abstracted aspects of nature from the whole: theirs was a protest on behalf of the organic view of nature. Indeed it has been regarded as a sweeping revolt against rationalism and the Enlightenment. The Romantics maintained that science was inadequate to explain all the phenomena with which humans are confronted. They regarded aspects of life which are apprehended through intuition, instinct and emotion as the most noble. While scientists denigrated them, Romantics elevated them. Subjective knowledge of, and oneness with, nature, as expressed particularly through art: this was a superior form of knowledge to that of objective, empirical and coldly calculating classical science and its Cartesian dualism.

Romantics exalted in fantasy, imagination and unrepressed depth of feeling. Spontaneity, inner truth and the extent to which the unique point of view of the artist was expressed: all were criteria by which they judged their work and life. A Romantic was also 'sensitive and eager for novelty and adventure', revelling in disorder and uncertainty (Edwards 1972). Romantics feared the dehumanisation which science and 'civilisation' appeared to bring. This fear was symbolised by Mary Shelley's (1818) dreadful product of science, Frankenstein's monster, which ceased to be society's slave and became its destroyer.

This dualism of the rational versus the romantic was not a new feature of Western thought. The eighteenth- and nineteenth-century material conditions of most people particularly encouraged a backlash against rationalism, but the two sides of the dualism have been there since classical times. They oppose, says Russell, passion and inspiration to method and discipline, content and colour in art to form and line, Dionysus to Apollo (the former was Bacchus, the god of intoxication, associated with instinct and passion; the latter was the sun god, who was said to have been the father of Pythagoras the mathematician).

Many modern ecocentrics worry about this dualism. Pirsig (1974) bemoaned how the romantic was divorced from the 'classical': 'What's wrong with technology is that it's not connected in any way with matters of the spirit and of the heart . . . and so it does blind, ugly things.' Capra (1982) considers that the dualism is a substantial obstacle to be overcome in shaping an ecological society. Skolimowski (1992) too blames environmental and social crises on the divorcing of rational knowledge from (romantic) values.

Two other tenets of Romanticism stemmed from its opposition to rationalism applied to nature and culture to produce the industrial, 'civilised' society – the epitome of 'progress'. First, such a society was complex and sophisticated, but Romantics revered *simplicity* – of form (in art), of action and of ideas. Simplicity equated with honesty, and nature was beautiful because it was simple and honest: 'Beauty is truth, truth beauty, – that is all Ye know on earth and all ye need to know', said Keats (*Ode on a Grecian Urn*).

Second, an interest in the folk societies of the past followed from this: they were imagined to have been closer to nature, simpler and more honest than modern, corrupt, urban society. This idea, which also runs through the political ideology of anarchism (some Romantics were also anarchists), was epitomised in Rousseau's 'noble savage' and in his dictum, from *The Social Contract* (1762), that 'Man is born free and everywhere he is in chains'. Inherently people were good, and were born so, but civilisation corrupted and degraded them.

This concept harked back to the myth which underlay Gilbert White's perception of Selborne, and which Romantics perpetuate, of the harmonious idyllic rural society of Arcadia in ancient Greece. It often amounted to idealising *the past*. This tendency in Romanticism was illustrated by the interest expressed by Keats, Walter Scott, William Morris, John Ruskin and others in aspects of medievalism: the chivalry and heroism, and the supposed unity of people and the land in its 'organic' – and hierarchical – society. From this belief (really a restatement of senescence theory) came the back-to-the-land ideal, subsequently a persistent feature of Western counterculture (Chapter 4.5). It reappears in the 1990s alternative health movement also, says Coward (1989).

The European movement

In rational thought, the Cartesian dualism meant that the kinds of qualities in nature – beauty, colour, majesty – which Romantics elevated were regarded as 'secondary': not objectively real, but the products of the human mind. However, Romantics denied this utterly; qualities like colour or beauty were not secondary but were inherent. This is important, for it meant that the Romantics ascribed a significance and integrity to nature which did not depend on humans. The intrinsic value concept is a key facet of modern ecocentrism's bioethic (Chapters 1.2, 2.1).

The move to respect nature in this non-utilitarian way was part of profound and broad changes in attitudes to wild nature which started around the Romantic period. For up to the eighteenth century it was generally thought that wild uncultivated nature was to be deplored, while the regular and symmetrical forms associated with ploughing, planting, hedging and other agricultural practices were a welcome mark of civilisation. When it became a symbol of progress to tame nature, and when colour, taste and emotion became secondary qualities, then the formal and mathematical triumphed over the wild. The gardens at 1660s Versailles were the apotheosis of this triumph, but still, over a century later, William Gilpin (cited by Thomas 1983) said that most people found wild country in its natural state totally unpleasing: 'there are few people who do not prefer the busy scenes of cultivation to the greatest of nature's rough productions'.

Gilpin was an eighteenth-century aesthete and theoretician of the 'picturesque'. This refers literally to what we have seen in pictures, and

192

during this century travellers valued scenery according to how much it reminded them of the carefully composed landscapes which they had previously seen in paintings. They were particularly influenced by the compositions of Claude Lorraine (1600–82), who 'first opened people's eyes to the sublime beauty of nature' (Gombrich 1989, 310). He, in turn, drew from Nicolas Poussin (1594–1665), whose paintings conveyed nostalgic Arcadian landscapes of innocence and calm repose, and from his first-hand studies of the Roman Campagna. Realistically as he represented natural objects, Lorraine, however, selected only such motifs as conformed to this dreamlike vision of the past. Rich Englishmen, influenced by this version of the 'picturesque', modelled their country estates, through the talents of designers like Lancelot Brown and Humphrey Repton, on Lorraine's Italianate dreams of what constituted beautiful country. By comparison with such models, then, unproductive mountains were traditionally regarded as physically unattractive: people complained of the 'desert, barren and very terrible' aspect of the Lake District, the 'hideous' Pennines and the 'hopeless sterility' of the Scottish Highlands (Thomas 1983, 257–8).

Wales, until the late eighteenth century, was ignored or similarly denigrated. The *Gentleman's Magazine* said in 1747 that Wales was acknowledged a 'dismal region, generally ten months buried in snow and eleven in clouds'. To eighteenth-century rational people, beauty was well-proportioned and cultivated land, and wilderness held no attraction.

But the mood changed dramatically at the end of the eighteenth century. This was partly a reaction by artists against the growing fact of their becoming mere producers for a middle-class market in picturesque images, rather than being commissioned by prestigious individuals or groups – they revolted against their appropriation by commercialism (Cosgrove 1984). 'Wild, barren landscape ceased to be an object of detestation and became instead a source of spiritual renewal.' The mountains which in the seventeenth century had been

> hated as barren 'deformities', 'warts', 'boils', 'monstrous excrescences', 'rubbish of the earth', 'nature's *pudenda*', had a century or so later become objects of the highest aesthetic admiration.
>
> (Thomas 1983, 258–9)

Definitions of what was 'picturesque' changed too, in the Romantic period. Natural (wild) objects were now preferred to artificial (human) ones, asymmetry to symmetry, irregular and curved lines to straight ones, rough surfaces to smooth, complexity to simplicity, and diversity to sameness (Hargrove 1979).

In line with this were shifts in the very meaning of the words 'nature' and 'wilderness': the former, from being identified with *human* nature and reason, acquired ideas of goodness, innocence and that which humans had *not* made; the latter, from carrying overtones of fear and hideousness, became associated with purity. The world 'sublime', which previously related to religious awe or dread, began to be used to describe mountain scenery: it inspired awe and wonder by its grandeur, nobility and extra-ordinariness. The 'harmful' as well as the pleasing became appreciated aesthetically. 'People began to be impressed with vastness, massiveness, chaos and disharmony' says Hargrove (pp. 220–1):

> a person who appreciated the sublime could not look at the world as a place created solely for man's use, or perhaps even for human purposes at all. In this sense, the sublime signified the end of the age in which natural objects were evaluated exclusively in terms of human uses and needs.

Romantic inspiration thus came from what was grand and remote. Snowdonia became the 'British Alps'. Zaring (1977) says that the 'rain-soaked uplands, sparsely populated and largely unploughed, were beautiful in the eyes of those who were reacting against the ideas of their father'. Merionethshire, formerly the rudest and roughest country of all Wales, had replaced the civilised county of Kent as a standard of ideal beauty. Coleridge, Shelley, Wordsworth, Southey: all experienced the solitude of North Wales in the 1810s.

Edmund Burke, in the *Philosophical Enquiry into the Origin of Our Ideas of the Sublime and the Beautiful* (1756), urged his readers to see the sublime in the vast, rugged, dark, gloomy and that which excited ideas of pain and danger. From such scenery 'came ideas elevating, awful and of a magnificent kind'. The rugged rocks, precipices and gloomy mountain torrents had become the epitome of aesthetic experience for some Romantics because their dark soaring masses expressed the immensity and infinity associated with God. The artistic category of the 'sublime', then, appreciated the 'harmful' as well as the safe, and vastness, massiveness, chaos and disharmony; that is, a world *not* created solely for humans and existing long before them (Hargrove 1979).

The apotheosis of the sublime vision lay in J. M. W. Turner's (1775–1851) avalanches, deluges and storms. In them he attempted to depict the *forces* in nature, rather than give a realistic picture of what was immediately apparent to the eye:

> He dissolved subjects into colour and light, much as science trans-formed matter into energy, and by conceiving of a universe composed of vortexes or fields of force he anticipated forces in physics and astronomy . . . To the romantics, a general theory of the unity of nature based on the transmutability of the elements appealed just as much as one based on the interdependence of natural phenomena.
>
> (Rees 1982, 264, 269)

Turner's approach of depicting the world beyond the immediate senses can be contrasted with John Constable's (1776–1837), to illustrate what Rees calls a basic division in nineteenth-century science. If Turner symbolised the search for general and universal forces, Constable symbolised naturalism: the world as it is or seems to be. He sought to make lifelike images of 'natural' objects (albeit in lowland agricultural landscapes rather than in the mountains, which he found oppressive). This required detailed knowledge of their form – hence the development of natural sciences and naturalism in landscape painting went hand in hand. As a naturalist, Constable partly broke with the neoclassical Arcadian tradition of Poussin and Lorraine, in not setting out to select idealised motifs. However, his depictions of labour might be considered idealised, and his images certainly did become the basis of a new English pastoral myth (see p. 204).

Wordsworth: a proto-green?

The notion of modern ecologism as a quintessentially neo-Romantic movement is reinforced by Bate's (1991) re-examination of William Wordsworth from a green perspective. Bate concludes that Wordsworth should be the virtual patron saint of ecologism, for he put himself squarely in the green tradition by teaching his readers to look at, dwell in and respect the natural world, and to be sceptical about material and economic 'progress'.

Bate sees Wordsworth as counterpart to Thoreau (see pp. 197–9). American romantic ecology was concerned with vast wildernesses, of course, while the British equivalent was about localness and 'small enclosed values'. Nonetheless they both, through Wordsworth and Thoreau particularly, were very holistic and preached a symbiosis between nature's economy and human activity. The *Guide to the Lakes*, a central part of Wordsworth's work, is 'an exemplar of romantic ecology' (Bate, p. 45); a supremely holistic work moving from nature to the Lakes' human inhabitants, and demonstrating the increasingly disruptive effect of the latter on the former. In *The Excursion*, Wordsworth wrote of the animation in, and unity of, all things in the universe. Everything is linked to everything else, and the human mind must be linked to nature and must accord it moral respect.

Besides such bioethical sentiments, Wordsworthian ecology entered into a broader and explicitly political tradition, via John Ruskin (1819–1900). This, says Bate, championed a form of labour harmonised with nature in artisan, cottage industry – contrasting with capitalist labour. Wordsworth and Ruskin both argued for, and worked to restore, local, small-scale industry. This 'proto bioregionalism' was reinforced by Wordsworth's attachment to place. Places became defined through his detailed description of their character. In some of his *Lyrical Ballads* his 'naming poems' contained local place names and features in profusion: they 'develop a highly original sense of specific place and the relationship between self and place' (Bate, p. 99). This tradition was continued by 'Wordsworth's successors', Walter Scott, John Clare, Thomas Hardy and A. E. Housman.

Bate is anxious to defend Wordsworth politically, from charges by some that he was a reactionary idealist; by others that he was a revolutionary materialist. Left critics like Raymond Williams have accused Romanticism and its pastoral poetry of masking the real conditions of oppression and exploitation in agrarian economies, feudal and capitalist. They therefore, like the painters of Arcadian scenes, promote a comforting, conservative, aristocratic fantasy, in a 'great pastoral con trick' (Bate, p. 18).

Others have placed Wordsworth and Romanticism, as represented by the likes of Ruskin, Morris and Shelley, squarely on the left because of its anti-capitalist, anti-materialist stance. Shelley, in *Queen Mab*, for instance, denounced money's contamination of human relationships. And in *The Masque of Anarchy* (1819) he castigated 'every rotten facet of capitalism, its law, its judiciary, its priests, its parasitic class and the foulness of its oppression': this is 'real socialist literature' (Montague 1992).

Bate, however, argues that both characterisations are inappropriate. Neither enraged radical nor conservative, escapist counter-revolutionary, Wordsworth's ideology was based on a harmonious relationship with nature, 'which goes deeper than the political model we have become used to thinking with'. (In this defence itself there is a familiar argument about green politics: in essence, 'neither left nor right but green' is the claim.) In Books 7 and 8 of *The Prelude*, Wordsworth laid out his fears about the city, and his ecological vision of the rural community of Grasmere. The latter was an ideal 'commonwealth', where labourers, like the shepherd in *Michael*, were not appropriated and alienated but worked for themselves. They were close to nature, but this was no neo-classical pastoral Arcadia. It was a harsh and hard-working paradise. Wordsworth aggrandised common people and wrote them into history, leading on, in Book 9, to a more generalised expression of love of humankind in keeping with the revolutionary spirit in France.

Persuasive as Bate's argument appears, it needs to be considered alongside Cosgrove's (1984) observations on the politics of Romanticism. Despite its apparent anti-capitalist radicalism, it in fact celebrates the 'central myth of capitalism': that is the nobility and 'naturalness of the isolated individual'. Certainly, elements in Romanticism that seek solutions in nature rather than in changing society, sit uneasily with any radical left political doctrine. Bate's revolutionary Wordsworth is more a champion of liberalism, the political ideology of capitalism and the French Revolution.

The American Romantics

As in Europe, the love of wilderness which the American Romantics such as Muir, Thoreau and Emerson expressed partly displaced earlier very opposite feelings towards nature. Nash shows that wilderness at first frustrated physically the attempts of early pioneers to find a second Garden of Eden in the West, and it 'acquired significance as a dark and sinister symbol'. The pioneers therefore shared the Western tradition of 'imagining wild country as a moral vacuum, a cursed and chaotic waste-land'. So the frontiersman 'in the name of nation, race and God' saw himself as civilising the New World by destroying wilderness and trans-forming it into cultivated landscape. This became 'the reward for his sacrifices, the definition of his achievement and the source of his pride' (Nash 1974, 24–5). Being pitted *against* nature in this way was one thing that united the diverse peoples who sought liberation in America.

By contrast, the American Romantics took on European ideas. Euro-peans such as Chateaubriand, de Tocqueville and Byron visited the USA and praised its wilderness. They were echoed by a gradual public percep-tion, even among frontiersmen such as Daniel Boone, that wilderness had aesthetic values. In the nineteenth century growing concern over the loss of wilderness 'necessarily preceded the first calls for its protection' (Nash, p. 96) by such as the ornithologist J. J. Audubon, who observed on his travels the destruction of the forest. These calls led eventually to the National Parks movement which was inspired by Henry David Thoreau's phrase 'In wildness is the preservation of the world'.

Thoreau (1817–62) has been thought to epitomise the values of the isolated individual living in nature and free of all social attachments. As an experiment he cast himself away from society for two years, and, taking virtually nothing with him, he built a cabin and won a living from nature by the side of Walden Pond in Massachussetts. The journals of his experience, published in 1854, rejected the materialism which saw only monetary worth in nature:

> I respect not his labors, his farm where everything has its price, who would carry the landscape, who would carry his God to market, if he could get anything for him . . . on whose farm nothing grows free . . . whose fruits are not ripe for him till they are turned to dollars. Give me the poverty that enjoys true wealth.
>
> (Thoreau 1974, 145)

Thoreau is constantly singled out as an inspiration for modern greens because, says Worster (1985, 58), his 'romantic approach to nature was fundamentally ecological'. As an anarchist, subverting classical science, capitalism and the anti-nature element in Christianity, Thoreau fits the radical green bill perfectly.

A self-educated naturalist, he inherited Gilbert White's arcadian legacy, says Worster, who considers that the 1850–61 journals (rather than *Walden*) represent his mature ecological philosophy. He sought out, and wanted to preserve, a pre-human wilderness, rejecting its antithesis in the Puritan 'civilisation' of Eastern USA. For this 'civilisation' marred nature. His critique of industrial society's standards and aspirations is cutting, pertinent and entirely apposite for the late twentieth century. For instance, his observation that

> We are eager to tunnel under the Atlantic and bring the Old World some weeks nearer to the New; but perchance the first news that will leak through into the broad, flapping American ear will be that Princess Adelaide has the whooping cough
>
> (Thoreau 1974, 48)

could well be transposed to the current technological 'miracle' of satellite TV. Its sentiment precisely foreshadows Schumacher's ecocentric anxieties about a civilisation which has 'know-how' galore but lacks moral guidelines, that is, 'know-what': that has techniques for saying, but nothing interesting to say.

Hence a document like *Walden* is not to be seen as a chronicle of Romantic escapism, full as it is of such incisive social comment. But then Thoreau's writings do also contain much about nature. They rejected the Linnaean synthesis of religious and scientific values, projecting instead a dynamic evolutionary picture of nature, and taking above all a holistic perspective. Thoreau constantly stressed interdependence in the community of plants, animals and humans, being much influenced by von Humboldt's writings. Rather than hunting squirrels, he asserted, we should honour them for the part they played in the economy of the universe (Worster, 70).

Thoreau pre-echoed permaculture (Chapter 6.3), advocating scientific forest management by replanting along natural lines. And he foreshadowed bioregionalism, advocating loyalty to one's birthplace, where one belongs and can find all. On travel, his holistic observations might have been written by a green economist, when he argued that it was faster to walk than take a train. For by the time one had worked the hours to earn the fare one could have gone by foot:

> And so if the railroad reached round the world, I think that I should keep ahead of you; and as for seeing the country and getting experience of that kind, I should have to cut your acquaintance altogether.
>
> (*Walden*, p. 48)

His down to earth prose often sang of 'sensuous attachment to earth and its vital currents . . . a visceral sense of belonging to earth and its organisms', and of vitalism: 'The earth I tread on is not a dead, inert mass, it is a body, has a spirit, is organic' (cited in Worster, pp. 77–9). Like a Gaianist, he spoke of a living earth with one 'great central life'; like a deep ecologist he was non-anthropocentric: 'The poet says that the proper study of mankind is man. I say, study to forget that' (cited in Worster, p. 85).

Despite such spiritualism, even mysticism, the empirical scientist would approve of Thoreau's detailed descriptions of nature. As Erisman (1973) notes: 'For all its intuitive insights and mystical meditations, the *Walden* journal abounds in raw bold data'. Thoreau strongly conveys the idea that there is pleasure and reward in the close and systematic observation of nature. Detailed observation will reveal the endless variety, complexity and fecundity of nature – its plenitude – and this was the quality in which the American Romantics revelled. 'I love to see that nature is so rife with life that myriads can be afforded to be sacrificed and offered to prey on one another', he said.

Bioethical sentiments were also evident in Ralph Waldo Emerson. In his essay *Nature* (1836) Emerson wrote:

> Such is the constitution of all things . . . that the primary forms, as the sky, the mountain, the tree, the animal, give us a delight *in and for themselves,*

And, as with some later ecocentrics, one gets the impression that so strong is the integrity and purity of wild nature that humans sully it by their very presence. 'This pond [Walden] has rarely been profaned by a boat, for there is little in it to tempt a fisherman' said Thoreau, suggesting the kind of anti-humanism betrayed in the 1970s ZPG (zero population growth) movement's phrase, 'people pollute'. And Romanticism generally *was* at odds with the ideals and assumptions about nature of much humanism. There is a sense in pure preservationist Romanticism that economic activity vandalises nature:

> This Sierra Reserve . . . is worth the most thoughtful care of the government for its own sake . . . Yet it gets no care at all . . . lumbermen are allowed to spoil it at their will, and sheep in unaccountable ravenous hordes to trample it and devour every green leaf within reach.
>
> (Muir 1898)

Transcendentalism

Contact with wild nature inspired Romantics to think beyond, or *transcend*, material life and to contemplate the spiritual plane. For Wordsworth the mountains inspired spirituality and sacredness without specifically referring to the Christian God. For Ruskin and others, nature inspired the kind of awe and reverence traditionally reserved for God, reminding them of him.

François de Chateaubriand wrote in 1802:

> I am nothing; I am only a solitary wanderer, and often I have heard men of science disputing on the subject of a supreme being. But I have invariably remarked that it is in the prospect of the sublime scenes of nature that the unknown being manifests himself to the human heart.

It seemed that to find God and come close to him, one must find 'unspoiled' nature, his creation.

Transcendentalist Romantics believed, then, that contact with wild nature purifies and refreshes people spiritually. For the religious all wild nature is a manifestation of God (pantheism). Thoreau wrote that 'we need the tonic of wilderness . . . we can never have enough of nature', and by Walden's banks he was 'affected as if in a peculiar sense I stood in the laboratory of the Artist who made me'. Emerson's *Nature* called the woods 'these plantations of God'; in them he saw all the

> currents of the Universal Being circulate through me. I am part or parcel of God . . . The greatest delights which fields and woods minister is the suggestion of an occult relation between man and the vegetable.

Emerson's reference here to vitalism – the currents of a universal being – strongly echoes both back to the medieval Great Chain of Being and forward to today's deep ecology. John Muir, who called the Grand Canyon 'this grandest of God's terrestrial cities' strongly promoted wilderness' tonic and spiritual values, in a famous passage which could be an anthem for middle-class America today:

> Thousands of tired, nerve-shaken, over-civilised people are beginning to find out that going to the mountains is going home; that wilderness is a necessity and that mountain parks and reservations are useful not only as fountains of timber and irrigating rivers but as fountains of life. Awakening from the stupefying effects of the vice of over-industry and the deadly apathy of luxury they are trying as best they can to mix and enrich their own little ongoings with those of Nature, and to get rid of rust and disease . . . some are washing off sins and cobweb cares of the devil's spinning in all-day storms on mountains.
>
> (Muir 1898)

Emerson evokes holism in his view that 'There is a property in the horizon which no man has but he whose eye can integrate all the parts, the poet.' We note that the integrator is the poet – the artist – not the scientific systems ecologist.

But as transcendentalists, Thoreau, Ruskin and Emerson did also revel in scientific study. This was, however, not merely for its own sake – to increase the store of 'objective' knowledge in subjects like geology, botany or meteorology. For unlike classical scientists they followed the tradition of Goethe and von Humboldt, considering that the subject (themselves) could not be divorced from the object of enquiry (nature). Furthermore, like Turner, they wanted to go beyond form and morphology to understand subjectively the *essence* of landscape. That is, they desired phenomenological knowledge of it through experiencing first hand, as individuals, nature's raw powers.

All of this quest was for a higher purpose; to discover nature's 'symbolising of God's intentions towards mankind'. And for Ruskin, the constant recurrence of certain forms in nature (e.g. the similarity between the curvature of a scree slope and that of a bird's wing) displayed the 'inward anatomy of creation and sure sign of the ideal form to which each object aspires and which is to be found in perfection only in the divinity' (Cosgrove 1984, 244–5).

Thoreau, too, made much of the repetitiveness of form in nature, in, for instance, his detailed description of mudflows in Walden Pond's banks, and how their lobate structure was replicated in the vital organs of animals, in vegetable leaves, lichen, coral, leopards' paws or birds' feet, brains or 'lungs or bowels, and excrements of all kinds'.

Ruskin considered that nature conveyed moral lessons about the proper conduct of individual and social life: there were literally 'sermons in stones', so that clouds, for instance, symbolised divine mercy and justice. Emerson, in *Nature*, similarly restates this pre-modern doctrine of signatures and nature as a book: 'What is a farm but a mute gospel? The chaff and the wheat, weeds and plants, blight, rain, insects, sun – it is a sacred emblem'.

Transcendentalists, then, thought that humans were created to live in nature and that nature was designed by God to teach them, through signs and symbols, how to live properly. Today's ecocentrics have secularised this message in the 'nature-knows-best' principle. Popularised by Barry Commoner, this 'third law of ecology' 'acknowledges that *any* major human change to an ecological system is likely to be detrimental to that system' (Button 1988, 288). Therefore it follows that we must study ecological principles (carrying capacity, strength in diversity) and base our lives on them. Ecology rather than God gives us messages in nature about how to live.

Worster detects in transcendentalism a contradiction, a tension, which he believes is fundamental to ecology (and therefore to ecocentrism). On the one hand, Romantics, including transcendentalists, revelled in the material fact of being *in* nature. For some Romantics this logically led to a pagan animism which required 'direct communion with the animated energy of the cosmos' in such as tree worship. The outcome was a heathen, non-Christian religion, 'God' being identified with the one life coursing through the ecological system (Worster, p. 86).

But on the other hand Romantic transcendentalists preferred to elevate spirituality above material, animal instincts (squarely in the Enlightenment tradition). Thus for them (Blake, Coleridge, Emerson, Fichte, for instance) nature was but a means to an end: a lower order *not of value in and for itself.* When Emerson's *Nature* placed the living world as a resource for human imagination, and when Ruskin re-stated the doctrine of signatures in nature, they both placed the *human mind* at the centre of things: it gave beauty and meaning to the world. This issue constitutes another version of the modern problem of whether there can be a Christian eco/biocentrism. And it raises again the central problem of whether biocentrism is a realistically attainable human position, let alone a desirable one (Chapter 2.1).

Country and city and the arcadian myth

For Romantics the excrescences of industrial capitalism were epitomised in the city. Anti-urbanism is a prime feature of romantic thought, as it persists in some ecocentrism today. The Romantic movement reflected a reversed perception of the city, as it did of wilderness. Tuan (1974) shows that the design of the ancient and medieval city made it not only a shrine to God but also an expression of society's highest cultural and technological achievements. This 'sacredness' contrasted with the 'profanity' of wild nature, as noted above. With the increased importance of industrial manufacturing in the city, however, these positions were reversed, and as the wilderness became sacred, so the city was regarded – especially in Romanticism – as profane (Tuan 1971).

Thus Emerson wrote of how 'The tradesman, the attorney, comes out of the din and craft of the street and sees the sky and the woods, and is a man again', and the poet nourished by nature can never altogether lose the benefit in the 'roar of cities or the broil of politics'. Thoreau considered that 'in nature, not in the pomp and parade of the town, the individual may walk with the Builder of the universe'.

Once again, the American Romantics were here echoing a series of attitudinal changes:

> In Renaissance times the city had been synonymous with civility, the country with rusticity and boorishness . . . Yet long before 1802, it had become commonplace to maintain that the countryside was more beautiful than the town.
>
> (Thomas 1983, 243–4)

This reversal was partly connected with the deterioration of urban environments. Dirt-laden London air had been the subject of complaints in the thirteenth and the sixteenth centuries, while in the seventeenth and eighteenth centuries Oxford, Newcastle and Sheffield also experienced atmospheric pollution. Thomas asserts, however, that concern was often more about the city's immorality than its physical dirtiness. He cites (p. 246) John Norris, who wrote:

> Their manners are polluted like the air
> From both unwholesome vapours rise
> And blacken with ungrateful steams the neighbouring skies.

Similarly, Ruskin, in *The Storm Cloud of the Nineteenth Century*, observed that climatic deterioration and pollution were creating a new form of cloud, a 'loathsome mass of sultry and foul fog, like smoke . . . a plague wind' (cited in Cosgrove 1984, 251). In what some consider a premonition about the greenhouse effect, Ruskin seemingly alluded to pollution from the blast furnaces at Barrow-in-Furness, apparent in the Lakeland atmosphere. What most seems to have concerned him about it was the symbolism: the cloud, for him, was the material expression of moral decline brought by industry and the market place.

This dualism of town and country often involved romanticising the latter. The arcadian tendency to do this, as opposed to romanticising wilderness, stretches back beyond the Romantic period, as noted (Introduction; Short 1991). Williams (1975, 9) observes:

> In the country has gathered the idea of a natural way of life: of peace, innocence, simple virtue. In the city has gathered the idea of an achieved centre of learning, communication, light. Powerful hostile associations have also developed: on the city as a place of noise, worldliness and ambition: on the country as a place of backwardness, ignorance, limitation. A contrast between city and country, as funda-mental ways of life, reaches back into classical times.

From time to time in history one aspect of this dualism surfaces while the other becomes relatively dormant, but the two strands are always there, in fundamental tension. Williams shows us that there is a powerful Western tendency to associate the country's more benign image with the *past*: a past where things were invariably better than they are now. In this, we emulate not just the Romantics but those artists from the classical poets onwards who have always promulgated the fiction of Arcadia: of this romanticised countryside of simplicity, virtue and society–nature harmony. As Leo Marx (1973) wrote, it is a (senescent) vision of a past 'Golden Age, of grazing flocks, unruffled waters and a calm, luminous sky, images of perfect harmony between man and nature'.

This mythical pastoral idyll has no room in it for the necessity of human labour, for the rich patrons of art did not want to be reminded of the more baleful aspects of the production from which their wealth came. 'In the pastoral economy nature supplies most of the herdsman's needs and, even better, nature does virtually all of the work'. In some Romantic paintings this 'magical extraction of the curse of labour' was achieved by simply removing any labourers from pastoral images:

> The actual men and women who rear the animals and drive them to the house and kill them and prepare them for meat . . . these are not present: their work is all done for them by a natural order.
>
> (Williams 1975, 45)

This was particularly true of the classical Euro-pastoral of Poussin and Lorraine, which continued the tradition of Virgil's *Eclogues*. It became modified after 1700, when there came a demand for a more specifically English pastoral in response to the nostalgia for an English rural golden age that was triggered when foreign competition unhinged the rural economy (Cox 1988).

This essentially political demand eventually assimilated Constable's work into the myth of a rural idyll, although his paintings, largely, were not romanticised images. As a naturalist he wanted to 'keep to the motif in front of him and explore it with insistence and honesty'; he was not interested in Lorraine's idyll (Gombrich 1989, 392–3). And, unlike the arcadians, aristocratic patrons of the English pastoral did now want labour to be shown in their countryside – albeit as pleasantly social, honest toil – to symbolise and legitimate the movement to 'improve' and manage nature for greater output. Constable, in his lowland England landscapes, represented labourers as distant and indistinct, to present, in keeping with his rural Tory background,

> a crafted image of harmony between labourer and landscape . . . an image redolent of the myth of the lost organic community . . . a time before distinctions between consumer and producer, the idle and industrious had emerged.
>
> (Cox, p. 26, citing the view of John Barrell 1980)

In this way, Constable represented the English countryside as a place of work, development and improvement – rejecting elements that were not human-made – but also one of refreshment and recreation by comparison with the city.

Hard labour was, of course, an important if under-recognised part of pastoral images, because in them what constituted 'nature' was not in fact wilderness. Even in Lakeland the landscapes bore the indelible imprint of sheep farming; lowland idealised landscapes were even more the product of farming. They both were, and are, 'middle landscapes' between city and true wilderness (Tuan 1971). This 'civilised' nature-as-a-garden has been romanticised not only in the arcadian myth but also in Christianity's Garden of Eden. When Adam and Eve committed original sin they had to leave the garden. Ever since, Christianity has invoked us to be deserving of its rediscovery through labour and moral living, or, in its most conservative form, through striving against the very forces of industry and materialism. William Blake wanted to recreate 'Jerusalem' in 'England's green and pleasant land'. Such romantic anti-industrialism remains a contemporary feature of the British psyche (Wiener 1981), and it is manifested in liberal ecocentrism of the sort voiced by Jonathan Porritt (1984).

The garden, or middle landscape, also featured strongly in American pioneer ideology. Thomas Jefferson said

Those who labour in the earth [to create the safe farmed landscape out of wilderness] are the chosen people of God . . . Corruption of morals in the mass of cultivators is a phenomenon of which no age or nation has furnished an example . . . [whereas] The mobs of great cities add just so much to the support of pure government, as sores do to the strength of the human body.

(cited in Marx 1973)

Romanticism thus attached social messages to environmental images of city and country, as ecocentrics are wont to do today.

4.5 UTOPIAN SOCIALISM

Definition and antecedents

When it comes to a social programme – views about an ideal society and how to get there – ecologism frequently reiterates utopian socialism. The utopian 'tendency' within socialism inherits from pre-industrial utopians: millenarian peasants, artisans and intellectuals from medieval times to the Civil War who wanted an egalitarian communalist society. It was then strongly developed between the Napoleonic wars and the 1848 revolutions. Three people are particularly associated with what Bottomore (1985, 504–6) calls this 'first stage' in the history of socialism: Claude Henri de Saint-Simon (1760–1825), Charles Fourier (1772–1837) and Robert Owen (1771–1858). Then there were the nineteenth-century anarchists, particularly those who developed anarcho-communism rather than liberal-individualist anarchism. Of these, Peter Kropotkin has most affinity with ecologism. And there were the Victorian utopian socialists, of whom William Morris (1834–96) was the most obviously 'green'. Throughout the nineteenth century and into the twentieth there were also writers and activists concerned with 'doing it': trying to establish socialist utopia in the here and now, in intentional communities and the back-to-the-land movement (a movement which is not, however, exclusively of the left, as Chapter 4.6 discusses).

Utopian socialists were and are 'utopian' in two ways. First, they created visions of their ideal society. Other socialists, particularly Marxists, disapproved of this on grounds that socialism was particularly about people being enabled to control and shape their *own* society – to 'make their own history'. Therefore it was for that society, not us, to say what it should be like.

In practice, however, Marxism itself inevitably contains utopian envisioning of the future, as does anarchism. It is partly because of this that liberals in turn attacked Marxism, considering that it tended to create blueprints for the future. Their rigidity lead to dissenters or 'deviants' being not tolerated.

In other words, it was a recipe for totalitarianism (Goodwin and Taylor 1982 – see Chapter 6.4).

Second, socialists could be 'utopian' in the sense of naive and unrealistic (as Marx and Engels saw it) about how to achieve the socialist society. They shunned conventional politics, imagining they could eliminate individualism, competition and private property without directly confronting capital in a class struggle where the proletariat played a crucial revolutionary role. And they were idealistic (Chapter 2.7), considering that the realm of moral ideas was the determining basis of all other human behaviour. Therefore they thought that their battle was particularly against pre-existing religious and political theories and ideas. They appealed to people to *think* differently – to change their values – as precursor to total social transformation, rather than overturning capitalist society's material base (its economic organisation). And they imagined that if they set an example, by trying to *pre-figure* the ideal society – living it out in intentional communities for instance – others, on observing them, would rush to follow it. I have argued elsewhere (Pepper 1993, Chapter 3.9) that these criticisms, if such they are, can be levelled at ecocentrism today.

Pre-industrial utopians

Merchant's account of pre-industrial utopians suggests that the egalitarian communalism which they regarded as the natural way to live was also 'ecologically sound'. In Tommaso Campanella's *City of the Sun* (1602) nature and society were an organic whole, and natural harmonies were maintained in people through regulated and diverse diets and natural medicines. The theme re-emerged in city planning and environmental design in Valentin Andreä's *Christianopolis* (1619), which saw the city as a replica of the larger cosmos. In these utopias nature was to be emulated, not altered and 'tortured', as Merchant calls the treatment accorded to it in Bacon's technocentric utopia *New Atlantis* (1627).

During the English Civil War sectarian movements arose against enclosure of the commons. Around Cromwell's time there were the Ranters. They were the only utopians then who did not see humans as the battleground between good and evil (Manuel and Manuel 1979). Their pantheistic beliefs sound positively New Ageist, and they rejected rationalism and the moral hypocrisy of the times.

The Levellers were active in Britain in the seventeenth and early eighteenth centuries. There was a Levellers rising in Galloway in the 1720s, where those dispossessed of their right to graze cattle denied that landowners had any right to evict them and tore down enclosure walls.

A small group of Levellers formed the Diggers, who in 1649 dug up common land on George Hill, near Walton in Surrey. They planted it and lived off the produce, building communal houses, until they were forced to stop in 1652. This action has reverberated down the centuries, repeatedly striking chords with radical left back-to-the-land movements and now with ecocentrics.

Its instigator was Gerrard Winstanley, a bankrupt clothing tradesman given to religious mysticism and utopian visions. In *The Law of Freedom in a Platform or True Magistracy Restored* (1650) he sketched a complete scheme for radically reconstructing society around the principle of community. Opposition to the poverty and drudgery known by ordinary people, and to buying and selling nature in private property were consistent features of his often contradictory writings (Manuel and Manuel 1979).

His utopianism was revealed also in an approach which abhorred violence and hoped that rich landowners would willingly share and work with their fellows when they saw the example of the Diggers. Furthermore he saw God as an internal rather than external agent; a spirit within people, among whom knowledge of Christ would create a mystical communion. They would form a world movement initiated by an elect few who had been given word of the divine light. The New Age movement today sees social change in very similar terms, with themselves as the enlightened harbingers of the ecologically sound Aquarian age.

Early nineteenth-century utopian socialists

Saint-Simon, Fourier and Owen tried to produce both a social evolutionary scheme and an account of what evolution's end product should include. This end seems highly compatible with ecologism, by Atkinson's (1991, 115–25) account. The visions of all three emphasised decentralising power to small-scale communities. Like ecocentrics, they were often ambiguous about industrialism and private property. Egalitarianism and common ownership of the means of production were frequent themes, as was the desire to abolish leaders, hierarchies and the town–country antagonism. Contrary to some later Marxist approaches, the *way* that the ideal society was to be achieved was seen as equally important as *what* was achieved, for example a peaceful society could not be got through violence. The needs of the individual and of the whole community rated equally, science and technology of an 'appropriate' nature was accepted, and work, it was said, should fulfil people's creative instincts. Furthermore, says Atkinson (p. 121)

> Nature and the things of the earth are treated with respect and modesty . . . whilst the elevation of 'ecology' to a key position may appear as a new aspect of Utopianism . . . in fact respect for nature has always constituted an important principle of Utopian thought and action.

The Manuels describe how Saint-Simon drafted blueprints for world organisation in which each of three different 'natural types' of people were represented: scientists, administrators, and artists and emotional moralists. Hence the good society was a harmonious cooperation of fundamentally dissimilar people: each had a role, although not all were equal. But each should be able to express their true natures so that misery and ignorance would be abnormal. This, then, was an organic, vital community, different from the atomised egalitarian societies championed by Enlightenment rationalists such as Condorcet. The vital element, thought Saint-Simon, was love and a sense of belonging: there would be no class conflict because everyone was doing what was natural to them. There would instead be only mutual aid between the classes: society would not be commanded by rulers but run by administrators in accordance with what the élite among scientists, artists and industrial classes thought best for all. Hence in this brotherly community there was no need for a political power, a state. Even though Saint-Simon greatly influenced Marx, the true heir of his vision was a society of bankers and capitalists, says Joll (1979, 37).

Saint-Simonianism was the doctrine of Saint-Simon's followers in the 1820s and 1830s. Its heirs might well include romantic ecocentrics, as a list of some of their beliefs suggests (Table 4.1). The list reflects Saint-Simonianism's domination by Barthélemy Enfantin, who championed emotionalism, free love, sensuality and the search for a female messiah in a quest to awaken the dormant feelings of love in humankind. This reaction against the eighteenth-century rationalists leads us not only towards

Table 4.1 Some beliefs and values of the Saint-Simonian movement

1 Emancipation of women
2 Opposition to inherited property
3 Horror of strife and revolution
4 Indictment of contemporary selfish, loveless generation
5 Indictment of *laissez-faire* as leading people from their true pacific, cooperative, associative nature
6 Need for overall industrial and scientific planning
7 Approval of the outburst of Romanticism among artists
8 In the good society there will be a balance of the material and spiritual, body and soul
9 Indictment of scientific thought with its lack of values
10 World history consists of the diffusion of love
11 Association and sociability are gaining ascendency over exploitation of workers and international strife
12 History is about to enter an organic age – such ages recur throughout history, punctuated by more critical, destructive epochs
13 The new moral order's main principle will be 'from each according to capacity'
14 Each individual is sacred and saintly
15 But because people are different they will be rewarded unequally

Source: Manuel and Manuel 1979, 616–35

Romanticism but also (in its idealist views of history and of the individual) to the 1960s and contemporary New Ageism.

Another of the utopian socialists, Fourier, takes us more towards an anarchistic, feminist, communalist and hedonist (based on pleasure as the chief aim) ecocentrism. He worked out a system of social organisation based on 'phalansteries', with no overarching state to regulate them. These ideal communities of up to 1700 people followed the Platonic tradition of seeing the good society as essentially small, say the Manuels (p. 667). In them, traditional family relationships would be dissolved and replaced by a wide spectrum of sexual relationships; capital, private property and inheritance would be preserved; material wealth disparities would remain, but all would be emotionally satisfied and spiritually rich. There would be much social mobility and no class alignments based on wealth. Creative talents would be expanded, meals and child care would be communal, and 'industrial armies' would carry out important environmental projects. Fourier held out phalansteries as the only alternative to the unnatural repression of 'civilised' society. Its gratuitous evil prevented material needs from being satisfied, artificially repressed pleasure, and, through industrialisation, squandered natural resources.

> Today, we can see Fourier as a source and inspiration for red green theory. His writings brilliantly illuminate basic green principles: small is beautiful; make love, not war; steady state economy; quality of life not accumulation of consumer durables; work as play; 'post patriarchal values'; abolish hierarchy; respect nature truly; and many others . . . His solution, entirely relevant to the contemporary world, was collective consumption, which he regarded as economically, environmentally and socially superior to the individual family system . . . Fourier can be seen as wanting to 'feminise' the world, so that peace, cooperation, environmental protection, nurturance, comfort, pleasure, decoration, the arts, relationships and food occupy the highest rungs. (That sounds suspiciously like the 'green' project.)
>
> (Roelofs 1993, 70 and 84)

Fourier preached the ultimate triumph of the passions, together with economic planning and social engineering by altering the form of society to conform with 'human nature'. Robert Owen, by contrast, thought that environment shaped behaviour: he therefore proposed to change people's natures by changing their educational, social, working and living environments. A successful businessman and mill-owner between 1800 and 1812, thereafter he began to criticise contemporary society and economy, denouncing the family and organised religion.

He reformed the conditions for his own mill workers at New Lanark, then proposed self-sustaining communities of about 500–3000 people each, without private property, organised in rectangular units close to manufacturing plants, and all surrounded by intensive agriculture. Such exemplary communities would lead to a cooperative socialism that would ultimately embrace the earth, formed by a new generation of rational beings. The condition of the working classes would therefore be permanently relieved.

Though Owen's utopia smacked of paternalism and anti-urbanism, it saw people as a clean slate at birth, without original sin. It also saw labour as the ultimate source of value. Hence much in it appealed to Marx, even though Owen was locked into the analytical perspective of individualist behavioural psychology, though he shunned revolution and political action and though his practical attempts to establish his ideal townships in New Harmony, Indiana, failed.

Goodwin and Taylor's (1982) review of early utopian socialism – Icarianism, Saint-Simonianism, Fourierism, Owenism, German artisan and early American socialism – shows that none was the direct expression of the modern working class. Rather, they were manifestations of concern by middle-class intellectuals and leaders, reacting to the plight of workers in early industrial capitalism. It is as such that Marx and Engels criticised them – as an élite, idealist vanguard rather than part of the spontaneous proletarian material revolution which Marxist purists demanded. Utopian socialism was thus a 'fantastic standing apart' as Marx and Engels dubbed it. But, say Goodwin and Taylor (p. 162) Marx and Engel's own socialism had all the elements of the earlier utopians; neither were *they* working class, and their visions of social harmony and an end to class antagonisms were utopian.

Anarchism and utopian socialism

I have suggested elsewhere that ecologism today often restates various forms of anarchist ideology (Pepper 1993, see also Eckersley 1992). Most ecocentrics would support what anarchism favours and oppose what it opposes (Table 4.2). Perhaps ecologism also shares with anarchism the tendency to resist neat categorisation, having shifting beliefs and, as a 'new social movement', embracing many groups who would appear to have little else in common (Scott 1990). Historically, anarchism has been pursued by rural European peasants, urban Australian unions, immigrant Americans and many other groups.

The Manuels note the links between the nineteenth-century anarchists and seventeenth-century Civil War literature of the Levellers, Diggers and Ranters. But the main body of anarchist theory was composed, from the 1840s, by Proudhon, Stirner, Bakunin and Kropotkin, and anarchism was also well-represented in the Romantic movement, especially by Southey, Coleridge and Shelley, who espoused animal rights and vegetarianism.

Shelley's father-in-law, William Godwin, set out anarchism's starting point in *Political Justice* (1793). This refuted the idea of social contracts

Table 4.2 Anarchism's principles

The features of social life which anarchists broadly *favour* include:

1 Individualism or Collectivism
2 Egalitarianism
3 Voluntarism
4 Federalism
5 Decentralism
6 Ruralism
7 Altruism/Mutual Aid

The social features which anarchists broadly *oppose* include:

1 Capitalism
2 Giantism
3 Hierarchies
4 Centralism
5 Urbanism
6 Specialism
7 Competitiveness

Source: Cook 1990

(proposed by Locke or Rousseau) overseen by a government (the state) and having force down the generations. For they infringed the individual's right of free action: they and the state created an *unnatural* society of hierarchies, competition and division. By contrast the *natural* society was one of harmonious cooperation and spontaneous association.

Kropotkin, in *Mutual Aid* (1902), attempted scientifically to show how the human and animal instinct towards mutual aid was a major factor in evolution, in contrast to the Malthus-Darwin model of aggressive competition (Chapter 5.6). However, anarchists shared with social Darwinists, and many ecocentrics today, the ideas that there *is* a fixed human nature and that the rest of nature provides a model for human society. Again like ecocentrics, anarchists also champion the individual, feeling that people can attain most control over their own lives, and greatest quality of life, in small-scale as opposed to state-centralised society.

But there was, and is, a fundamental cleavage in anarchist thought between those, like Max Stirner or Pierre-Joseph Proudhon, who tended to prioritise the individual over the community and those, like Kropotkin or Michael Bakunin, who saw people's individual fulfilment only in terms of how closely they could relate to the community. This distinction underlay Woodcock's (1975) classification of anarchism (Table 1.10, p. 45) and it separates more liberal (e.g. Proudhon's mutualism) from socialist forms (particularly anarchist-communism and anarcho-syndicalism). 'Social ecology' today, which is really green anarchism, tends to waver between liberal and socialist political philosophies and social prescriptions – its socialist content is clearly utopian.

211

Proudhon's vision did not exclude some features of libertarian capitalism and it was quite 'green' too. Free individual producers would enter into mutual contracts with others as and when they deemed it to be beneficial. All would have equal access to credit, perhaps with labour cheques as the exchange medium. Large-scale monopolistic capitalism and the state fetter the individual's voluntary transactions, along with vast accumulations of property – they should all be abolished.

For Kropotkin, anarchism led to communism, of the utopian sort that Marx also was ultimately seduced into visualising – the free association of producers, without class division, wage slavery or even money, where each individual contributed work according to ability and took goods according to need. Kropotkin initially conceived of his communes as a specific spatial form (based on the Russian *mir*), but later, like Marx, he saw them as groupings of equals in thousands of towns and villages.

The differences between Marx and Engels on the one hand and Bakunin and Kropotkin on the other illustrate how the latter could be seen as utopian socialists in the naive sense scorned by Marxists. For most anarchists opposed confrontational political struggle, especially when it meant struggle to control the state (Table 1.5, p. 32). Believing that you could not use means which were incompatible with ends, they rejected the socialist idea that violence could help to create a peaceful world, or that the state could be taken over on behalf of the proletariat and turned against ruling classes, ultimately to abolish the state and class society altogether and achieve the 'higher stage' of communism. Political power corrupts, they felt, no matter who held it and for what purpose. Hence one kind of oppression would replace another: therefore the 'higher' stateless stage had to be achieved without ever enlisting the state, on the very morning after the revolution – or even before, in exemplary communes. Thus

> The true anarchist stands aside in his pristine virtue until the moment of apocalyptic revolution [by spontaneous mass uprising rather than initiated by an élite vanguard of leaders], or, at most, in the prerevolutionary period he may form voluntary groups among his fellows for mutual aid.
>
> (Manuel and Manuel 1979, 740)

This argument still features in the sometimes uneasy dialogue between 'reds' (Marxists) and radical greens (utopian socialists and other anarchists). The latter usually point to the former Soviet Union as an example of the pitfalls of Marxism.

The reds, however, disown Soviet Marxism (i.e. Marxism-Leninism) as state capitalism, not socialism. They favour that tradition in Marxism which was initiated by William Morris, 'the first English Marxist' (Morton 1979). While he painted a utopia, he did not doubt the need to overthrow capitalism in proletarian revolution to achieve his small-scale, decentralised, truly democratic and ecologically benign society.

His picture of an ideal socialist landscape, in *News from Nowhere* (1890), was very close to Kropotkin's in *Fields, Factories and Workshops* (1899) (see Pepper 1988), and also to that of today's 'bioregionalists'. Sale's (1985, 85–6) list of features of the bioregional society make it virtually indistinguishable from their utopias (Chapter 6.3).

William Morris

O'Sullivan (1990) considers that Morris applied Marxism practically to show how we could rid ourselves of certain social evils. And

> What he also achieved, by no means incidentally, is to provide radical environmentalists with a document setting out many of their basic ideas in plain English . . . [making] an unrivalled contribution to both revolutionary thought and environmentalism.

Wall (1994, 9–10) agrees, and Morton (1979, 30) similarly underlines Morris' ecological credentials:

> The working out of a truly self-reviewing ecological basis for the earth may well be the next task before humanity, a task impossible for capitalism, possible though still not easy for socialism. The profound wisdom of William Morris can be of immense value to us in attempting it.

Indeed, it is sometimes impossible to distinguish Morris from a Thoreau-esque ecocentric:

> What brings about luxury but a sickly discontent with the simple joys of the lovely earth? . . . shall I tell you what luxury has done for you in Modern Europe? It has covered the merry green fields with the hovels of slaves, and blighted the flowers and trees with poisonous gases, and turned the rivers into sewers . . . Free men, I am sure, must lead simple lives and have simple pleasures.

> (Morris 1887a)

Morris also feared restoration of architecture purely

> for the benefit of hunters of the picturesque, who, hopeless themselves, are incapable of understanding the hopes of past days, or the expression of them. The beauty of the landscape will be exploited and artificialised for the sake of villa-dwellers' purses where it is striking enough to touch their jaded appetites; but in quiet places like this it will vanish year by year (as indeed it is now doing) under the attacks of the most grovelling commercialism.

> (Morris 1889)

The first quotation demonstrates Morris' sensitivity to pollution as a product of consumerism. But, neither a Malthusian nor an ascetic, his economics started from a premise of potential abundance for all people. This, however, depended on earth's resources being employed to meet *use* value rather than value as commodities in a market place: production for need rather than 'artificially induced wants' as the green epithet now has it. In the second quotation Morris demonstrates his eco-*socialist* credentials by identifying commercialisation and commodification of nature *in capitalism* as the source of its impoverishment. Morris did not oppose the fruits of industry *per se*, particularly in small workshops as craft production, nor did he abhor any change which humans make to nature. In fact nature was glorified and improved by contact with humanity, provided that contact was not aggressive, as it is in capitalism.

Morris thus elaborated virtually all of the themes 'discovered' by radical environmentalists over the past quarter-century about a century before they did. And he did it within a Marxist framework which in no sense could be seen as paving the way for the ecologically disastrous centralised state bureaucracies of twentieth-century Eastern Europe.

Morris inveighed against pollution, urbanisation of the countryside by mass housing, commodification of nature and of human labour, alienation of the labourer from his/her creativity, and consumerism. He rejected material luxury but advocated a luxurious quality of life, in education, enjoyment of nature, community, and creative and 'useful work' as against 'useless toil'.

His preoccupation with art and craft was not élitist: rather he saw them as the way that all people could express their individuality and relate to others. O'Sullivan considers that this perspective on work is also his most important *ecological* perspective. In it Morris (1885) emphasises that our labour should not be appropriated by others, that it should produce useful and beautiful things, should bring us mental and physical pleasure, should always be creative and should not be surplus to real needs. The workplace should be small, with less specialisation and more rotation of tasks. Machinery and all products of science should be reserved to save us from truly arduous and unpleasant work, and for

teaching Manchester how to consume her own smoke, or Leeds how to get rid of its superfluous black dye without turning it into the river.
(*Collected Works*, cited by O'Sullivan, p. 171)

However, in his *News from Nowhere*, a utopia also acceptable to most anarchists, places like Manchester and Leeds are gone – converted, like London, into a series of villages set in woodland. The huge manufacturing districts are broken up, for nature to heal the scars they made, and we have ceased 'disgracing the earth with filth and squalor'. Despite this, and curtailed machinery and division of labour, there is such an abundance

214

of ordinary necessities that no exchange is needed. There is no private property as of right, hence no hierarchy based on it.

Though the bulk of Morris' work, especially his political writings, display an analytical edge and an insight in advance of much ecocentrism, *Nowhere* tends to emulate a common ecocentric failing (from a humanist perspective at least). Its utopian medievalism idealises the countryside and the past, as the setting for harmonious social relations. But it slides over the problem of how to improve the environment of the masses, only deriding the rash of 'cockney villas' which Victorian London spread over the countryside. *Nowhere*'s hero falls asleep by the Thames in that London and wakes up in the twenty-first century, when capitalism, ugly urban environments, commercialism, wage-slavery and pollution are gone. *Nowhere*, say the Manuels, goes alongside Edward Bellamy's Fourierist *Looking Backward 2000–1887* (1888) in a tradition of nostalgic socialist utopias which by-pass the dynamism of industrial–scientific civilisation. They envisage simple, genteel communities: a 'golden age of the future' which really recaptures an imagined past of freedom from fear. In *Nowhere* Morris idealised the medieval artisan, like other pre-Raphaelites. But as a whole his work went further than they did in rejecting capitalism, so that he is ultimately the prophet of an eco-socialist future rather than a romantic ecotopian past.

Communes and back-to-the-land socialism

Many of the communes and community experiments of the nineteenth and early twentieth centuries were associated with utopian socialism and anarchism. They also displayed many principles and practices relevant to modern ecocentrism. Robert Owen's ideal communes embodied such 'green' principles as self-sufficiency, an agricultural base with small industry mixed in, public kitchens and communal eating and child care, mixing town and country and producing to meet social need. Hardy's (1979) description of the Owenite commune at Harmony Hill (1839–45) refers to careful and systematic organic farming, while Concordium in Surrey (1838–48) is described as a centre for health reform that prohibited salt, sugar and tea. Concordium was also mystical, believing in the power of the 'love spirit' to change society, and publishing in 1843–4 a journal called *New Age*, with articles on mysticism and vegetarianism. George Ripley, who founded the non-religious Owenite Brook Farm commune near Boston in 1841, is described by Mercer (1984) as a leader of the transcendentalists, the romantic followers of nature mysticism.

Hardy goes on to describe agrarian socialism, which in some superficial ways, though not because of a root concern for nature, was ecologically sound. Communal working and control of the land was seen as the source of all economic power and therefore the key to establishing socialism. Its idyll of peasant fulfilment and village cooperation contrasted with the general reality of long-established rural enclosures and alienation of common people from the land against which the Diggers had fought.

Agrarian socialism's first strand, Chartism, involved Feargus O'Connor's attempt to establish many families in self-sufficiency on two- to three-acre plots practising intensive organic husbandry, with potatoes, pigs and sheep the staples of each colony. New colonists at Charterville were promised 40 tons of dung each.

The second strand was founded on John Ruskin's romantic socialism, which often yearned for feudal communities. Lost values were to be restored through craft production; the land was to be labour-intensively farmed to restore its full potential, to save people from the alienation of industrialism and to narrow the distance between humans and nature. Ruskin's medieval-sounding Guild of St George made several community experiments, for instance at Totley, Sheffield. But the ecological principles on which it was founded foundered on the rock of some very unecological disagreements among colonists and trustees.

The third strand of agrarian socialism was in the home colonies/back-to-the-land movement of the 1880s and 1890s, which Gould (1988) calls

> the most fecund and important period of green politics before 1980
> . . . During that period the philosophy of industrialism, the relationship between the individual and the social and physical environment, and the degree of functions and successes of the city received an extra-ordinary degree of critical examination.

Faced with rising unemployment, the decline of rural society and concern for Britain's political and economic world role (a context broadly paralleling that of the 1970s and 1980s rise of green concerns), people turned to the natural world and the countryside to solve individual and social problems.

The back-to-nature and back-to-the-land themes were adopted by socialists as part of a recipe for radical social change. They invoked the simple life, an alternative to urbanism, harmony with nature, liberal sexual and social relations, a sensitive approach to animals, and hankering for a past golden age of freedom to work on the land of one's own choice and enjoy its produce. The socialist Edward Carpenter advocated 'small communities of limited wants and needs' to cover the whole nation, while industry, too, should be run on communal or cooperative lines (Gould, p. 24). Carpenter's concerns for inner and world consciousness, and organic unity between humans, animals, mountains, the earth and constellations, paralleled the present concerns of 'deep ecology'.

Robert Blatchford, editor of the socialist newspaper *The Clarion*, wanted, says Gould (p. 39), to 'see socialism and nature as established institutions'. Like today's greens, Blatchford deprecated the environmental consequences of liberal, *laissez-faire* economics, and thought that material betterment should come second to improved quality of life (e.g. simple pleasures and nature enjoyment) for the working classes. He supported and visited Starnthwaite Home Colony (1892–1901) in Westmorland, one of many set up to accommodate the urban poor and unemployed. He described it as a 'small utopia of green beauty'. Hardy says it practised real communal living and eating, and carried out fruit and mixed farming amid the beauty of nature.

Gould points out that creating new low-population settlements in the country to replace cities was the ultimate aim of the back-to-the-land socialists, and was entirely compatible with Marx's call in the *Communist Manifesto* to abolish town–country antagonisms as one of the first conditions of communal life. However, it also had counter-revolutionary, implications, being a way to cope with unemployment and defuse the discontent which capitalism spawned.

One of the largest colonies of the period was at Purleigh in Essex. Tolstoy was its dominant inspiration, and its 75 members in 1898 aimed to organise a million people into a voluntary cooperative commonwealth. A breakaway group was set up at Whiteway (Gloucestershire) in 1898 and it lasted into the 1920s, although steadily forsaking collectivism and adopting private ownership. But at the beginning its members embraced feminism, vegetarianism and non-aggression, working communally and sharing all possessions.

This very green pedigree seems to be common to other 1890s anarchist communes which Hardy describes. Norton Colony in Sheffield, for example, was based on a 'return to nature', practical horticulture and crafts (sandal making). It was vegetarian, teetotal, non-smoking, and against salt, chemicals, drugs, minerals and fermenting and decomposing foods. Unlike some of the socialist communes these had no leaders, were non-hierarchical, without majority voting, and favoured small group cooperative relationships. Perhaps more than any other communes of the period, these anarchists most closely approximated in their social, economic and political relationships to the green lifestyle.

4.6 INTO THE TWENTIETH CENTURY

Aesthetics and science

There were three distinct phases of environmental concern in Europe and America: from the mid-1880s to the turn of the century, in the middle inter-war years and from the late 1950s to the present (Lowe and Goyder 1983). There are some very direct links between the nineteenth century and today's environmental movement. For example in 1892 John Muir (who had been influenced by Thoreau) and friends formed the Sierra Club to defend America's wildernesses. Like many other Victorian environmental conservation organisations, the Sierra Club survives today. Furthermore one of its leading figures, David Brower, led a splinter group in 1969, aiming for more direct action. That group, Friends of the Earth, now operates internationally.

The divisions already noted between 'arcadian' and 'imperialist' strands in ecology, romantic and classical scientific attitudes, ecocentrism and technocentrism, appeared in all three periods. Those who wanted change in the name of material progress, so taking a managerial conservationist stance, met opposition from resisters of change, adopting defensive preservationist attitudes.

And this 'division between a fundamentally aesthetic appreciation of landscape and a primarily scientific understanding of nature conservation . . . is still evident in British culture', says Cox (1988). It persisted between the post-war British government-sponsored bodies, English Nature (scientific) and the Countryside Commission (aesthetic). Its nineteenth-century antecedents were the differences between Constable (naturalistic) and the Lakeland artists (romantic) in how they interpreted nature. Constable's depiction of landscapes, as we have noted, was naturalistic and therefore scientific (rigorously describing things as they 'objectively' were), although later appropriated into the romanticised mythical English pastoral. This approach to 'reading' nature complemented the eighteenth-century drive to 'improve' it: to create a tamed, inhabited and more economically productive landscape of neatness, symmetry and regularity. This arranged design was all part of a countryside ordered for producing goods, besides being an aesthetic object to please the rising bourgeoisie.

But later, and partly alongside this movement, came the other values, expressed in different landscapes portrayed by Wordsworth, Ruskin and others. They were tied to an emerging preservation ethic that smacked of élitism. Wordsworth's opposition in 1844 to the proposed Kendal–Windemere railway, for instance, was based on his view that access to natural beauty such as Lakeland's should be only for those cultivated people with an 'eye to perceive and the heart to enjoy' (p. 92 of *The Guide*), not artisans, labourers and shopkeepers with 'common minds'. 'These tourists, Heaven preserve us,' said Wordsworth in *The Brothers*. (Ironically he also pioneered 'the kind of tourist guide genre of writing that invites nature to be consumed in ultimately destructive ways' (Harvey 1993)).

In 1883, Ruskin, William Morris, Thomas Carlyle, Octavia Hill, Hardwick Rawnsley and Robert Hunter helped to form the Lake District Defence Society, after earlier unsuccessful attempts to prevent Manchester Corporation from turning Lake Thirlmere into a reservoir. This led to the formation of the National Trust in 1895, the concept of which had arisen from the work of the oldest national environmental group, the Commons, Open Spaces and Footpaths Preservation Society (Evans 1992). Parliament passed a bill in 1907 giving the Trust the right to declare its property inalienable against development.

Cox and Lowe and Goyder all agree that preservation groups like these shared the anti-urbanism and anti-industrialism that characterise British culture (Wiener 1981), and are part of the rejection of economic liberalism and materialism noted in Chapter 4.4. Lowe (1983) believes that this

> reflected the absorption of the urban bourgeoisie into the upper reaches of British society and its genteel value system . . . which disdained trade and industry.

The preservationist tradition re-asserted itself in the 1920s with the Council for the Preservation (now Protection) of Rural England, formed in 1926 particularly to counter 'threats' to the countryside from urbanisation and afforestation. CPRE joined other groups to press for establishing National Parks. Cox again notes elements of cultural élitism in this movement, despite the broadening of its class base to include socialistic mass-access groups such as the Ramblers Association (1935). It smacked of preserving countryside from 'the nation' in the form of 'the public' who wanted to 'discover' a bucolic rural England 'almost totally at variance with the surrounding agricultural realities which prevailed at this period' (Newby 1987).

Such preservationism contrasted with a more scientific, managerial approach to wildlife and landscape conservation stimulated by inter-war developments in ecological research (the British Ecological Society was founded in 1913). After World War II ecology became institutionalised: politicians became ostensibly willing to treat naturalists' concerns about habitat destruction seriously. The Nature Conservancy (later English Nature) was established in 1948 to notify governments of sites that should be designated of 'special scientific interest' (SSSIs). This organisation was to be administered by scientists, and was sympathetic to changing landscapes through 'positive management'.

The romantic reading of landscape was enshrined in the National Parks Commission (1949), which became the Countryside Commission in 1968. Its role was advisory and supervisory, and many of its objectives were circumvented, as were SSSIs destroyed, by mining, agriculture, tourism and other commercial interests. Cox regards the Commission as defensive, protective and attempting to sustain 'traditional' rural ways of life (see p. 223).

Conservation and preservation

Rationalism and romanticism intertwine in American environmental history too: the former manifest in a 'resource conservationist' approach; the latter in 'nature preservation' (Petulla 1988).

Interest in conservation followed early soil exhaustion in the East in the eighteenth century, and the US Department of Agriculture was established in 1862. George Perkins Marsh's seminal work, *Man and Nature* (1864) documented human-induced soil erosion and argued, holistically, the need to replant forests for soil and water conservation's sake. Fears of a timber famine led to some forest reserves being established in 1897, but at the same time Congress ruled that no land was to be used as reserves which could be better suited for mining or agriculture. Such legislation reflected the early political clout of commercial interests against nature preservationists such as John Muir.

Pressure from the latter had led to sixteen National Parks being established in America as early as 1916. But Congress' attitude to public set-aside land was none the less parsimonious and cavalier. In a running battle between preservationists and conservationists the values of the latter generally prevailed. They were epitomised by Muir's former Sierra Club colleague, Gifford Pinchot, head of the Forest Service at the turn of the century. He advocated multiple land use and scientific management rather than set-aside. In 1910 he wrote:

> The first great fact about conservation is that it stands for development . . . Conservation does mean provision for the future but it means also and first of all the recognition of the right of the present generation to the fullest necessary use of all the resources with which this country is so abundantly blessed.
>
> (cited in Opie 1971)

Opie notes that this point of view 'still dominates today's federal land use programs although the ecology movement in the 1960s encouraged the "wilderness-for-wilderness'-sake" preservationists'.

This 'utiliser–preservationist' split was dramatised in the struggle over the beautiful Hetch Hetchy valley in Yosemite National Park, which San Francisco applied to dam in 1912. The Sierra Club preservationists lost the day to Pinchot's 'wise use' philosophy in 1914, when Congress finally authorised the reservoir, and now 'The Hetch Hetchy scenario has been replayed dozens, perhaps hundreds, of times since it first unfolded' (Petulla 1988, 321).

Pinchot's conservation ideals of 'rational', 'efficient' management constituted a major plank of the Progressive political movement. President Theodore Roosevelt (1901–9) was its major spokesman. Its tradition of placing the nation's economy above that of nature stretched back, says Worster (1985) beyond the 'improvement' philosophy of the British agricultural revolution, ultimately to the Baconian project of managing nature for utilitarian gain.

Ironically this 'conservation' movement was responsible for massive wildlife extermination, in the name of efficient agriculture. 'Pests' and predators like wolves, bears and coyotes, which impeded profitable farming, were depicted in government propaganda campaigns as craven monsters. Their extermination had become a government responsibility.

Meanwhile, attempts to maintain game such as deer for the hunting lobby were often grossly mismanaged. In the mid-1920s, thousands of deer died because of overpopulation and an absence of predators.

Nonetheless the 'Progressive' approach persisted, although scientific societies from the 1920s (e.g. the American Society of Mammalogists) increasingly warned about the effects of disappearing species in the food chain. Their arguments, Worster notes, were mainly about the 'balance of nature' and how much it could be tampered with: they were pragmatic.

But in the 1940s a more moral argument, concerning the rights of animals to exist, was introduced. Its proponents wanted to balance human interests with those of other creatures, in what they regarded as a whole community. Aldo Leopold, a follower of Progressivist environmental management and member of Pinchot's Forest Service, shifted his own position from the 1930s to the 1940s. In 1933 he wrote *Game Management*, detailing a precise, calculating approach to conservation. But in the 1940s he wrote an essay on 'The Land Ethic', as the last chapter of *A Sand County Almanack*, a series of rural natural history sketches published post-humously in 1949. Merchant (1992) calls this the first formulation of modern ecocentric ethics. Worster says it signalled the 'arrival of the age of ecology', and indeed all the ecocentric catechism is there: anti-materialism; love and respect for the land; the land as one organism; the extension of 'natural rights' from humans to the rest of nature; the need for an ecological conscience rather than mere agronomic management; the plea to return to an outdoor holistic science of natural history. Most memorably Leopold wrote that predators should be spared not just for pragmatic reasons but because they were members of a community, of which humans were just plain members and citizens – a circle of cooperative communal relationships extending to all beings.

Persistence of the pastoral myth

Concern about British countryside, wildlife and ecology extended from upper to middle classes during the second and third environmentalist periods described by Lowe and Goyder. But in the first, Victorian, period there was, as noted above, more élitist reaction against the influx of tourists to wild places. Their desire to visit landscapes had partly reflected the growing popularity of naturalist and field societies. But, more than this, tourists had sought specific images of nature and countryside in pictur-esque landscapes. These images had power to command people's emotions, and this quality, Daniels (1993) reminds us, generated political potential to make diverse people come together to defend countryside and wilderness, and values associated with them.

Already, in the 1890s, Thomas Cook ran tours to 'Constable country' in the Stour Valley in eastern England, and Constable landscapes were soon taken to represent quintessential English countryside, not necessarily in any specific location:

> Constable country, in the form of the Hay-wain, pictures the metaphor of the 'south country' as the essential England, a metaphor which became compelling across the political spectrum from the later nineteenth century.
>
> (Daniels, p. 214)

Such strong feelings for landscape and 'nature' can be politically appropriated – most obviously, by the cause of nationalism. For instance, as Daniels describes, the *Haywain* was lumped together with Shakespeare, Spencer, Keats and Wordsworth in a 1916 *Country Life* article about the love that made men die for England's green and pleasant land. People in the ugly and fearsome landscape of the trenches dreamed of what Daniels calls a 'languid little England pastoral'.

This pastoral was again enlisted in World War II:

> Constable has set down for all time . . . the countryside which it is our privilege to defend against the gates of hell infernal itself

wrote S. P. Mais in a 1942 volume in Batsford's *Face of Britain* series. Constable, says Daniels, symbolised the sturdy vernacular culture, rooted in Tudor England, thought to epitomise the nation. As ornithologist Peter Scott expressed it in a radio broadcast in 1943 (cited in Wright 1985, 83):

> 'England' means something slightly different to each of us. You may, for example, think of the white cliffs of Dover, or you may think of a game of bowls on Plymouth Hoe, or perhaps a game of cricket at Old Trafford or a game of rugger at Twickenham. But probably for most of us it brings a picture of a certain kind of countryside, the English countryside. If you spend much time at sea, that particular combination of fields and hedges and woods that is so essentially England seems to have a new meaning.

Even the Constable sky-scapes were commandeered for the war effort, in a series of paintings showing vapour-trailing Spitfires swooping among the clouds over rural country. The English sky was held to reflect national character in what Gruffudd (1991) calls this 'RAF pastoral' style.

Radical patriotism also appropriated Constable in a much-reproduced 1980s anti-nuclear photomontage by Peter Kennard depicting cruise missiles set in the *Haywain* picture. Here, says Daniels, it was the American style of the Thatcher government which both left and right critics resisted: the Americanisation, or as many see it, the 'modernisation' of British culture.

Indeed the rhetoric of the English pastoral was used from the turn of the century to resist invasive rashes of modern development. A *Punch* cartoon depicted the bitter irony of British soldiers going off to defend their land, their countryside, then returning to find that land destroyed – concreted over by development (see Figure 4.1).

The class interests defended consciously or unconsciously by this kind of rural chauvinist environmentalism were those of landed aristocracy, the farming lobby and, as the century advanced, increasingly the rich middle classes who had gained their stake in this countryside and now did not want it 'spoiled' by further development.

The main inter-war fears concerned the very rapid growth of suburbia and accompanying roads and advertising hoardings, particularly in unordered, chaotic development:

> In rows of cheerful villas and bungalows the dormitories of contemporary stockbrokers sprawl over the Surrey hills; the city man approaches and the earth breaks out in an angry red rash, for all the world as if it had caught measles . . . coming upon Worthing as I did recently in the middle of a hot day in July, I was struck anew by its purulent beastliness, its utter horror . . . Machinery is, however, the real villain of the piece . . . wherever there is a main road in England today, there over an area two miles wide along the whole length of the road the peace of the countryside is shattered.
>
> (Joad 1933, 190, 193, 196, 199)

Despite Joad's familiar-sounding irritable outbursts against 'the motoring classes' and 'the new industrial revolution', most inter-war environmentalism was not anti-modernist, says Daniels. Rational, modern, planned development as exemplified by the national electricity grid, arterial roads and National Parks was in keeping with the spirit of the time.

However, with the appropriation of modernism in art, architecture and planning by the Third Reich, this changed. After World War II industrialism, planning and the centralised state increasingly became the villain of the piece, held responsible for despoiling rural England and 'natural' ecosystems. Hoskins (1955) bemoaned:

> England of the arterial by-pass, treeless and stinking of diesel oil, murderous with lorries . . . England of the bombing range . . . Barbaric England of the scientists, the military men and the politicians.
>
> (cited in Daniels, p. 224)

Such phrases strikingly pre-echo contemporary rural preservationist sentiment, as expressed in Shoard's (1980) effective anti-agribusiness defence of 'traditional' landscapes, for instance, and in many specific environmental campaigns. A 1970 *Country Life* article (cited in Daniels) supporting the *ad hoc* groups trying to defend rural Suffolk and Dedham Vale against development described it as 'the physical realisation of the ideal rural scene of every Englishman's dreams'. Again, giving evidence for a north Buckinghamshire anti-airport group at the 1969 Roskill inquiry into London's third

1914. MR. WILLIAM SMITH ANSWERS THE CALL TO PRESERVE
HIS NATIVE SOIL INVIOLATE.

1919. MR. WILLIAM SMITH COMES BACK AGAIN TO SEE HOW
WELL HE HAS DONE IT.

Figure 4.1 Patriotism and rural preservation
Source: *Punch*

airport, historian Arthur Bryant said of the area: 'all English history – its strength, its sleeping fires, its patient consistency – are contained in its sleeping silence'. An airport would destroy 'an organic part of the permanent heritage of England, as near to perfection as anything of its kind at present existing'. Poet John Betjeman declared that the area was 'what one means by England' (Pepper 1980).

Despite this appeal to heritage, however, the membership of such environmental groups is overwhelmingly adventitious – urbanites or suburbanites who have enough money to move to a rural village (where the resultant rise in house prices coupled with unemployment prohibits the families of the 'original' population from staying on). This new population is really searching for the Arcadian English pastoral:

> a peaceful, if mythical, rural idyll out beyond the high rise flats and the Chinese takeaways which, if not quite inhabited by merrie rustics, is at least populated by a race which is supposedly attuned to verities more eternal than the floating pound and the balance-of-payments crisis . . . Somewhere . . . at the far end of the M4 or the A12, there are 'real' country folk living in the midst of 'real' English countryside . . . 'real' English countryside, in this idyllic sense, is located only in the minds of those engaged in the search for it, on a few calendars and chocolate-box lids – and in the wholly misleading paintings of John Constable.
>
> (Newby 1985, 13–14)

This search is encouraged by everyday commercialism, through fabric designs, 'country' fresh food, 'everyday stories of country folk' on radio, and countless popular romances in books, films and TV series. Romantic escapism, anti-urbanism and anti-modernism persist: intensified, it seems, during national crisis and unhappiness such as economic depression and war. This is not just a European phenomenon. The romanticised polarities between countryside, nature and wilderness and urban civilisation are replicated in another popular inter- and post-war cultural manifestation, the western film from America (Table 4.3).

Of course, rural escapism, as well as the transcendentalist desire for spiritual regeneration, was partly what inspired the national parks movement in both Britain and America. Shoard (1982) describes this movement in Britain, noting that no lowland parks were created. This was partly because of a 'romantic legacy' which helped to shape the perceptions of key figures in the conservationist movements pressing for the parks. Vaughan Cornish, for example, a 'typical child of the romantic movement', thought that mountains and sea-lapped headlands constituted the supreme landscape. And John Dower, whose 1945 Government-commissioned report on potential national parks influenced what countryside was to be included in them, was essentially an 'upland man' – raised in Northumberland and the Yorkshire Dales.

225

Table 4.3 Polarities of the Western film

The wilderness	Civilization
The individual	**The community**
Freedom	Restriction
Honour	Institutions
Self-knowledge	Illusion
Integrity	Compromise
Self-interest	Social responsibility
Solipsism	Democracy
Nature	**Culture**
Purity	Corruption
Experience	Knowledge
Empiricism	Legalism
Pragmatism	Idealism
Brutalisation	Refinement
Savagery	Humanity
The west	**The east**
America	Europe
The frontier	America
Equality	Class
Agrarianism	Industrialism
Tradition	Change
The past	The future

Source: After Kitses 1969, cited in Short 1991

Shoard discovered what moorlands enthusiasts consider to be the attractive features of these uplands. They are nearly all qualities of the wilderness which the Romantics also eulogised. There is *wildness*, the antithesis of domestication, and *naturalness*, an apparent absence of human handiwork (even though the enthusiasts know full well that moorlands were human-created four thousand years ago). There is their height, perhaps fulfilling a desire for aloofness, and the *freedom* to wander at will, both satisfying the desire for *solitude* and *individuality*. And there is that much-praised quality *simplicity* in the homogeneity of form. The openness and grand vistas, in which the sky dominates, facilitate communion with the creator, providing 'almost a religious experience'.

The legacy of right-wing Romanticism

Radical ecocentrics might claim that today's ecologism is not about rural chauvinist preservationism. This may be true, but preservationism does constitute an ever-present *tendency* to which ecocentrics may be prone, or a movement for which they may become enlisted.

Again, ecocentrics criticise Anna Bramwell for describing the inter-war group of intellectuals whom she studied as 'ecologists' – on the grounds that this misleads as to the 'true' nature of ecologism. Yet it is useful to consider such people in contemplating the *potentialities* of contemporary ecologism. In particular, Bramwell's studies show, first, just how many of the ideas of current ecocentrism were popular then among 'chattering classes'. Second, Bramwell demonstrates how such ideas, by their eclecticism, can lend themselves to radical right- as well as left-wing thinking. This may well be an object lesson for today. As Allison (1991, 21) puts it, green

> campaigns and organisations often juxtapose very orthodox and respectable ideas and personnel with those which are much more eccentric or even deeply unrespectable.

Like many greens, Allison appears to consider this inconsistency as some kind of strength: others however may regard it as a weakness.

Bramwell (1989, 104) insists that the '1930s saw the full development of the group of ideas we call ecologism today'. They flowered among people whom she calls anti-capitalist 'soft alternative right: Tory anarchists'. Theirs was a back-to-the-land lobby, among which John Hargraves' alternative to the Scouts, the Kibbo Kith Kin, was foremost. Hargrave's beliefs – in pantheism, socialism, Quakerism, eugenics, Eastern religions and Anglo-Saxon nationalism – represent 'a confusing synthesis of political attitudes'. So did the group surrounding the Kin: for instance Patrick Geddes, H. G. Wells, Havelock Ellis, Frederick Soddy and Rolf Gardiner. Their interests in folk roots and oriental religions resembled the alternative counterculture developed in Germany between 1890 and 1933. There was a current of nationalism running through it – a 'teutonic élitism'.

Hargrave merged the KKK with unemployed groups in 1931 to form a Legion of Unemployed who eventually called themselves 'Green Shirts'. Like the Fascists with whom they had affinities, they sought a 'third way' (a familiar ecocentric phrase) between capitalism and state socialism. This involved ideas like social credit (where the state issued credit to working people to keep demand going in the economy) and, crucially, pro-rural anti-urbanism. Amid the 'traditional', rural folk anti-capitalist values, they believed, could thrive. A fresh start could be made. As demonstrated in Chapter 4.5, similar sentiments motivated utopian socialists.

Rolf Gardiner, co-founder in 1945 of the pro-organic Soil Association, was another politically eclectic proto-ecocentric according to Bramwell. Among his influences was D. H. Lawrence (himself influenced by Haeckel), who was a nature worshipper. Bramwell characterises Gardiner as a guild socialist, pro-German ecologist, supporter of Nazi rural policies and of paganism, organic farmer and enthusiast for social communality, and for harmony and balance in nature and society, English folk dancing and personal, spiritual rebirth! Gardiner's dream of a third way involved a united, pagan, England and Germany. The pro-German, pro-Nazi

227

sympathies, which were still fairly respectable in English intellectual circles, were, says Bramwell, not spurious: they were fundamental to his beliefs.

He was attached to another Anglo-Saxon nationalist Lord Lymington, a Nazi sympathiser whose programme was based on healthy food, healthy soil and, again, rural regeneration. His Kinship in Husbandry organisation embraced Rudolph Steiner's bio-dynamic agriculture. This advocated self-sufficient farms, tilled in accord with Steiner's vitalist, New Age, vision of a living earth, of which the soil was an actual organ imbued with life forces. Lymington attributed every social evil to urbanism: his suspicion of trade and finance (with its inevitable anti-semitic undertones at that time) is in the 'broad green tradition' thinks Allison (p. 17). Lymington was 'true green, far right and very shrewd'. Allison (pp. 18–21) makes a point-by-point comparison of Lymington's ecological and social beliefs with those of Jonathon Porritt today. 'The agreement is considerable,' he concludes, 'despite the two writers being from opposite ends of the orthodox political continuum.' Allison, incidentally, prefers Lymington to Porritt's 'oily indecision'.

Some conservative proto-greens were anti-Nazi, such as H. J. Massing-ham (1888–1952), another Kinship in Husbandry member. His brand of rural resettlement envisaged millions of smallholdings and a Europe of small regional groupings. Though he hated machinery and urbanism and wanted to preserve rural craft industry, he also disliked the idea of the country as a picturesque museum piece. He sought to recreate the English rural tradition in a form compatible with some of the modern develop-ments which he thought more salutary.

However, others, like Henry Williamson (1896–1977), leaned towards more extreme right views. Williamson, whose books all 'hammer out an environmental point of view' (Bramwell 1989, 136), espoused Oswald Mosley's British Union of Fascists, and Hitler's rural programme as an example for 'our England – the great mother of our race' to follow. Bramwell (p. 150) says that

> Put into his correct category, as an ecologist of his era, these ideas of Williamson's which are intolerable to us today slot into place as painful but entwined with his real beliefs, a society oriented around the natural man on the natural earth.

This diagnosis points up the slipperiness of the slope that some ecological beliefs can lead to.

H. J. Fleure, the geographer, and George Stapledon, the plant breeder, for example, both celebrated rural life for its supposed sense of community, innate nature wisdom, and localised non-alienating production. But they also advocated national racial planning – a eugenic programme that would draw on this pure country stock and 'improve' the race. Stapledon, a prophet of the New Age, looked for new teachers who would take Europe into this age; leaders such as Mussolini!

This brings us to Bramwell's controversial characterisation of the Third Reich's leadership itself as 'green'. Hitler and Himmel, she reminds us, were vegetarians and animal rights supporters. Hess, Hitler's deputy, was a homeopath and promoted Steiner's bio-dynamic farming. And Walther Darre, the Agriculture Minister, advocated ecologically sound, organic land-use planning.

They all subscribed to German naturist philosophy, whose general features included the critique of technological culture and excessive rationality; acceptance of nature's laws as the bedrock of social structures and of nature as a teacher; Gandhian and Tolstoyan anarchy; and a penchant for ruralism. This last, says Bramwell, saw peasants as the repository of all honest and eternal values – the nation's life blood. If this was not the same sensuous arcadian reading of countryside as the British view, none the less it was still romanticised fiction, glorying in folk myths, paganism and mysticism. One of its most mystical concepts was that of 'blood and soil', a concept which subsumed both the fancied indivisible, historical bonds between a race and their nation (see Chapter 4.3) – or a people and their countryside – and the legitimacy of spilling blood on behalf of territory, the state and aggressive territorial claims.

If these are not exactly the immediate historical antecedents of today's 'deep ecology' biocentrism, bioregionalism, anti-urbanism and the cult of the natural, they nevertheless serve as a salutary warning to greens. Overindulging in romanticism and naturalism, and neglecting to prioritise social justice and to contemplate rationally the nature of social, political and economic structures could lead in this 'eco-Fascist' direction. Haeckel may be cited as the 'first ecologist', but in some views Haeckel does not embody rationalism, progress, liberalism and socialism, as has been traditionally thought:

> A close investigation of the major ideas of Haeckel and his followers reveals a romantic rather than a materialistic approach to biology and a striking affinity not with liberalism of socialism but with the ideology of National Socialism.
>
> (Gasman 1971, xiv)

The Darwinist movement that Haeckel created synthesised evolutionary science with romantic volkism. As such, thinks Gasman, it was a 'prelude to National Socialism'.

Others have noted these affinities between neo-fascism and certain ecocentric ideas, and how, today, 'Ecology has been particularly targeted as an ideological Trojan Horse to carry fascist ideas into the citadel of orthodox political debate' (Griffin 1991, 171, see also Wall 1989). Biehl (1993) documents how this has happened. Like the volkische movement of the 1930s, today's American New Right and European neo-Fascist groups denounce modernism, with its appeal to a universal homogenous culture of egalitarian democracy and racial mixing. Instead, they advocate 'ethnopluralism', where all cultures would have sovereignty over themselves and their immediate environment, in their 'own' country. Biehl considers that this dovetails neatly into elements of ecologism and bioregionalism. She also details several German neo-Fascist parties with strong 'deep ecology' orientations. Like volkism before them, they are attracted to Steiner's 'Anthroposophy' movement and bio-dynamic agriculture. Nowadays, says Biehl (pp. 149–62), even Rudolph Bahro preaches 'spiritual fascism', calling for a 'Green Adolph' to lead Germans to ecological salvation.

4.7 ECOLOGICAL ECONOMICS AND THE SYSTEMS VIEW

Energy economics

Ecocentrism wants to bring economics and ecology together. This is a specific aspect of the general desire to apply ecological concepts to society. The ecosystem concept, developed by A. G. Tansley in the 1940s and 1950s, studies nature not as isolated components, but in terms of the relationships between its components. These are often expressed in terms of material and energy exchanges between them.

Ecocentrism suggests applying this analytical framework to society, and to economics in particular. It sees the use and dissipation of energy as fundamental to production. Indeed, the energy implications of economic processes are so crucial that some have suggested that units of energy should be substituted for money in accounting, and that economic efficiency should be assessed primarily in energy terms (Georgescu-Roegen 1971, or Chapman's (1975) account of economics on the 'Isle of Erg').

Such perspectives have been advocated since at least the mid-nineteenth century. Bramwell (1989) and Martinez-Alier (1989; 1990), on whom the following account particularly draws, describe the advocates as mainly socialists. They started from a critique of how both conventional economic theory and orthodox Marxism arrived at measures of value. Serge Podolinski (1850–91), for example, sought an energy rather than a labour theory of value, trying to model production efficiency in terms of human intake of calories compared with their expenditure in work. He considered that physical and biological phenomena could be regarded as transformations of solar energy, and that human labour's task was to increase solar energy accumulation on earth. Eduard Sacher, too, conceived of humans as thermic machines, calculating in 1891 that individuals could do at most 1000 kilocalories of work per day and needed 3000 kilocalories of food intake to do it.

Frederick Soddy (1877–1956) developed an energy critique of capitalism. Energy, not 'capital', was the fundamental means of production, he thought, but conventional economic language obscured this real truth about wealth, supposing that capital represented accumulated and stored wealth. This, however, was irrational, for real wealth, the sun's energy, could not be stored; it could only flow. A sack of wheat, a machine, a building: all would decay in line with the second law of thermodynamics if merely stored *ad infinitum*. For them to continue to be valuable, energy had continuously to be expended on maintaining them. Hence the conventional economic view, that wealth could be stored as finance capital and that its value could *increase* as compound interest, was a nonsense. Those who had shares and bonds actually shared something of stagnant, even falling, value. Living off investment, therefore, was tantamount to living off the destruction of non-renewable resources. Soddy reiterated, here, Ruskin's view that in exchange there is no real profit, and that therefore the shadow in economic transactions is mistaken for the substance.

Underlying such perspectives was, and is, a perception of the second law of thermodynamics as a governing universal principle. The tendency it identifies, towards increasing disorder (Chapter 2.3), randomness and coldness means that free energy, which dissipates when used to do work, will decrease as a whole. There will eventually be a 'heat death' of the universe, said Henry Adams (1838–1918), where no energy will be available for work.

This scientific revival of senescence theory, by extension, pessimistically implied the ultimate degradation of society. Rudolf Clausius (1822–88) had drawn attention to a direct corollary of the second law, in *Energy Stocks in Nature* (1885). This was, as ecocentrics today constantly repeat, that energy should at all times be *conserved*: recycling of energy from states of lower to higher organisation is not possible. But Clausius saw that capitalist economics treated resources like coal or soil fertility as if they *were* inexhaustible, whereas the second law dramatised their exhaustibility – their inevitable decay. Arguing similarly, Adams forecast that the world would 'go smash' in 1950 through lack of coal. As Martinez-Alier (1990, 117) puts it, such arguments contributed less in the way of firm scientific analysis based on energy flows than a 'metaphysics' of doom-laden threats to humanity arising from the second law. Ecologism enthusiastically endorses this metaphysics, building its politics and economics on limits to growth premises and evoking the grim-reaper spectre of the second law, for example

Today's inflation is tied directly to the depletion of our nonrenewable energy base . . . inflation . . . is ultimately a measure of the entropy state of our environment.

(Rifkin 1980, cited in Button 1988)

231

Most of these writers were concerned not just about capitalism's wasteful-ness of energy, but energy inequality. Adams' work showed that some economies and social groups used vastly more energy than others.

Leopold Pfaundler (1839–1920) also studied flow and distribution of energy within and between different societies. His concern was carrying capacity, and he thought that mutual aid (and trade) could increase a region's carrying capacity, but at a price. This would be loss of energy in transporting goods – a 'frictional coefficient of transport' which had to be overcome (compare Schumacher's (1973) strictures about the energy-wastefulness of today's international exchanges of virtually identical goods like cars).

Socialists consider that their own economics would, unlike capitalism, be rational enough to overcome such failings. One distributional injustice which capitalist economics perpetrates is towards future generations, who cannot register their preferences in today's market, while discounting means that future resource consumption is seen as less rewarding than consump-tion now (Chapter 2.3).

Such intergenerational problems featured prominently in the critique of economics developed by Patrick Geddes (1854–1932), biologist, town planner, back-to-the-land decentralist, Kibbo Kith Kin member and arche-typal proto-green, who greatly influenced another proto-green, Lewis Mumford (Chapter 5.2).

Geddes, like Soddy, strove to make energy accounting part of a holistic economics. He argued for bringing biology, physics and psychology to bear on conventional economics, criticising the narrowness of contemporary definitions of utility and wealth. His ecological critique of industrial urban-isation owed much to Soddy's rational debunking of the notion of capital, as well as to the more romantic Ruskin: the market, he said, could not answer people's need for beauty. Value was not therefore a matter of narrow material utility: it should include the aesthetic. Economics should be dedicated to improving quality of life. And industrial wealth was not merely a question of present returns on capital invested but had to take into account the effect of present use on future reserves of resources. Geddes (whose ideas on town planning were 'proto-green') saw history as a story of growing production, but fossil fuel depletion would inevitably slow growth, hence renewables should be developed.

Joseph Popper-Lynkeus (1838–1921) proposed an economics based on only moderate non-renewable resource use. His anarchistic social arrange-ments would provide basic needs 'free' for all, through state-organised national conscription where everyone would give up to twelve years' labour. The rest of the economy could be organised along market lines if people wished.

Karl Ballod-Atlanticus (1864–1933) envisaged land reallocation into 500-hectare mechanised farms, fertilised by returning city waste to the land – all to feed communes of up to 2000 people each. In 1927 he also pointed up the resource implications of the private car (the greenhouse effect having been hypothesised as early as 1903 by Svante Arrhenius). Such visions were quite technocratic, being more compatible with Kropotkin than with William Morris' romanticism. Many of their elements feature in today's 'ecotopias' (Callenbach 1978, 1981), as well as in green political programmes, like the basic income scheme.

Reduction and mechanism

These attempts to derive a holistic economics of energy flows could produce just the reverse effects to those ecocentrics desire: not of holism, but of *reducing* everything to energy – as bad as reducing everything to money in capitalism. Writers like Geddes and Soddy were not reductionist, for, following Ruskin, they recognised many different kinds of values. Others, however, were reductionist.

Wilhelm Ostwald (1853–1932), for instance, was an 'extreme "monist" reductionist; he went so far as to believe that "mental" or "psychic" energy could be measured in physical units' (Martinez-Alier 1990, 183). He defined cultural progress as increase in the availability of energy and substitution of human by other forms of energy. Martinez-Alier notes the absurdity here: if the theory were correct then twentieth-century Austin, Texas would be a more 'cultured' city than Renaissance Florence!

Thus perspectives derived from ecology can be just as defective as the classical science which ecocentrics disdain. Indeed, twentieth-century ecology did take on the mechanist, analytical, reductionist mantle of classical science. Worster says that Raymond Lindeman's (1942) work on the 'trophic-dynamic aspect of ecology' was significant here. It attempted to reduce all interrelated biological events to energy terms, amenable to abstract analysis. It saw ecosystems as assemblages of energy producers and consumers. As energy passed up to higher trophic levels much of it was lost. In fact there was only 10 per cent conversion efficiency, so that of every 100 calories of net plant production only ten are expected at herbivore level and one at carnivore level (ecocentrics often regard this as the strongest argument for vegetarianism).

This 'bioeconomics paradigm' now rules, says Worster, so that the 'hard' science of ecology speaks of energy 'income', nutrient 'capital' and energy 'budgets', reflecting the values of wider society. The idea of energy coursing through the system superficially resembles Eastern mystical and romantic perspectives, but Tansley's ecosystem model has nothing really in common with Goethe's and Thoreau's life force. Nor, says Worster, was it really born of classical mechanistic science: it came from modern thermodynamic physics. This allowed ecology to merge with economics – to become a quantitative, reductionist science, serving the impersonal, manipulative agronomic management philosophy – and it is not really connected with ecologism's rise in popular consciousness:

> Tansley hoped to purge from ecology all that was not subject to quantification and analysis, all those obscurities that had been part of its baggage at least since the Romantic period.
>
> (Worster 1985, 301)

It is significant that in the 1960s, when social scientists such as geographers wanted to establish their discipline as a 'hard' science in keeping with the spirit of the times (and with funding opportunities) some of them proposed adopting the 'ecosystems approach'. Its supposed advantages were that

> The study of the ecosystem . . . requires the explicit elucidation of the structure and functions of a community and its environment, with *the ultimate aim of the quantification of the links between the components* . . . partaking in general system theory, the *ecosystem is potentially capable of precise mathematical structuring* . . . Ecosystems are *ordered arrangements of matter in which energy inputs carry out work.*
>
> (Stoddart 1965; emphases added)

But it is difficult for a social science to view human societies as 'ordered arrangements of matter in which energy inputs carry out work'. A related 'cybernetic' approach to society, by the ecocentric Goldsmith (1978), seems equally unsatisfactory, recalling as it does structural functionalism's deterministic, and reactionary tendencies. Societies, he says, are exactly analogous to biological organisms such as cells or other ecosystems – 'units of behaviour within the biosphere'. Cybernetics shows that systems are controlled by *control mechanisms*, which detect data 'essential to the maintenance of the system's stable relationship with its environment'. The data can be resolved into a 'model or set of instructions' ensuring the normal daily 'behaviour of a biological organism such as a dog or man'. For society, such a model or template is the consensus world view, and the control mechanism for this, says Goldsmith approvingly, is *religion*. Religion can ensure that society's basic structure is maintained: religion 'admirably satisfies cybernetic requirements'. It encourages order, giving the stability which the social system needs. In India, the caste system 'supplies a religious basis for inequality' and has helped maintain 'a stable plural society in the face of overwhelming odds'.

Social ecologists might say that this is bad rather than good. They would be uneasy with these sorts of manipulative approaches to human society: approaches which could logically follow from the current paradigm of ecological science. Bookchin (1980, 88) scathingly summarises the fundamental objections:

If energy becomes a device for interpreting reality . . . we will then have succumbed to a mechanism that is no less inadequate than Newton's image of the world as a clock . . . Both reduce quality to quantity . . . both tend towards a shallow scientism that regards mere motion as development, changes as growth . . . systems analysis reduces the ecosystem to an analytic category for dealing with energy flows as though life forms were mere reservoirs and conduits for calories, not variegated organisms that exist as ends in themselves and in vital developmental relationships with each other.

Organic systems

However, as well as the mechanical tradition, organicism has also worked forward into ecological thinking, albeit as a minority current (Table 1.6, p. 36). Organicism holds that complex wholes have

the kind of systematic unity characteristic of organisms. An organism differs from mere mechanism of aggregates because the *nature* and *existence* of the parts depend on their position in the whole. A hand is a hand only if united to the living body . . . [Furthermore] the whole has a characteristic life cycle as organisms typically do.
(Bullock and Stallybrass 1988, 613; emphases added)

Capra (1982, 37) wants systems science to incorporate this organic world view, 'characterised by the interdependence of spiritual and material phenomena and the subordination of individual needs to those of the community'. Organic systems form (p. 303)

multilevelled structures whose levels differ in their complexity . . . At each level of complexity we encounter systems that are integrated, self-organising wholes consisting of smaller parts and, at the same time, acting as parts of larger wholes [see Figure 4.2].

Capra's *Turning Point* restates the organicism and the critique of classical mechanistic science which, in its twentieth-century form, was most memorably put by Alfred North Whitehead, Jan Smuts (1926) and Lewis Mumford (1938 – his concise summary of 'the organic outlook' is reproduced in Wall 1994, 100–101) (see Chapter 5.2).

Contemporaneously, and up to the 1970s, Ludwig von Bertalanffy (1968) developed his 'organismic conception' of general system theory (see Figure 4.2). This wanted to discern and describe general principles governing how all kinds of systems worked and developed throughout the universe. It saw that universe not as a deterministic machine but having design and purpose. It showed *self*-regulation, *self*-direction and *self*-orientation. Thus, for instance, feedback, servo-mechanisms and circular systems and processes all help to maintain homeostasis in systems – a condition of dynamic

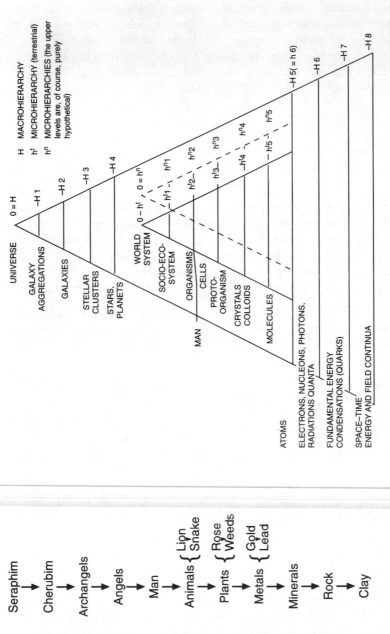

Figure 4.2 Chains of Being. On the left is a sketch of a medieval Chain of Being, a cosmology of the living and non-living world arranged in a pyramid and hierarchical order. Such an arrangement is also seen in the cosmology of twentieth-century general systems theory on the right.

Source: Laszlo 1983

balance by which stability is maintained although there is constant fluctuation and oscillation around a moving mean.

Various approaches help to discover these general principles, such as classical systems theory, set and graph theories, cybernetics, information, queuing and games theories:

> The point to be reiterated is that problems previously not envisaged, not manageable, or considered as being beyond science or purely philosophical are progressively explained.
>
> (von Bertalanffy 1968, 23)

These are problems which an understanding of systems as just local does not adequately explain. They are problems of order, for instance, whereby disparate structures (atoms, proteins, cities) exhibit similar organisational principles and patterns. Thus

> an exponential law of growth applies to certain bacterial cells, to populations of bacteria, of animals or humans, and to the progress of scientific research measured by the number of publications . . . The entities . . . are completely different and so are the causal mechanisms involved . . . nevertheless the mathematical law is the same.
>
> (von Bertalanffy, p. 33)

Or again, biology and economics both share the same 'laws of competition'.

Again, there is the obvious danger that such holism can, perversely, really be reductionist. By seeking universal principles it could treat a bacterial colony and a city as essentially similar. Von Bertalanffy shows awareness of this by stressing the importance of differences between those phenomena which he also links via universal principles. Organicism has been an important theme in the minds of some interpreters of twentieth-century science, to which we now turn.

Chapter 5

POSTMODERN SCIENCE AND ECOCENTRISM

Subjectivity, ideology and the critique of classical science

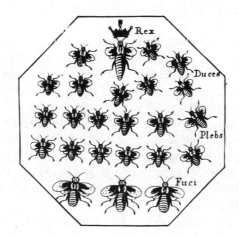

POSTMODERN SCIENCE AND ECOCENTRISM
Subjectivity, ideology and the critique of classical science

5.1 ECOCENTRIC AMBIGUITIES TOWARDS SCIENCE

All radical environmentalists should think about the nature and role of science in society. For, as we have noted, science constitutes a principal component of the 'cultural filter' through which Western society views nature and environment. Many people tend to regard science as the most 'respectable' way to know nature: to show irrational or emotive attitudes towards an environmental issue runs the risk of not being taken seriously. By contrast, people often see science as the source of absolute, monolithic 'truth' about the environment: the route to the objective 'facts' on which we can base decisions.

But straight away this presents us with problems. We have noted that ecological science tells us as a 'fact' that *humans are part of natural systems. They must be subject to the same basic laws and must live in harmony with nature. They must not exploit nature.* Yet Baconian, 'classical', science seems to say something quite different. It asserts that *scientific knowledge equals power over nature, and if that knowledge is managed properly we can and should use science for social progress, via transforming nature.* A science of monolithic 'truth' surely cannot tell us both contradictory things at the same time. To understand the apparent paradox we must consider some scientific history, as Chapters 3 and 4 have done, and some social and philosophical aspects of modern science, as this chapter is about to do.

Ecocentrics are ambiguous about science. On one hand they claim that the science of *ecology* (especially its 'arcadian' strand) indeed conveys absolute wisdom and truths, and universal principles about nature and how all society ought to behave towards it. On the other hand they claim that romantic – non-rational, emotive and intuitive – knowledge of nature is as legitimate as scientific knowledge. Again, on the one hand they attack classical science for its view of nature as separate, machine-like and reducible to basic components that can be objectively known and predicted. They criticise, too, much development associated with the modern period that enlisted the aid of big science and technology (e.g. conurba-

240

tions, pesticides, nuclear energy). But on the other hand they are not averse to enlisting notions of progress, detachment and nobility of purpose that are associated with the stereotypical image of classical science in order to legitimate their own aspirations and interests. If environmentalists can show that their cause is supported by scientific evidence and research, and scientific experts, following scientific method, then the public will be more likely to see them as above sectional interest, legitimate, respectable and worth supporting.

For some ecocentrics these ambiguities can be reconciled by recourse to twentieth-century science, especially physics. They consider that this science totally legitimates their world view – which is organic, monist, subjective, mystical, and believes in design and purpose rather than chance in evolution. But subjectivism would suggest no 'objective' facts about nature, that is, properties of nature are not separate from our interpretations of them. However if this were so, alleged ecological 'facts', say, about how industrial society degrades ecosystems would have no authority, being merely a function of the observer's interpretations.

Such ambiguities may lead us to regard science as an 'unreliable friend' to the environmental movement, as Stephen Yearley (1991) puts it. Yearley, a sociologist, is well aware of how sociologists, historians and philosophers of science have revealed that the authority commonly associated with scientific experts may have been unjustly exaggerated. In the Baconian tradition this authority partly derived from the belief that scientists were 'unbiased' in the sense of not being partial to specific vested interest groups, social, political or economic. They were engaged in an intergenerational and international search for truth, and were not subject to fashion or fad or undue pressure from society. Yearley, however, emphasises that *science is a social activity* and that it rarely conforms to this Baconian ideal. The very questions that scientists ask, and the answers they reach, cannot be fully understood without understanding the social context in which the scientists work. Economic and political power relations within science and in society influence this work, laying doubt on claims to absolute truth and authority.

Scientists sometimes attack sociologists for overstating this relationship of science to its social context. In this argument scientists may adopt *'direct realist'* and *naturalist* philosophical positions (see Martell 1994, 3–4, 121–4 and Dickens 1992). This means they consider that nature has objective properties which can be studied and known directly as free-standing truths about the world. Such material properties of nature embrace and condition human thought and behaviour.

By contrast, sociologists may take a *social constructionist, relativist* view, implying that the findings of science are never totally objective in that they mirror or even depend on influences from society. In social constructivism, scientific findings, 'facts' and 'truths' are not, then, absolute and universal, but are relative to the society from which the scientists come.

Underlying social constructivism may be the assumption that we cannot ever know nature's 'objective' qualities directly, and 'objectively', because they will inevitably be filtered and mediated through human perception. This could lead further: to an idealist view claiming that nature *has* no independent, objective properties – they are all the creation of the mind (see Chapter 2.7).

The ecocentric interpretation of twentieth-century science tends towards such idealism (implying a very intimate bond indeed between us and nature). Again, though, this ecocentric tendency to embrace subjective knowledge and relativist positions sits uneasily and ambiguously with the much-loved concept of nature having intrinsic (objective) qualities, giving it value independent of humans (see Chapter 2.1).

5.2 TWENTIETH-CENTURY PRECURSORS OF THE GREEN CRITIQUE

Ecocentrism's view of science borrows from some writers in the first half of the twentieth century: for instance, Alfred North Whitehead (1861–1947), Jan Smuts (1870–1950), Henri Bergson (1859–1941), Ludwig von Bertalanffy (1901–72) and Lewis Mumford (1895–1990, see Mumford 1934; 1938). Indeed, Griffin (1993) considers that Whitehead's view was 'deeply ecological', because his cosmology supports biological egalitarianism, the ecological nature of the self, the notion of intrinsic value, and the idea that each individual thing arises out of and is constituted by its relations with other things – this last being a tenet of organicism.

Organicism

Whitehead (1926) proposed organicism (see Table 1.6, p. 36, and Chapters 3.4, 4.3 and 4.7) to replace classical science's machine metaphor. The latter's view of nature as consisting of matter with primary qualities, such as location and size (fundamental solid particles) was, thought Whitehead, 'quite unbelievable'. For these 'objective' qualities were really partial qualities detached from their context. That context is a 'flux', or whole continuum of the universe, of which we are intimately a part. As Smuts (1926) put it:

> Both matter and life consists of unit structures whose ordered grouping produces natural wholes which we call bodies or organisms. This character of 'wholeness' meets us everywhere and points to something fundamental in our universe. Holism . . . is the term here coined for this fundamental factor.
>
> (cited in Clarke 1993, 178)

Whitehead thus thought that while the concept of objectivity, suggesting separation of humans from the rest of nature, had been astoundingly useful, fundamentally it represented a 'one-eyed reason, deficient in depth'. From the eighteenth century on, this reason had derided everything that did not fit its schemes, and resisted subsequent attempts in biology to impress organic thinking onto philosophy.

Whitehead's whole outline of his 'theory of organic mechanism' greatly paralleled the interpretations of modern physics which contemporary eocentrics make (see Chapter 5.3). What we saw as 'things', he said, were really 'events' (e.g. a series of interactions among, say, fundamental particles). And every event incorporated aspects of other events into its own unity, while other events took aspects of it into their unities. Hence the organic philosophy of nature by definition hinges on the concept of interaction, so important to ecocentrics.

Like some later physicists (e.g. David Bohm), Whitehead conceived of 'objects' as temporary structures in a dynamic flowing universe, like vortexes in a stream (where the vortexes are the 'objects' and the stream is the general flux of energy and matter). Furthermore

> One and the same pattern is realised in the total event and is exhibited by each of these various parts grasped into the togetherness of the total event.

That is, as Bohm now says, that each little bit of the universe reflects in its structure the whole. So 'objects', or entities, are what they are partly because of their own history and partly because they derive from aspects of other entities (including us, the observers) forming their environments. It follows from this that the observer in some way helps to shape the nature of the things being observed.

Organic science and technology, and holistic knowledge

Mumford (1934) shared Whitehead's reservations about classical science's world view. However, in a quasi-technocentric way, he was at least conditionally optimistic about the potential of science and technology (less so in later writings, see Guha 1991). They could provide adequate material needs for everyone, and were already extending the richness of non-material life (creating enhanced communication, a beautiful and stimulating human environment and a 'machine aesthetic' to go with it).

Unlike Whitehead, who tended to conceive of history as principally a battle of ideas, Mumford was a materialist, so to him what militated against science and technology realising this humanistic potential was the prevailing economic system – that is, the material mode of production of capitalism. But this mode of production was the outcome of a specific historical period, and Mumford looked forward to a future when a different mode would replace it. This would be 'communist' – characterised by common ownership – and it would be 'organic'.

For instead of simplifying the organic to make it intelligible (after reductionist classical science and capitalism) it would complicate the mechanical to make it more 'organic', and therefore more effective because more harmonious with our living environment. In this coming 'fresh gathering of forces on the side of life' (Mumford 1934, 117) the Victorian myth of a struggle for life in a blind and meaningless universe would be replaced by a picture of partnership in mutual aid (cf. Kropotkin 1902).

And more holistic perspectives would reduce the total need for machines. For example in medicine better knowledge of the body, better nutrition, healthy housing and sounder recreation would cut the amount of high technology surgery necessary. Ecocentrism proffers exactly this argument today (see for instance Kemp and Wall, 1990, chapter 10) for 'greening the National Health'.

Mumford's future communism pre-empts 'ecotopia', as it also echoes William Morris. The pace of change will slow:

> The emphasis in the future must be, not upon speed and immediate practical conquest, but upon exhaustiveness, inter-relationship and integration [p. 121] . . . [In the organic age] our success in life will not be judged by the size of the rubbish heaps we have produced: it will be judged by the immaterial and non-consumable goods we have learned to enjoy . . . handsome bodies, fine minds, plain living, high thinking, keen perceptions, sensitive emotional responses and a group life keyed to make these things possible.
>
> (p. 148)

Against scientific materialism

Rather like deep ecologists today, Whitehead thought that classical 'scientific materialism' was actually *anti-rational*, despite its claims to rationalism, whereas the pre-modern religious thought that preceded it was truly rational. For proper 'faith in reason is the trust that the ultimate natures of things lie together in a harmony which excludes mere arbitrariness' (1926, 23). That is, it is rational to think that the world has design and purpose something classical science denied. Furthermore true knowledge comes not from inductive generalisation but by directly inspecting the 'nature of things as disclosed in our own immediate present experience'. Our experience is partly *intuitional* and *metaphysical*, hence true rationality must include a metaphysical analysis of the nature of things, showing fully how they function. Rationality is not just studying the 'empirical' fact of antecedents and consequences' (isolated chains of cause-effect relationships). Here, Whitehead anticipated ecocentrism's call for *wisdom* – uniting metaphysical 'know-what' with technological 'know-how' (Schumacher) and reuniting fact with value after their separation in the modern period (Skolimowski).

Design and emergent evolution

The organism is fundamental, said Whitehead, and the evolution of organisms 'depends on a selective activity which is akin to purpose' (1926, 135). Here, Whitehead like Smuts, Bergson and modern eco-centrics, rejected the classical scientific idea that chance and randomness alone makes the world what it is. Bergson (1911) argued that the tendency for species to develop structures like the eye – complex structures shared by many species – was not the result of random mutations, as neo-Darwinians maintained. It reflected continual change in a definite direction, towards greater complexity. An 'original impetus' of life, regularly passing from generation to generation in a purposeful way, was what led to the emergence of new and more complex organisms.

Smuts (1926) held that 'There is behind Evolution no mere vague creative impulse ... but something quite definite and specific in its operation '. This produced not merely more complex organisms, it was what was responsible for the emergence of life from non-life. There was an evolutionary design producing complex chemical compounds from mere physical mixtures, organisms from these compounds, minds and conscious-ness from simple organisms, and from consciousness:

> Personality, which is the most evolved whole among the structures of the universe, and becomes a new orientative, originative centre of reality . . . Holism is not only creative but self-creative.
>
> (cited in Clarke 1993, 178)

As Chapter 6.1 suggests, this has become the basis of deep ecological and New Age thinking about social change (Russell 1991).

Furthermore, Whitehead's view that

> the key to the mechanism of evolution is the necessity for the evolution of a favourable environment ... Any physical object which by its influence deteriorates its environment commits suicide
>
> (1926, p. 138)

advances Gaian co-evolutionary theory fifty years ahead of Prigogine and Lovelock. They follow him in proposing that the optimum environment is produced by each organism making the best conditions for other organisms, while 'the environment automatically develops with the species and the species with the environment'.

Ludwig von Bertalanffy also believed in 'emergent' properties in nature, which general systems theory could help to uncover. These are properties not definable by simply totalling the properties of the individual compo-nents in the system. They are 'synergetic', that is, more than the sum of their parts.

Theories of *emergent evolution* came from the Chicago School of Sociology, operating between the 1920s and 1940s. C. Lloyd Morgan proposed that nature evolves by sudden leaps. As Smuts proposed, inorganic matter produced life, and life in turn produced mind. Each new, superior, level was an irreducible whole greater than the sum of the lower levels; it cannot be reduced to those lower levels. As Bergson's theory of vitalism stated, life cannot be reduced merely to non-living things.

Such theories challenge classical scientific premises like causality, by arguing that what followed from given evolutionary processes might differ fundamentally from what preceded them. They also challenge entropy, since evolution produces *greater,* not lesser, organisation and structure with time: something which it is now recognised that open systems can do (as opposed to closed systems, which show increased disorganisation).

In all this, 'non-scientific' concepts revived from pre-modern times now become the legitimate concern of scientific ways of seeing and enquiry. These include the notions of design, and of nature struggling forward to final perfection, and the perspective of 'the one' incorporating the many within a common body (cf. the Great Chain of Being – the living and the non-living world arranged in pyramidal and hierarchical order (see Figure 4.2, p. 236 and Laszlo 1983)).

Having noted all these precedents, we can now consider how they re-emerge in the ecocentric interpretation of the findings of late twentieth-century, 'postmodern' science.

5.3 TWENTIETH-CENTURY SCIENCE

The challenge summarised

Classical science tells us that nature is a machine, whose parts are related but discrete. Its fundamental particles, such as atoms, electrons or quarks, are solid bodies in empty space. We, as observers of nature (subjects) are separate from it, the object, so we can be objective about it. Such dualism, as between mind and matter, sets the paradigm for understanding most of Western culture. The Western view of nature as a machine is essentially deterministic and fatalistic. For if nature's laws determine events, then given sufficient knowledge of these laws we could have exactly predicted the present. Our future is already set, and free will is an illusion (Zukav 1980, 51–2).

With such perspectives classical science meshes with the 'hard headed' philosophy of *materialism*, which Whitehead so criticised. This says that the physical world consists of independent material objects each with independent properties; all events are determined by prior physical causes acting according to invariable laws; the behaviour of a complex whole is explained in terms of its constituents (Powers 1985, 2).

But some scientists (by no means all) now consider that twentieth-century science challenges all this. They say that relativity theory destroys the notion of objectivity, for instance. What we regard as independent particles are really inseparable from the greater undivided whole of the rest of the universe. Separateness is an artificial construct imposed by our own minds.

To some interpreters, twentieth-century physics indicates that the observer is the observed, everything affects everything else, the universe as a whole is flowing movement, reality is multidimensional and there is a changeableness in nature's laws (Briggs and Peat 1985). Such principles have also become part of the ecocentric (particularly Gaianist) perspective. They have been taken also to support notions of panpsychism – that is, all matter, living and non-living, has rudimentary consciousness – and idealism – the ideas in the mind constitute the primary reality of the way the world is, and how it changes.

Subatomic physics

Capra (1975) and Zukav (1980) each discuss the most important discoveries in subatomic physics in the twentieth century. These, they say, suggest that atoms are not indivisible, solid building blocks of matter. They are vast regions of 'space' in which subatomic particles like electrons move at high speeds (e.g. 600 miles/second) in specific orbits ('shells') around a nucleus. Subatomic particles can be further split into other particles, by accelerating them and causing them to collide.

It is important to realise that the resultant traces made on recording apparatus in a particle accelerator register not a series of 'things' (i.e. solid particles) but that a particular *event* – a collision – has happened. The traces register what took place at the end of a process where humans made specific arrangements in order to record and discover something in a particular way. This premeditated process of interacting with nature observes what happens at the beginning and end of an experiment, but it does not thereby provide us with direct, objective experience of what nature *is*, only those events which happen when we set up certain apparatus and initiate specific processes.

Since such apparatus is but an extension of our bodily senses of sight, touch, etc., the obvious but important corollary is that we never do perceive nature – 'reality' – directly, but only as mediated through our senses and *minds*: they determine *what* we observe and how we interpret and structure it. So, as Zukav puts it (reminding us of phenomenology), whether there is an absolute truth 'out there' is irrelevant since we never directly perceive it: the mind can only ponder its own sense-transmitted ideas about reality, never reality itself directly.

Yet we still want to ask what *is* the reality with which our minds, via the senses, interact? A subatomic particle is a 'quantum', which means a quantity of *something*. Nonetheless, what that something is is a matter of speculation. Some say that it may not be meaningful even to search for the ultimate 'stuff' of the universe. Zukav considers that an important conference of physicists in Copenhagen in 1927 declared that it does not *matter* what it is that quantum mechanics is describing the behaviour of, for as discussed above it can never be known 'objectively', that is, in a way unmediated by our minds. Therefore what is most important to describe is what our minds make of 'it'. What follows, for some, is an excursion into subjectivity which denies Descartes' conception of a separateness between us and the rest of nature. (Again it should be recognised that others vigorously repudiate the validity of such excursions, as an 'ontological extravagance that all properties are observer-dependent' which the Copenhagen Interpretation simply does not justify. In fact the Interpretation *denies* the possibility of replacing the concepts of classical physics by any others, says O'Neill (1993a, 184).)

Capra says that matter's apparent solidness is an effect, similar to a whirling aeroplane propeller, of the high speed at which electrons whizz around a nucleus. But what we are really sensing is not solid particles but standing wave patterns, confined to a finite region, like the waves on a vibrating guitar string. These waves, though, are not in three dimensions like airwaves or water waves. They are 'probability waves', the shapes of which tell us where, at a particular instant, the 'particle' is most likely to be. This reflects the fact that 'particles' do not exist anywhere with certainty, but rather show tendencies to exist and turn up in a given space. Hence, Capra concludes, at subatomic level the solid material objects of classical physics dissolve into wavelike patterns, or clouds, of probabilities. They are not probabilities of *things* existing so much as of *interconnections* between the start of an event and the subsequent measurement of something, for example as in a scientific experiment.

All this is hard to visualise, especially to minds like ours, which have been socialised into classical science norms. 'Strange' also is how electrons move from one orbit to another around an atomic nucleus when energy is added to or taken from the electron. But it is not possible to trace a path, between one orbit and another, which the electron takes when moving position. At one instant it is in one position in one orbit, at the next in another orbit. There has been no intermediate position: it has taken a 'quantum leap'.

In this and other ways twentieth-century scientists discovered that their questions about the behaviour of nature at subatomic level were answered by seeming paradoxes – contradictions of 'common sense'. They might be more understandable, however, if they are regarded as outward manifestations of a deeper and grander multidimensional reality, which we are only now beginning to apprehend through science, but which we have always been able to glimpse intuitively on occasions. Capra, particularly, is keen to suggest that mystical thinking has provided insights into a reality which science, throwing off its classical conventions, is only now perceiving.

One set of 'paradoxes' involves the way that 'particles' can display dual properties that seem to contradict each other (what Niels Bohr (1885–1962) called the principle of 'complementarity'). For instance, sub-atomic units (quanta) of 'matter', such as light, can appear as particles (photons) or waves. Experiments to discover 'the nature' of light can show that it 'is' composed of particles, or it 'is' waves or that it can display properties of both at the same time. Yet this seems impossible, for a particle is something confined to a small, discrete volume of space, while a wave, in opposite fashion, is spread over a comparatively large area.

Similarly, according to Heisenberg's uncertainty principle, a subatomic particle's position and its momentum (mass times velocity) cannot be known simultaneously with precision in the way that we could know these things about, say, a football or a planet. We can either precisely know position and be ignorant about velocity and momentum or vice versa, or we can know roughly about each. The interpretation which some physicists make of this paradox is startling. Capra (1975, 145) says:

> The important point now is that this limitation has nothing to do with the imperfection of our measuring techniques. It is a principle limitation which is inherent in the atomic reality. If we decide to measure the particle's position precisely, the particle simply *does not have* a well-defined momentum, and if we decide to measure the momentum, it *does not have* a well-defined position.

Similarly, the answer to the question: 'Is light made of particles or waves?' is: 'It *is* particles if you arrange for it to be recorded as particles, and it *is* waves if you decide to record it as waves'!

Implications

Capra (1975) continues: 'In atomic physics, then, the scientist cannot play the role of a detached objective observer':

> Nothing is more important about the quantum principle than this, that it destroys the concept of the world as 'sitting out there', with the observer safely separated from it by a slab of plate glass.

For in observing the electron (nature) the scientist becomes a 'participator', making and altering its properties; what it is.

Zukav (1980, 117–18) spells this out. The wave-like behaviour which we observe if we set up what is called the 'double slit' experiment, the particle-like characteristics which we detect with the 'photoelectric effect', or the dual properties which the 'Compton scattering' experiment reveals – none of these are properties *of light itself*. They are properties *of our interaction with light*. And since we cannot know simultaneously all about these different properties of our interaction with nature, we must choose which, at a given time, we wish to know best. Now, he asserts, this is not just a matter of our influencing reality; we partly are *creating* it — we create properties when we decide to measure them. In a similar way, when we observe bacteria under a microscope we have to bombard them with light to see them. But this changes their properties. Hence we cannot eliminate ourselves from the picture. We cannot know objectively any external world. The distinction between the 'in here' and the 'out there' is an illusion. Even 'the point of view that we can be without a point of view is a point of view' (p. 56).

From this position, Capra and Zukav go on to argue idealistically that twentieth-century physics tells us that scientific 'laws' are really creations of the human mind: properties of our conceptual map of reality rather than reality itself. Similarly, distinctions between 'empty' and 'full', 'something' and 'nothing' are *false* distinctions that we have created — abstractions from the whole cosmos. Again:

> One of the most profound by-products of the general theory of relativity is the discovery that gravitational 'force', which we had so long taken to be a real and independently existing thing is actually our mental creation.
>
> (Zukav, p. 206)

Ecocentrics may seize on such interpretations of twentieth-century science. For they legitimate deep ecology's holism:

> Quantum theory thus reveals a basic oneness of the universe . . . nature . . . appears as a complicated web of relations . . . [which] always include the observer in an essential way . . . we can never speak about nature without at the same time speaking about ourselves.
>
> (Capra, p. 71)

And if *we* create the structures which our senses perceive, out of a greater cosmic reality, then there is no significant division between us and our environment:

> Ecology does not know an encapsulated ego over and against his or her environment . . . The human vascular system includes arteries, veins, rivers, oceans and air currents. Cleaning a dump is not different in kind from filling a tooth. The self metabolically, if metaphorically, interpenetrates the ecosystem. The world is my body.
>
> (Rolston 1989, 23, cited in O'Neill 1993a, 150)

These interpretations of science also seem to make New Age idealism, mysticism and irrationalism respectable:

250

the structures and phenomena we observe in nature are nothing but creations of our measuring and categorising mind. That this is so is one of the fundamental tenets of Eastern philosophy. The Eastern mystics tell us again and again that all things and events we perceive are creations of the mind, arising from a particular state of consciousness and dissolving again if this state is transcended.

(Capra, p. 292)

In similar vein, Zukav (pp. 117–18) draws out 'psychedelic' implications from quantum theory:

Since particle-like behaviour and wave-like behaviour are the only properties that we ascribe to light and since these properties now are recognised to belong not to light itself but to our interaction with light, then it appears that light has no properties independent of us! To say that something has no properties is the same as saying that it does not exist. The next step in the logic is inescapable. Without us, light does not exist . . . This remarkable conclusion is only half the story. The other half is that, in a similar manner, without light, or, by implication, anything else to interact with, *we do not exist!*

To put it another way (p. 95):

'Correlation' is a concept. Subatomic particles are correlations. If we weren't here to make them, there would not be any concepts, including the concept of 'correlation'. In short, if we weren't here to make them, there wouldn't be any particles.

Such ideas resonate not only with ecocentrism and Eastern mysticism, but also with phenomenology. They emphasise Gaian holism – tending towards monism. Physics today, say Briggs and Peat (1985, 69) take us beyond the view that everything consists of indissoluble units of matter at subatomic level. For it makes of the whole universe a unified field. Mass equals energy (according to Einstein), gravity equals acceleration, space equals time.

The idea that the universe is a moving stream of energy, constantly changing into matter and back again into energy – in a 'cosmic dance' – immensely appeals to ecocentrics. Coward (1989, 25) points out how it has become the basis of alternative, 'holistic' medicine, where '"Energy" is assumed to be the force behind nature's vigorous renewal. It is the common factor between humans and "natural" substances', that is, between humans and the rest of nature. 'Everything is ultimately equivalent to and transforms into everything else', says Capra (p. 71).

But such messages are not unalloyed good news for deep ecologists. For Zukav's analysis also clearly undermines any 'Gaian' concept of a nature with *intrinsic* (i.e. independently existing) value, which could continue in any meaningful way without the presence of humans. It implies that the 'rights', 'interests' and other measures of value attributed to all members of the biotic community, including animals, have no meaning apart from the presence and intentionality of humans.

Implications for cosmography

Postmodern subjectivist conceptions of the universe have several elements in common with pre-modern ones in the West, including the idea of interdependence, and the notion that humans as well as other 'parts' of nature constitute a microcosm of the larger order (Cosgrove 1990).

Scientists such as David Bohm have presented a cosmology of the universe as a dynamic web of interrelated events, where no properties of any part of the web are fundamental. All the properties of any part follow from the properties of the other parts, so everything must be defined in terms of everything else.

'Things' like particles are merely abstractions – something withdrawn from a larger whole. In reality 'What is, is always a totality of ensembles, all present together' (Bohm 1983, 184). This means that what our senses normally experience (what is explicit, or 'explicate') is but part of a larger universe which we do not regularly apprehend. The larger universe, says Bohm, is implied, or 'implicate'. And where we (or our instruments) experience seeming paradoxes (light as particles and waves) in what is explicate, this can be understood by reference to the larger, implicate universe.

Thus there is a deeper order hidden in events that appear superficially random, fragmented or confused. Bohm illustrates this idea of implicate order by reference to three principles, which can partly be grasped by analogy.

One analogy is the hologram. When an ordinary photographic negative has a piece torn off, all that this piece could show, if developed, would be a *part* of the original image. By contrast, a small piece broken from a larger holographic plate will, if projected by light, display the *whole* of the original image (Figure 5.1). The plate contains the interference patterns produced by the interaction of a laser beam and the object being holographed. And all of these patterns are contained on every part of the plate.

By extension, since (after Einstein) matter is also energy, the perception of the matter of which objects in general are composed is the perception of interference patterns between the energy of the objects and the energy, including light, which streams through the cosmos. And the patterns of energy in any 'object' (bit of the whole holographic plate) reflect wider encoding patterns of the matter and energy which spreads through the universe: 'each region of space containing the whole, including the past and with the implications for the future' (Briggs and Peat, 119).

Figure 5.1 The hologram analogy. The holographic plate records, coded in its interference patterns of concentric rings, a three-dimensional image of the rabbit. The interference patterns have been produced by light bouncing off different features of the object. The patterns encode those features. By shining a laser beam through the plate, the encoding can be retrieved so the rabbit seems to appear in space. This can also be done by shining a beam through only a piece of the plate!
Source: Briggs and Peat 1985

Here is the pre-modern microcosm-macrocosm concept revived, in Bohm's principle that says that *everything in the universe mirrors everything else.*

Bohm's second principle, that *wholeness is flowing movement* is illustrated by the analogy of two cylinders of different size. The smaller one is placed inside the larger, and some glycerine is poured between them. Then a drop of insoluble dye is inserted into the liquid. Since its colour and form is clearly seen, the dye drop is 'explicate' within the liquid. But when the outer cylinder is rotated, the dye drop becomes elongated, dispersed and eventually invisible (Figure 5.2). It has become 'implicate', or enfolded, in the fluid. However, if, then, the rotation is reversed, eventually the initial dye drop *reappears* (becomes explicate again) in its original place! This can actually be explained via the 'old-fashioned' causal explanations of classical science, and Bohm uses it merely as an illustration of how matter and structures become, in his schema, explicate from the flowing movement of the larger implicate order in which they had become enfolded.

253

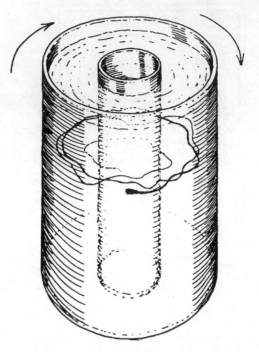

Figure 5.2 The dye drop analogy
Source: Briggs and Peat 1985

We can now imagine several dye drops placed, one after another, in
positions next to each other at the same time as the cylinder was being
continuously rotated. In succession they would each disappear. However, if
after this the rotation were continuously reversed, then each drop would
reappear in its original place and then disappear in the opposite direction.
But to the observer it would be as if one drop was continuously present,
and was actually moving (like the early cinematographic 'flicks').

The implication of this is that all discrete objects ('subtotalities') in our
apparent three-dimensional reality – stones, particles, us – are really 'dye
drops'. They are parts of the larger implicate whole, made temporarily
explicate. They are experiencing a relatively brief period of stability in the
moving whole of the (not fully seen or apprehended) universe. They are like
currents, eddies, or whirlpools in a flowing stream. They are semiperma-
nent structures formed from constantly moving 'stuff', such as energy. All
that classical scientific laws describe is apparent relationships between the
'subtotalities', for example the 'eddies' and the 'vortexes'. But the *real*
relationships cannot be fully understood by reference only to such structures. The structures have to be seen as parts of the stream in which they
appear. Living and non-living things alike are mere 'subtotalities', and since

they are structures in the *same* stream they are intimately related – the energy/matter which is a part of (i.e. explicate in) one vortex/subtotality in one instant is part of another in the next instant. This idea, incidentally, is also suggestive of the pre-modern view that no hard distinction can be made between living and non-living things: in a way, that everything is 'alive'.

Bohm's third principle is that the *universe has countless dimensions*, which 'embody its wholeness'. This means that what cannot be understood by reference to a lower order of dimensions can be by reference to a higher order. For instance, subatomic particles which are spatially separated from each other sometimes display apparently related behaviour (as do larger entities, including people). In this phenomenon, called 'nonlocal causation', the two apparently independent particles are not really separate according to Bohm: they are merely a three-dimensional projection of a six-dimensional reality. It helps here to think of two converging railway lines in a photograph. The photograph does not reflect the real truth. Its two-dimensional projection has the lines getting closer, whereas in the three-dimensional 'reality' the lines maintain their separation but get further away. Similarly, time and space themselves, Bohm proposes, are actually 'sub-totalities' from a higher dimensional reality. We thus inhabit a world of 'shadows', very different from the 'hard', 'factual', 'concrete' world of classical science. We know that when we see the two-dimensional shadow of a building on the ground the length of the shadow is not necessarily the true length of the building. To understand that we must look at the building itself in three dimensions (but if we cannot do that directly we can deduce the true length by considering the shadow's dimensions in their *context*, that is, the height of the sun above the horizon and the angle it makes with the ground).

Bohm's picture of the cosmography of contemporary physics presents us, then, with 'space' as an enormous sea of energy, in which matter is only a 'small quantised wave-like excitation . . . rather like a tiny ripple'. The 'nothingness' (energy) and the matter are as inseparable as the whirlpool and the river – at a distance they seem separate, but a close viewer cannot decide where one begins and another ends. Similarly, 'facts' are abstractions of things out of the total flow. They 'are' what the instruments and experiments that detect them structure and determine them to be: made by how a given particular scientific theory orders the universe. But there is no randomness about all this: what seems random is in fact ordered when seen at the higher, implicate level.

Reality, then, is flux, flow, or 'holomovement' in Bohm's terminology. Everything is constantly changing, so what 'is' is also 'becoming'. Elementary particles, like knowledge, are only *relatively* constant forms: 'vortices', abstracted from flux. But there are no solid objects really; neither is there a series of basic fixed truths. Bohm wants our collective consciousness to develop these new notions of order. Whereas in Cartesian thought we start with the parts and see how they relate to form a whole system, in the 'new' notions we would start with the whole and derive parts through abstraction from it. To gain such insights will involve intuitive thought, but not mysticism, which Bohm dismisses.

Some other physicists, in turn, dismiss Bohm, whose cosmography is at odds with big bang theory and lacks experimental evidence. Nonetheless his theories do attract the New Age/mystical tendency in green thought (Chapter 1.2). This is partly because he considers that the implicate order of the 'holomovement' enfolds consciousness as well as matter, that the two come from common ground and are essentially different aspects of the one general order (Bohm, p. 208). This of course appeals to New Age idealism and monism. Bohm sees thought as basically a material process; mind and matter have the same origin, and 'matter is the unfolding of consciousness'. Hence the content of our immediate perception (e.g. a 'moving object') is really an intersection between two orders of movement. First, there is light and the response of our nervous system to it (matter), and, second, there are those things that determine how we order those responses (mind). An event thus consists of a series of orders of movement made explicit together.

Morphogenesis

This link between consciousness and matter meshes with Rupert Sheldrake's (1982) theories of 'formative causation'. Sheldrake is another scientist – a biologist – considered heretical by mainstream opinion. He is interested in how living and non-living things attain, maintain and pass on their forms and functions. He considers that mechanical DNA theory and neo-Darwinian evolution do not adequately explain this process. Rather like Jan Smuts before him, he argues that there are hidden 'fields' (analogous to the elusive 'life force') governing the development of species.

Reminding us that patterns of electricity have been detected around people, Sheldrake proposes that all 'living' and 'non-living' entities possess 'morphogenetic fields'. Their growth and form is determined by the action of these fields on the 'matter' of which they are composed. The fields therefore constitute a channel – and a blueprint – via which form and growth are passed on. The fields of individuals interlock, and larger fields orchestrate smaller ones – there are for instance fields governing the development of whole species, as well as of individuals within species. Some New Age gardeners who embrace Sheldrake's theories enthusiastically maintain that the essential principle in growing healthy plants is to communicate with the 'devas' (spirits or morphogenetic fields) governing the growth of those plants (Hawken 1975, Pepper 1991). To grow prize carrots you need to strike up a rapport with the 'carrot devas'.

The bigger morphogenetic fields are simultaneously made up from the fields of the entities that they are forming. So each individual's experience is constantly transmitted to, and influences the nature of, the larger field. This in turn alters what that field 'tells' other and future individuals. Thus the laws of evolution are affected by the cumulative experiences and habits of individual entities, forming a collective 'memory'.

In this process also, fields in different *times* and *places* affect each other. Individuals and groups in one place and time can learn from the experiences of individuals and groups in other times and places, through communication between their morphogenetic fields. 'Morphic resonance' is Sheldrake's term for it – and this causes quicker adaptation than random mutation as in Darwinian theory. According to formative causation, Westerners who have never learned traditional morse code would learn it more quickly than they would learn an entirely new form of morse code, because traditional morse code has already been learned by others in the culture. Experiments are said to confirm that this is so (despite the difficulty of isolating the relevant variables: Mahlberg 1987, cited in Fox 1990, 255). Similarly, youngsters would be likely to learn to use computers more quickly than their parents and grandparents: a familiar occurrence, though there could be many reasons for it.

Sheldrake's fields enable transmission of behaviour and instinct, and it would appear that they are a form of consciousness affecting matter. In Bohm's terms, a higher-order collective consciousness which is 'implicate' affects the 'explicate' appearance and behaviour of individuals and groups.

The deep ecology appeal of such theories is that they constitute 'scientific' basis for idealistic and 'personal-is-political' theories of social change. They apparently substantiate New Age notions of scattered minority groups of people in different parts of the globe, who, by thinking new thoughts, meditating and living new lifestyles, would ultimately exert a collective influence on the course of world history out of all proportion to their numbers. Eventually, it is anticipated, when a 'critical mass' of people has attained the new consciousness, the majority of the world's people will spontaneously join in (Chapter 6.1).

Sheldrake's evidence for his theories is thin, however. Lyell Watson's (1980) 'hundredth monkey' anecdote is frequently retold in support of morphic resonance:

> There were these Japanese scientists in the '50s who left potatoes out every day for these wild monkeys on Koshima Island and then watched what they did. One of these monkeys learned to wash the potatoes and began teaching this to the others. Then, when a certain number had learned, maybe a hundred – scientists called this a 'critical mass' – an amazing thing happened. Suddenly all the monkeys knew how to wash potatoes, even monkeys on other islands hundreds of miles away!

257

Scientists consider this to be conclusive proof of a telepathic 'group mind'.

(Ken Keyes, cited in Coleman 1994, 36, see also Keyes 1982)

Briggs and Peat (1985, 251) point out that none of this is actually borne out by the scientific paper that Watson cites — there was no sudden or remarkable leap in monkey learning. Coleman (pp. 36–7) confirms that Keyes' and Watson's position is

> based on a complete misinterpretation of the facts. If anything, the popularity of the hundredth-monkey story proves, not the power of a group's mind, but the power of wishful thinking. The popularity of such stories is an indication of the extreme passivity and disempowerment of people in contemporary society, who, unaccustomed to effective, organized action, instead turn hopefully to the putative power of good thoughts. At the same time, the story of the hundredth monkey resonates with the Christian emphasis on salvation through faith.

Evolution

However, this lack of evidence for Sheldrake's and Bohm's ideas does not mean that all the problems of the evolution of forms, behaviours and order can be solved by more conventional theories. For one thing, there is a basic contradiction between the second law of thermodynamics and evolutionary law. The former holds that the universe must tend towards increasing disorder and simplicity (entropy), while the latter suggests the reverse — evolution producing, over time, forms of increasing order and complexity.

The resolution here lies in the fact that it is closed systems which degenerate towards maximum entropy — that is a stable equilibrium comprising maximum disorder and simplicity. By contrast, living systems are 'open', with inputs and outputs of energy and matter. They can therefore repair and develop themselves, maintaining themselves in highly ordered and complex forms and structures. The work of Ilya Prigogine (1980) has shown that 'Life and nonlife both appear in nonequilibrium situations and such situations are everywhere' (cited in Briggs and Peat, p 179).

Prigogine considers that new and higher structures evolve out of the 'fluctuating motion' of entities and forms which already exist. First the random fluctuations of individuals and groups reach a critical point; then, beyond it, larger numbers start fluctuating. Fluctuations are amplified, until they go over a threshold, when new, higher orders are established which are relatively resistant to further fluctuation. They are now relatively stable, but are subject to further change if the system experiences more intense fluctuation. Then a new phase of 'chaos' occurs and new and higher orders, more complex still, evolve (Briggs and Peat, p. 179). A parallel may be drawn here with the emergence of new social orders out of the chaos of political conflict and crisis, as in, for instance, Marxian historical change theory.

Living structures, then, are 'autopoietic', that is, they are self-producing, self-organising, relatively stable, and yet change all the time. Their apparent separateness and autonomy actually mask an interdependence with everything in their environment, like the whirlpool in the stream. We have the image of relatively stable but temporary structures within the environment of a flowing universe. Changes in the environment can lead to processes whereby these structures either maintain themselves as 'lower order', or mutate to 'higher order' structures. These terms, for Prigogine, do not, however, imply a hierarchy of structures, for (as in the pre-modern chain of being) lower and higher levels *depend* on each other for continued existence: there is a dialectic between them.

This leads on to the theory of *co-evolution* fashioned by Erich Jantsch (1980) out of the ideas of Prigogine and Ludwig von Bertalanffy:

> Coevolution says that changes which take place on the micro scale instantly effect changes which take place on the macro scale and vice versa.
>
> (Briggs and Peat, p. 207)

So low-level entities do not just build up in steps to high level, both levels evolve simultaneously. Hence the far-from-equilibrium conditions of the early Earth led to structures which could reproduce themselves via template molecules, and could pass on mutations. They did not compete, but flowed into each other, sharing their chemistry information. As they evolved they changed the Earth's whole chemistry, allowing further life forms to evolve (Lovelock 1989).

Jantsch thus sees evolution as a holistic unfolding, whereby an ever-expanding complexity of forms interpenetrates one another, not an interaction of separated parts. Earth, stars and galaxies are also co-evolving. Such a vision of development, hinging on mutualism, holism, vitalism, and elevating subjectivity – all validated by science – is immensely appealing to ecocentrism. It resonates with (a) anarchist social ecology and (b) Gaian deep ecology. For it emphasises (a) the essential role which *cooperation* plays, and (b) how micro- and macro-structures evolve *together as a whole*. It therefore informs the social change theory of many in the green movement, as Chapter 6.1 discusses further.

5.4 SCIENCE AND SOCIETY: INFLUENCING THE QUESTIONS ASKED

How science develops and changes

Most Western education preaches the supremacy of (classical) science as the study of 'real', empirically observable phenomena, established through common and repeated methods, and drawn together in law-like general-isations – an incontrovertible basic set of scientific truths, valid throughout space and time.

As such, science represents for many people a monolithic accumulation of inter-generationally established facts, as an anonymous writer in *The Economist* (1981) suggests:

> What should be beyond argument is that there is an accretion of known facts. On the whole science is 'true'. To deny that man knows more about the workings of nature now than he did in the Middle Ages is perverse . . . today's scientists stand on the shoulders of their prede-cessors to place new bricks on the pyramid of knowledge.

This popular image of scientific development recalls its Baconian roots. *What* scientists work on – *what* they consider to be the next significant problems to be solved – depends on what previous scientists have dis-covered. The image has it that, detached from society, scientists, singly or in groups, progress logically from simple to complex problems in an unerring search through the generations for ultimate truth (e.g. fundamental par-ticles), impelled by their insatiable curiosity.

An alternative view would follow from the conclusions of Karl Popper (1965), that no amount of verification of a theory (observations of it working out in practice) can ever prove it absolutely correct. However *one* instance where its predictions are shown not to be valid (falsification) will bring it crashing to the ground. Falsified theories must then be abandoned. The resultant picture of scientific development is discontinu-ous – proceeding via a series of traumatic upheavals, abandoning knowl-edge which has been falsified.

Paradigms

Thomas Kuhn (1962) provided a 'compromise' model of scientific devel-opment. Although sometimes disputed, his concept of *paradigms* has been widely used by historians of science and of particular disciplines. The paradigm model proposes that there are 'long periods in the development of scientific work during which scientists take for granted and are com-mitted to a particular view of the world' (Albury and Schwartz 1982). This view determines what problems are to be studied by most people in a discipline. It represents a consensus among the workers in the discipline, so that if asked what their subject was about they would all give similar answers. This consensus is the 'paradigm'. It

> represents the totality of the background information, the laws and theories which are taught to the aspiring scientist *as if* they were true,

and which must be accepted by him if he in turn is to be accepted into the scientific community.

(Richards 1983, 61)

Briggs and Peat (1985, 24–34) describe paradigms as like 'spectacles', which scientists put on. Once donned, the spectacles mediate and condition the scientist's world view: they filter in some things and filter out others. The spectacles are made from particular theories (e.g. quantum theory, relativity), together with the presuppositions surrounding the theories. They constitute a lens through which scientists discern what is worthwhile studying about nature, an object of scientific study.

Science students learn these world views. They don the spectacles as they progress in their disciplines. Hence they are socialised into what constitutes *acceptable*, worthwhile experiments and areas of study – 'normal science'. Most scientists solve puzzles within a paradigm, without considering or questioning for most of the time the basic assumptions or theories behind the paradigm – what their spectacles are made of. Without paradigms, scientists would not know where to look, or how to plan experiments or what data to collect. For, contrary to Bacon, one cannot simply go out and try to collect all data about the world before attempting to induce some structure and order from it in the way of law-like generalisations. *What* to look at and sample is pre-conditioned by making provisional hypotheses about what laws might be operating and deducing what field or experimental observations would follow if the hypotheses were correct.

As a small example, consider soil surveyors. They rarely attempt to sample a landscape's soils totally without preconception. First, by examining geological maps and air photographs, they provisionally draw boundaries around expected soil types on the base map – setting in a sense the 'paradigm' for the area. Much of the actual survey work will then simply confirm and refine what is broadly expected.

But sometimes (usually when the surveyor is tired, hungry and fed up at the end of a day's mapping!) the sampling encounters something entirely unexpected which undermines the whole schema that has been developing. As a good Baconian and inductivist, consistent with the public view of the scientist, the surveyor will immediately rethink and set out to change radically or revise the schema. But as a human being, he or she may be more inclined simply to ignore that sample and go home to tea, or, at best, to try to rationalise it into what has already been worked out.

Eventually, however, when just too many inconsistencies arise, whole schemas may have to be radically modified or replaced by new ones. Anomalies have caused a crisis in the paradigm, and the 'spectacles' are broken. However, a new generation of scientists grow up wearing new glasses – the new map of normal science – as 'true'.

One can think about paradigms and paradigm shifts at various levels.

261

There are overarching world views governing what constitutes 'science' (literally 'knowledge'). From the medieval physico-theological cosmology described in Chapter 3.1, where the physical world behaved according to God's will, there was a transition to classical science.

If a paradigm is an outline map of a discipline, Newton's *Principia* provided the outline for classical science. But by the nineteenth century, just when 'normal' science appeared to have filled most of the outline in, some workers found bits (like the particle–wave paradox) which did not fit. Ecocentrics often claim that the discoveries of twentieth-century physics were accompanied by a profound shift in scientific perspective. Cini (1992) summarises this as a change from concern with simplicity, order and rule (as in objectivity and reductionism) to concern with and acceptance of complexity, disorder and contingency in science (as in subjectivity and holism). While ecocentrics might represent this shift as the culmination of a logical progression of scientific enquiry since the Enlightenment, towards a more accurate representation of 'the truth' about the way the universe is, Cini takes a Kuhnian perspective. A new paradigm is supplanting an older one. And, most significantly, Cini points out that this shift within science roughly parallels a more general shift *within society* – from values of the modern period to those of postmodernism. Here is an example of how the paradigms of science are not determined merely by what happens within science: the directions of scientific enquiry are conditioned by wider society.

Coming down slightly in level, there are many more examples of paradigm changes. That from catastrophism to uniformitarianism in natural, especially earth, sciences was matched by the move from creationism to evolution in biology. Again, such changes may be better understood partly in relation to their social context, where the new philosophy of liberalism, part of a new economic order, was looking to imagined principles of nature on which to base its laws rather than looking, as in the past, to God's law.

Then there are those changes associated with specific disciplines (again, nearly always related to broader extra-disciplinary influences). In geology, for instance, continental drift gave way to tectonic plate theory; diluvial theories of superficial deposits gave way to glacial theories, monoglacial theories were replaced by multiglaciation. In biology, Darwinian evolution has been modified from considering adaptation and development to be continuous and gradual to perceiving a series of relatively sudden 'jumps', where new species appear, separated by more gradual transitions. This paradigm, of 'punctuated equilibrium' (Gould 1982) is still only a potential one, being disputed. In climatology, the theory that a new global ice age is on the way determined much normal research in the 1970s: today it is global warming that captures climatologists' attention.

In all this, as Briggs and Peat emphasise, scientists *do not* always objectively choose the 'best' paradigm: defined as that which is most accurate, consistent, simple, and fitting 'the facts' best. Indeed, they think it is less a question of evaluating what facts and data tell us 'in themselves' and more a question of examining the same data from nature's raw material in different ways. Hence, usually (p. 28)

> the parties in the paradigm debate speak different languages and 'talk through' each other . . . even if they should read the compass they wouldn't agree on the needle's direction.

There are usually two sorts of reaction to proposed paradigm changes. One is for workers in the established paradigm vigorously to defend it. This can be done by refusing to publish competing theories in the 'respectable' press. In the 1930s, geologist R. G. Carruthers (1939) could still advance his monoglacial theory in a mainstream geology journal. When he wanted to do so in the 1950s, however, he had to publish his paper privately at his own expense (Carruthers 1953).

Of course, there are less subtle and more aggressive ways of repressing competing views. The trial of Galileo for proposing that the sun and not the earth is the centre of the solar system is well known and dramatised (Berthold Brecht's play *Life of Galileo*), as is the trial and conviction of schoolteacher John Scopes in Tennessee in 1925 for teaching Darwininan evolution rather than creationism (Stanley Kramer's film *Inherit the Wind*). Magner (1979) noted how, in the year before President Regan was elected, anti-evolutionists had campaigned either to ban the teaching of evolution altogether, or to have equal time devoted to 'special creation' as documented in Genesis. These groups had considerable funds and political power, which they used to help Regan into office in return for his sympathy with their aims.

Creationism, as proposed in the nineteenth century by Bishop Ussher, held that the Earth and everything in it were simultaneously created in 4004 BC. This theory's clash with evolutionism presents us with examples of another sort of reaction to an emerging paradigm: attempting, usually unsuccessfully, to accommodate it within the existing one.

Some dedicated Christians who were also amateur naturalists felt increasingly uncomfortable with their position as the nineteenth century progressed. Darwin's theory appeared to be borne out by the empirical work being done on the fossil record. Yet

> Here was a dilemma! Geology [e.g. sedimentary strata, varves, tree rings] certainly *seemed* to be true, but the Bible, which was God's word, *was* true. If the Bible said that all things in Heaven and Earth were created in six days, created in six days they were – in six literal days of 24 hours each. The evidences of spontaneous variations of form, action, over an immense space of time, upon ever modifying organic

structures, *seemed* overwhelming, but they must either be brought into line with the six day labour of creation, or they must be rejected.

The writer is Edmund Gosse (1907), describing his father's anguish as a Plymouth Brethren religious zealot on the one hand and an enthusiastic naturalist on the other. To Phillip Gosse the problem was distilled into the question of whether Adam had a navel – if he had been created spontaneously by God there was no 'need' for one any more than rocks dating from only 4004 BC need have well-developed multiple strata. Gosse's (1857) bizarre accommodation declared that God had indeed created Adam, rocks and all simultaneously, but had made them *with* navel, strata, tree rings, to *look* as if they had a past (Oldroyd 1980, 259, note 15).

Equally bizarre was an attempt made as late as 1925 by the religious naturalist and Fellow of the Royal Geological Society, Inkerman Rogers, to reconcile scriptural and scientific paradigms (see Figure 5.3). His startling – and racist – solution was to have *two* origins for the human race. One branch, the 'inferior', was, as science told, as old as the Pleistocene deposits which buried its remains. The other, in which Adam was the 'firstborn of a superior race' was created in 4000 BC. To group Adam with the rest would throw doubt on scripture chronology: 'but Scripture chronology cannot be rejected'.

To Briggs and Peat (pp.. 30–3), because science is thus influenced by social fashions, mores and events, then it is essentially an extension of human subjectivity:

> Kuhn has, in fact, exposed an undreamed of crack in the long-held conviction that science is leading us towards the ultimate truth about nature's laws . . . Kuhn shows us that the observer, his theory and his apparatus are all essentially expressions of a point of view – and the results of that experimental test must be a point of view as well. Kuhn's analysis of scientific history strips us of the traditional preconception that science is objective . . . the *very laws* of nature are protean, changing with each new paradigm . . . What scientists in the Middle Ages knew about the universe, quantum scientists no longer know. Instead we know a *different* universe.

Such views smack of the extremes of social constructivism which many scientists dislike, because they imply, counter-intuitively, that there is no progress in science. And Briggs and Peat do question whether, despite technological advance, we really understand nature better now than in the fourteenth century. Indeed we could be *less* in tune with it now – less advanced today than we were, or American Indians are: we may have reached 'new levels of ignorance'. Here is a very ecocentric sentiment.

Social influences on science

Popular ideas about science often forget about the relationship between the scientists and the wider world. Not least, that relationship is one of *material* dependence. Scientists cannot live on thin air. But there is very little funding which is not tied to specific interest groups, particularly economic interests:

> Developments within the ivory towers of the academic world have long been conditioned by a gentle yet prolonged interaction with the economy

says Barnes (1985), observing that one-third of the £100 billion annually invested in research and development was, in the 1980s, spent on military and space projects. This investment was necessary to neutralise the effects of previous scientific R and D on weapons!

Most of the questions asked by scientists thus depend on 'society's' wants, which in practice are mainly the wants of the dominant economic (and therefore political) groups in it. Albury and Schwartz (1982) trace the history of several different scientific and technological developments, to demonstrate how they served the particular needs and ideologies of capitalists who funded them. Capitalism developed partly by expanding its markets through the territorial extension of trading. Two things followed from this. First, fast communication had to be developed between the industrial heartlands and their outposts: thus a large telecommunications industry was founded and developed. Second, these outposts had to be defended against those who would limit their right to 'free' trade with the heartland, so a 'defence' industry was fostered. Now, of course, the two industries serve each other as well as the interests of other industrial corporations.

Another way of expanding markets lies in product innovation, and it can be clearly seen that microelectronics has served this particular need. There has recently been a plethora of luxury consumer goods incorporating the microprocessor, a development of the telecommunications industry. To increase productivity and accumulation industry can also try to replace the workforce by machines wherever possible. Here again, research into the necessary robots, computers and world processors has been prioritised.

One powerful private funding body is the Rockefeller Institute, and Albury and Schwartz document this organisation's involvement in developments to foster a 'green revolution' in Third World agriculture. In particular, the Institute consciously decided in 1932 to fund biological research into plant genetics, producing uniformly high wheat and rice plants which could be machine harvested and would respond particularly to high artificial fertiliser inputs. All these characteristics suit plantation agriculture, which benefits multinational agribusiness corporations but not

THE ADVENT OF MAN AND HIS PROGRESS ON THE EARTH.

GEOLOGY supplies us with the evidence of the first dawn of humanity on our planet. ARCHÆOLOGY follows with her collections of prehistoric memorials of mankind. HISTORY has recorded and brought before us the various races of Man which have peopled the earth. LANGUAGE traces them back to their different beginnings. ETHNOLOGY assigns to each race its proper place in the procession of human life on the earth. SCRIPTURE RECORDS manifest the harmony of divine revelation and scientific research, and proclaim that God was the Author of Genesis, and in spirit with the holy men of old when the history of the first Adam and his descendants was written for our learning.

LATE PLEISTOCENE PERIOD.

PALÆOLITHIC MAN appears on the earth—how, when, and where, we cannot tell. Traces of his existence found in Pleistocene deposits; and of his presence long anterior to the first vestiges of civilization on the earth. These men lived in the west of Europe ages before the commonly received human era of six thousand years. Remains found in plateaux gravels and high-level river gravels. Cave dwellers; wild hunters; large, roughly chipped and unpolished stone implements; bone implements. Carvings and etchings. Shell-mounds. Dead consigned to rivers or ocean, or burnt. Cave lion, mammoth, woolly rhinoceros, hyena.

PALÆOLITHIC MEN (Old Stone Age).
    Acheulian.
    Chellian.
    Mousterian.
    Solutrean.
    Madelenian.

HIATUS.
Elevation of the Continental Shelf 300 feet above the present sea level.

NEOLITHIC MAN. Long ages elapsed between the date of the Old Stone men and the New Stone men, during which, the now extinct animals were slowly passing away from the world. General disappearance of glaciation. A rude, uncultivated people appear on the earth, inhabiting the north and west of Europe, and coming from the east. Cave dwellers. Pile dwellings. Hunting and fishing. Stone implements chipped, ground and polished. Bone and horn implements. Food: vegetable and animal. Mound builders. Carvings and etchings. Dead burnt.

NEOLITHIC MEN (New Stone Age).
PRE-HISTORIC MAN (Late Neolithic).

LATER NEOLITHIC MAN. A different race from the earlier stone men, but more advanced. Cave dwellers. Pile dwellings. Hunting and fishing. Polished and artistically chipped stone implements of various types. Mound Builders. Barrows, graves, tumuli, and rude megalithic and monolithic monuments. Burial urns. Arrow heads of quartz and jade. Weaving, and rude pottery. No traces of grain, nor any indication of the knowledge of agriculture. No trace of civilization; no work of art; no progress. No metals.

MONGOL RACES. } Descendants of
NEGROES. } Neolithic and
AMERICANS. } Prehistoric races.

MONGOLS wholly incompetent to advance either themselves or others to a higher position. NEGROES wholly uncivilized and incapable of self civilization. AMERICANS uncivilized savages. Descendants of these races surround us on every side. Existing races of men had no closer union with the Creator than the untutored savages of Africa. Mongoloid races always have been, and ever will be savages, until some external influence is brought to bear on them. Marked physical and moral distinctions between Mongol and Caucasian.

BASQUES, or Pyrenees. } European
MAGYARS Bulgars. } descendants of
OSMANLI Turks. } Neolithic and
INHABITANTS of gorges } Prehistoric races.
    of the Caucase. }

URAL-ALTAIC
(TURANIAN)

Subsidence of the Continental Shelf. Distribution of land and water nearly the same as at present.

HISTORICAL MAN. Continents and Islands inhabited from the remotest period to which history and tradition can reach. Different state of things appear. Races more advanced in manufacturing skill, who worked in bronze for centuries before the art of smelting and manufacture of iron. Pile dwellings. Grain; domestic animals. Monoliths and Cromlechs culminating in Egyptian obelisks and temples.

ADAM, firstborn of a superior race, 4,000 BC. Adam not the progenitor of all human beings that have inhabited the globe from the beginning. If it be assumed that he was the progenitor of all humanity, and that the Mongol and Negro, and all the other inferior races were his lineal descendants, then Scripture chronology for that period cannot be relied upon. It is either an impossibility, or the antiquity for Adam must be incalculably higher and more remote than that recorded in the Bible; but Scripture chronology cannot be rejected.

CAIN moves eastward. His descendants absorbed among Mongoloid peoples. Type lost in central Asia.

NOAH, 2,300 BC Descendants at time of dispersion in advanced stage of civilization. Cities built. Metals known and used. Tubal Cain an instructor in the art of metallurgy. Introduction of the arts of civilized life. Stone weapons still used, but bronze implements were added to them. Agricultural and pastoral pursuits. Cereals, comesticated oxen, pigs, and sheep. Civilization spreading. Pile dwellings. Neolithic and prehistoric men give place before the approach of men of the bronze period. The dead burnt and ashes collected into funeral urns.

JAPHETIC or Aryan family, gradually peopled the whole of Europe, north-westward, and eastward to Hindustan.

KELTS. An aboriginal race, descendants of Neolithic men, present in Europe on arrival of Kelts. Aborigines subjected, absorbed or destroyed. Aborigines disappear before advancing Caucasian.

Caucasians alone of all the inhabitants of the globe, have always been in a state of active and progressive improvement. Arts and sciences developed. Civilization highly progressive. Caucasians, identified as members of the family of Noah and descendants of Adam and Eve, have peopled all the lands which they now occupy throughout Europe and Asia; and the time requisite for such emigrations and settlements does no exceed the Scripture date of the Noachian deluge. The endeavor to account for the existence of the first man by searching for a link to connect his being with some pre-existing organism, must fail. Such a link never has, and never will be discovered.

Adam is the first man, but only in relation to Christ, the second man. The first man Adam a living soul, the last Adam a quickening spirit. The first man of the earth; the second man from heaven.

Death the primary casue of redemption to life—not death in general, which we know has reigned in the world before Adam, but death, the penalty incurred by him.

SUPERIOR MAN, advent of, on the earth—future.

INKERMAN ROGERS, F.G.S., F.R.A.S., 1925

Copper
Bronze

Iron

ADAM.
CAIN.

NOAH. { Shem.
        Ham.
        Japhet.

CAUCASIC
(IRANIAN)

SEMITIC FAMILY.
Chaldeans
Arabs
Hebrews

JAPHETIC FAMILY.
(*Common origin as a distinct family proved beyond question*).
Hindoos
Persians
Greeks
Latins
Slavs
Kelts (of Eastern origin)
Teutons

On all the languages from Sanscrit to English there is one common stamp of individuality.

Figure 5.3 The advent of man and his progress on the earth

most Third World people. Albury and Schwartz argue therefore that such research was 'geared to the aims of transnational companies seeking to create attractive investment opportunities for overseas capital'.

Neither is apparently 'pure', 'academic' research in universities immune from economic pressures. Social sciences tried to establish themselves as positivistic and 'socially relevant' subjects in the 1960s and 1980s. During both these decades, government had declared itself unequivocally most willing (or least unwilling) to fund science and technology – to aid the 'white heat of the technological revolution' (a Prime Minister's phrase) in the 1960s, and to make Britain commercially competitive in the 1980s.

Even the recent rapid shift in climatological paradigms noted earlier might be related to funding. This is certainly the view of the climatologists interviewed by Hilary Lawson in a Channel Four 'Equinox' TV programme in 1990, entitled 'The Greenhouse Conspiracy'. They systematically denied that there was any convincing evidence for the global warming theory: no evidence for temperature or sea-level rises, or for CO_2 as an efficacious greenhouse gas, or that the models on which predictions are based are at all reliable. Lawson asked why, then, imminent global warming has so forcefully replaced long-term cooling as the primary preoccupation. 'It is easier to get funding if you can show evidence for impending climatic disasters,' said Dr Roy Spencer of NASA. Most of the other climatologists broadly agreed, and believed that funding directly or indirectly affects the nature of the research and the results (Albury and Schwartz's conclusions also). Indeed one MIT climatologist whose work demurred from most model predictions had his funding cut.

Since science became a profession in the nineteenth century, rather than a pastime of the rich and leisured, few scientists can follow the advice of 'maverick' ex-NASA scientist James Lovelock. He says that in order to be able to ask the questions which they think important scientists should try to be self-funded.

> You may think of the academic scientist as the analogue of the independent artist. In fact, nearly all scientists are employed by some large organization, such as a governmental department, a university, or a multinational company. Only rarely are they free to express their science as a personal view. They may think they are free, but in reality they are, nearly all of them, employees: they have traded freedom of thought for good working conditions, a steady income, tenure, and a pension. They are also constrained by an army of bureaucratic forces, from the funding agencies to the health and safety organisations. Scientists are also constrained by the tribal rules of the discipline to which they belong . . . As a university scientist I would have found it nearly impossible to do full-time research on the Earth as a living planet . . . I would have been summoned . . . and warned that my work

endangered the reputation of the department and the director's own career.

(Lovelock 1989, xiv)

Such considerations apply also to the answers which science delivers, as we now discuss.

5.5 SCIENCE AND SOCIETY: INFLUENCING THE ANSWERS OBTAINED

The classical legacy of scientific authority

The principles of modern scientific doctrine were formalised by a group of scientists and philosophers known as the 'Vienna Circle' in the 1930s, who developed the philosophy of *logical positivism*. This held that intuitively, spiritually or emotionally derived knowledge was less valid and meaningful than knowledge verifiable by observation and experiment. It followed that empiricism and reason should form the basis for social action, because such action would be based on objective and not subjective judgements. Values, emotions, intuition, ideology – all those facets of life which could not be 'proved' through observation, measurement and logical argument – should not hold sway in decision making. European society was being permeated, in the 1930s, by 'anti-rational and anti-scientific' fascism, and this influenced members of the Circle such as Bertrand Russell (1946, 789), who wrote:

> In the welter of conflicting fanaticisms, one of the few unifying forces is scientific truthfulness, by which I mean the habit of basing our beliefs upon observations and inferences as impersonal, and as much divested of local and temperamental bias, as is possible for human beings.

This model still influences today's public perception of science. Despite disillusionment with some products of science, scientific knowledge and method have commanded respect and authority which is slow to be shaken. Scientists are often perceived as *detached* observers (after Descartes) of nature and society. Therefore they are regarded as ethically and politically neutral, with no axe to grind about what their results should be used for, that being for politicians and the public to decide. Jungk (1982, 13–14) strikingly conveys this notion:

> His face was wreathed in a smile of almost angelic beauty. He looked as though his inner gaze were fixed upon a world of harmonies. But in fact, as he told me later, he was thinking about a mathematical problem, whose solution was essential to the construction of a new type of H-bomb . . . this man had never watched the trial explosion of any of the

bombs which he had helped to devise. He had never visited Hiroshima and Nagasaki, even though he had been invited . . . To him research for nuclear weapons was just pure higher mathematics untrammelled by blood, poison and destruction. All that, he said, was none of his business.

But *because* of this perceived detachment the notion is also popular that scientifically literate people should have a special place in decision making, as experts promoting rational, 'common-sense' solutions according to objective truth. As Richards (1983) observes, science and scientists here become 'truth institutionalised'; usurping the territory formerly held by religion and priests. Scientific laws describe 'relations and regularities that are invariable' (like 'God's law' used to be). Empirical science therefore surely must be uninfluenced by cultural factors such as dogma or mere opinion: it emphasises what can be agreed by *all* observers.

Hence many people imagine that to win an argument, to be intellectually respectable and legitimate, they have to use scientific evidence, logic and method. Nothing is worse for your argument than having it dubbed 'unscientific' or 'just' 'emotional'.

By extension, to affix the word 'scientific' to one's product, one's message or one's evidence is to give it weight and authority of a special kind, encouraging the view that it is altogether respectable and must be taken seriously. For Barnes (1985, 72–79), there is no more powerful demonstration of this potency of scientific authority than Milgram's psychological experiments in the 1960s. Milgram showed that 60 per cent of a sample of ordinary people were prepared, albeit reluctantly, to administer (what they had been fooled into thinking were) progressively more agonising electric shocks to subjects who got answers wrong in a test. They had been told by the 'scientist' in charge that what they were doing was for a scientific experiment. But when this person in charge was presented as a layman, only 20 per cent were willing to go on administering 'torment' in this way.

This legitimating authority of science is today sought by all sorts of interests, religious, paranormal, flat earth and creationist enthusiasts. And it is used to further commercial ends, for instance as a tool in selling. We are familiar with this use in advertising. Products are dubbed 'scientifically tested'. They are said to contain impressive-sounding synthetic substances mystifyingly identified by formulae – 'The wonder ingredient, WM7' and other such gobbledegook (although nowadays concerns about chemical additives make this a ploy to be judiciously used). Products may actually be promoted by actors wearing the priestly robes of the white laboratory coat. The imagined intellectual integrity and detachment from interest groups of such characters may be amplified by making them appear super-intelligent (e.g. with ludicrously high foreheads) and/or eccentric or 'mad' (i.e. separate from the concerns of ordinary life). The general effect is that 'if they say that this product is good it *must* be true'. Any product which is the 'appliance of science' must be worth buying.

Legitimating environmental actions

In a similar way:

> It is their access to technical expertise which gives political elites their power. They are able to use that expertise both to arrive at decisions, and perhaps even more importantly to legitimate and justify them once made.
>
> (Barnes 1985, 100)

Blowers and Lowry (1987, 136–8) demonstrate how this can happen in environmental matters. Pro-nuclear government bureaucrats in Britain tried to legitimate their proposed plan of action – increased dumping of nuclear waste at sea – by scientific research in such a way as actually to pervert scientific method. An official review of waste disposal policy (HMSO 1979) said

> We believe that there can be quantitative justification for an increased sea dumping programme and we recommend urgent research to build up a body of knowledge which will demonstrate this.
>
> (p. 118 para. 7.23)

In other words: 'we have made our decision, now let us procure scientists to create research findings that will justify it'. It also nakedly says

> the international climate is such that it will be necessary to justify any substantial increase in sea disposal with more scientific evidence than is currently available. We recommend that the appropriate research and measurements to build up the necessary body of knowledge to do this should be pursued urgently.
>
> (p. 109 para. 6.32)

Ecocentrics have been quick to point out such perversions. For instance they have made accusations about the scientific method used by such a 'respectable' body as the US World Resources Institute. This Institute's 1990–1 annual report made the surprising claim that industrialised and non-industrialised countries share equal responsibility for global greenhouse gas emissions. However, said Patrick McCully (1991, 157) in *The Ecologist*,

> a close look at the raw data used by WRI, and the way in which they have interpreted it, reveals that the institute has used highly question-able estimates for the release of greenhouse gases from developing

countries and that their methodology contains some very dubious science.

He cites Anil Agarwal, Director of the Centre for Science and Environment in New Delhi, who says 'The WRI conclusions are based on patently unfair mathematical jugglery, where politics are masquerading in the name of science'.

It should not be imagined that ecocentrics are immune to this sort of perversion either. For instance, a BBC Radio Four programme alleged in 1989 ('The Zero Option') that Greenpeace was so anxious to claim scientific legitimacy for its view that toxic waste dumping might kill all life in the North Sea that it took the results of a Dutch scientific study on the matter and represented them in their publicity material in exactly the opposite way to the author's intentions.

Legitimating environmental combatants

Thus ecocentric pressure groups, while decrying the classical scientific world view and questioning the authority of science and scientific experts, at the same time are mindful of the legitimating potential of being 'scientific'. Hence they have enlisted scientific (including economic) experts to help them to defend environmental causes. Weston's (1989) history of Friends of the Earth documents how its Board members wanted to reinforce 'heartfelt views with sound scientific research' (p. 44), to balance 'fact and feeling' (p. 105) and to 'appear professional' (p. 106). FoE and Greenpeace have used informed expertise at public enquiries into road, airport, nuclear power station and other development proposals. Generally, however, they have been outgunned by the sheer financial capacity of their opponents – usually large commercial interests or government departments – to buy more expertise.

And experts are willing to be bought. Barnes (pp. 107–8) considers that scientists are so diverse that it is difficult to organise them all to further any particular political end. But they are united about the need to be paid (well) for their work, hence 'perhaps most important of all, experts are divided according to who employs or finances them'. There is no shortage of experts happy to prepare a case favouring a new project as 'safe, environmentally benign and profitable . . . Quite literally, many experts of this kind are killers' (p. 109).

Nelkin (1975) has shown how in environmental disputes available data are used by each side's 'experts' to make different predictions about the impact of the proposed developments. The 'facts' and the 'details of the technical dispute' are in this sense irrelevant, for each side uses them to embody 'their subjective construction of reality'.

But such technical disputes, as Wynne (1982) has shown, do have a function. They constitute a ritual which helps to secure public acceptance for potentially damaging projects, such as the thermal oxide nuclear reprocessing plant (THORP) approved of after the long Windscale Public Enquiry of 1977.

Wynne considers that decision-making institutions like public enquiries use rituals to defend their own credibility. These rituals define the problems and issues in such a way that their credibility is not questioned. For instance, the pro-nuclear lobby repeatedly claims possession of rationality, demanding debate of the 'hard facts' alone.

Yet to do this actually limits the amount and types of criticisms and objections that can be made. 'The demand for "hard facts" alone excludes debate about the interpretative social frameworks within which these facts have meaning' (Wynne, p. 3). So certain taken-for-granted assumptions are not challengeable, for instance that a given level (however small) of risk of exposure to radioactivity is acceptable: any claim that no risk at all should be allowed would be dismissed as 'irrational', therefore not worth discussing.

In fact the Windscale Enquiry, through its ritual of technical wrangling, projected a 'particular model of democracy – one that required only the expert discovery of objective facts about a narrowly defined question' (p. 10). This implicit, not-to-be-challenged (but in essence undemocratic) model was that 'the experts' rather than laypeople should dominate decision making. And since the nuclear industry had access to more experts then their opponents the outcome of the Enquiry was predictable.

All this, Wynne maintains, actually goes against the rationality which public enquiries claim to have, because 'a rational technological assessment would examine all hidden interests and agendas buried in claims and counter claims' (p. ix).

And repeated claims to rationality by the pro-nuclear lobby disguise its willingness to use emotion as a weapon. An ex-Chairman of the UK Atomic Energy Authority, for instance, said (being emotional about being non-emotional) that the nuclear industry would be judged 'upon the facts and upon our achievements and not upon the plaintive cries of the faint-hearted who have lost the courage and ambitions of our forefathers, which made mankind masters of the earth' (p. 166).

Wynne points out that the Windscale objectors were caught up in this ritual, whereby the boundaries of permissible discourse were defined by the nuclear and political establishment. Attempts to extend these boundaries would bring down contemptuous accusations of ignorance, political motivation or irrationality – the tactic of discrediting the opposition which Del Sesto (1980) also observed in clashes between pro- and anti-nuclear lobbies in the USA.

Yearley (1989) similarly notes that in a public enquiry about development on a raised peat bog in Northern Ireland the developer's tactic was to show that opposition witnesses were not specialised experts and that the scientific validity of their testimony was suspect. Again, Yearley notes a lack of real scientific rationality in this public claim to want 'proper' scientific debate – for proper scientific debate does quite legitimately involve subjective factors. It is about interpreting as well as gathering evidence, and interpretations are often fallible. But the developers here traded on an exaggerated image and expectation of science to impress the public.

The erosion of science's authority

To leave the argument here would be simplistic. For while the findings of science are often influenced and appropriated by groups who want science's powerful legitimating authority, that authority is not infallibly present. For example, after the Chernobyl incident in 1986, Welsh farmers began to mistrust government scientists who claimed to have an objective perspective on the effects of the discharge (Wynne 1989). Grove-White and Michael (1993) cite this in discussing how the Nature Conservancy Council in England also failed to gain the confidence of certain people (in this case, the government) by projecting an image of itself as a group practising value-free, empirical and universalistic science. The public authority of science, they believe, depends on social circumstances, and those circumstances may be changing. In the postmodern period science in the classical mode is perhaps being relegated from its position as a great arbiter and purveyor of truth (Lyotard 1984).

With the rise of environmentalism, all sorts of cracks now appear in the formerly respectable public edifice of classical science. A survey commissioned by the Edinburgh Science Festival in 1991 showed that 57 per cent of Scots believed science was *responsible* for environmental problems, with only 31 per cent disagreeing (*Guardian*, 28 February). And scientific experts in environmental issues have often been discredited. A 1992 BBC-TV series entitled *Pandora's Box* graphically detailed how this happened in two case studies.

First, in the 1940s and 1950s, agro-chemists were national heroes in Britain and America. As they developed herbicides and pesticides, they used their science ostensibly to pursue 'national security' (food self sufficiency), 'economic prosperity' (agricultural productivity) and 'to help the poor' (the green revolution in the Third World). But from the 1960s onwards they increasingly became villains. Widespread and serious agrochemical pollution was exposed by books which grabbed public attention (Carson 1962, Shoard 1980) and stimulated effective pressure group activity. Today public suspicion of scientific agribusiness and its food products is rife – at least among the well-heeled middle classes in the West.

Second, in the nuclear power industry assurances in the 1950s and 1960s that scientists were about to provide safe energy 'too cheap to meter' gave way in the 1970s and 1980s to defensiveness in the face of public antipathy about the industry. No amount of clumsy public relations by it could undermine the impact of real events, of which Three Mile Island and Chernobyl were but the tip of an iceberg. They caused the credibility of nuclear science to be irretrievably damaged (see Blowers and Pepper 1987, 1–35) – this against the background of the horrific destruction of two world wars, aided by science and technology, including nuclear weapons.

5.6 SCIENCE AND SOCIETY: LEGITIMATING SOCIAL GROUPS AND POLITICAL IDEOLOGIES

Legitimating political groups

If the questions asked and answers obtained by science are often influenced by social trends and group interests, the same might also be said of scientific theory. There is a long tradition of enlisting scientific theories, often about nature (including human nature), to support the political aspirations and ideologies of particular groups, contrary to the Baconian ideal. Wynne (1982, 161, 165) says:

> Political activity not only expresses pre-existent values but it also creates them, using even (perhaps especially) scientific knowledge as a medium for such covert moral persuasion . . . In Britain . . . the whole political culture depends on images of expert authority.

This started to happen when classical science first developed. There were groups who sought to use the image of science to bolster their social and political pretensions. Thackray (1974) studied a group of factory-owners and businessmen – emerging industrial capitalists – who were acquiring riches in Manchester between 1750 and 1850. But for all their 'brass' they did not have social status or political representation (Manchester had no MP). Such distinctions still lay with the landed aristocracy. As Anthony Trollope's Palliser novels demonstrate, it was exceedingly difficult to marry into this class and become part of it.

One path that was open, however, was to join scientific institutes, like the Manchester Literary and Philosophical Society. Thackray shows that this group of industrial capitalists saw science – doing it and being seen to be associated with people doing it – as a way to gain social status.

For scientific knowledge and activity were regarded as particularly appropriate for a group of progressive dynamic people with new ideas, building tomorrow. Science and technology had thus already taken on the association of 'progress'. (Significantly, in the 1990s a TV programme about scientific discoveries *today* is called *'Tomorrow's* World'.)

Furthermore science was generally regarded as 'polite knowledge' – gentlemanly and free from vulgarity – not the pursuit of lower orders. It was not seen as political – it was 'value free'. And science was democratic (unlike the old social order); for anyone who followed scientific method could be a scientist. Science was also new, and associated with the power to change nature and to provide material comforts. In all, science therefore ratified a new world order – of which the industrialists and businessmen were leaders. They joined scientific societies, in order to appropriate to themselves this image.

Interestingly, for these people this was not the end of this process of legitimation via science. Thackray's studies follow their social progress. Two generations later they had 'made it', having consolidated their riches with property and acquired social status and political power. They were still keen advocates of scientific activity and values – they advised all young people to do science and to learn about it. However, ironically, their underlying purpose was *exactly the reverse* of that which it had been two generations earlier. Instead of allying with science in order to break into and destroy an existing social order, now they saw it as something which could help to preserve an existing social order: the one they themselves now benefited from.

In essence they thought that if the younger generation studied science they would recognise the truth and inevitability of *laws* which 'governed' their existence, and their society: laws, for example, of economics and of the survival of the fittest. These 'laws', being universal, were inescapable, so obviously there was little point in trying to stir up and change a society bound together and determined by them. In other words, it would be futile to go into the taverns and alehouses trying to ferment social revolution among the lower orders – who were lower because this was their place in the natural way of things. Hence, dissemination of scientific ideas and practice would defuse social discontent, and legitimate an order in which the old entrepreneurs now had leading status.

Legitimating political ideology: Malthusianism

One scientific law, or 'principle', at the heart of ecocentrism is that of Malthus. Neo-Malthusians consider that humans tend to outbreed the Earth's productive capacity, and that this is today leading us towards disaster, pushing against nature. 'Overpopulation is the single biggest pressure forcing the planet's life support systems beyond their tolerance limits' (Stein 1993), green activists still say. Meanwhile the Ehrlichs still bemoan a *Population Explosion* (1990) and Garret Hardin (1993) continues to write poems on 'Carrying Capacity'.

Neo-Malthusians also doggedly claim that their arguments are above right–left politics, being concerned with the good of all rather than specific groups. However, this has never been true.

For Malthus himself and his seemingly scientific work have ideological significance, again, in defence of a particular social order. Malthus had affinities with the aristocracy, who in early nineteenth-century Britain feared that widespread social revolution would be imported from Europe.

Malthus' work – especially his *Essay on the Principle of Population* – can be seen as an argument against the state giving away 'taxpayers' money' (money substantially from the aristocracy) in the form of universal poor relief. It said that to do this would tamper with *natural* law, that is, the principle of population.

For to operate a 'welfare state' (poor laws) would increase survival rates among poor people – therefore hindering the operation of positive checks (famine etc.) on population increase. Malthus pitied the poor and would have preferred them to operate the preventive check of abstinence from 'early marriage'. But, he feared, ignorance, moral laxity and the encouragement of state assistance would mean that they would not do so. Instead they would 'convert' poor-relief money into more children, so ultimately more people would have to share food supplies (which could not keep up with population increase, Chapter 4.2). So, the good intentions behind widespread poor relief would actually increase the sum total of human misery. Too much attempted social reform was self-defeating.

At the same time he argued that the *aristocracy* were educated and morally enlightened enough not to convert *their* wealth into more children. Hence he was literally arguing one law for the rich and one for the poor.

So social improvement could not come from blanket poor laws. It did not thus lie in institutions or laws, or revolutionary changes. The solution to poverty ultimately lay in the hands of the poor, not government. The former should learn about the need for moral restraint, to ease their own plight: though it was good that relief should be given selectively to the deserving poor through charities.

Alan Chase says (1980) that this is 'scientism': 'using the language, symbols, findings and other attributes of science to advance unproved preconceptions and dogmas'. Marx called Malthus' work a 'libel on the human race' because it proposed limits to how much humans could improve their society and political economy. Malthus argued

> that the principal and most permanent cause of poverty has little or no *direct* relation to forms of government, or the unequal division of property; and that, as the rich do not in reality possess the *power* of finding employment and maintenance for the poor, the poor cannot, in the nature of things, possess the *right* to demand them; are important truths flowing from the principle of population, which, when properly explained, would by no means be above the most ordinary comprehensions. And it is evident that every man in the lower classes of society who became acquainted with these truths, would be disposed to bear the distresses in which he might be involved with more patience; would

feel less discontent and irritation at the government and the higher classes of society, on account of his poverty; would be on all occasions less disposed to insubordination and turbulence; and if he received assistance, either from any public institution or the hand of private charity, he would receive it with more thankfulness, and more justly appreciate its value.

If these truths were by degrees more generally known . . . the lower classes of people, as a body, would become more peaceable and orderly, would be less inclined to tumultuous proceedings in seasons of scarcity, and would at all times be less influenced by inflammatory and seditious publications, from knowing how little the price of labour and the means of supporting a family depend upon a revolution. The mere knowledge of these truths . . . would still have a most beneficial effect on their conduct in a political light; and undoubtedly, one of the most valuable of these effects would be the power that would result to the higher and middle classes of society, of gradually improving their governments, without the apprehension of those revolutionary excesses, the fear of which, at present, threatens to deprive Europe even of that degree of liberty which she had before experienced to be practicable.

(Malthus 1872, 260–1)

there is one right which man has generally been thought to possess, which I am confident he neither does nor can possess – a right to subsistence when his labour will not fairly purchase it. Our laws say that he has this right, and bind the society to furnish employment and food to those who cannot get them in the regular market, but in so doing they attempt to reverse the laws of nature . . .

If the great truths on these subjects were more generally circulated, and the lower classes of people could be convinced that by the laws of nature, independently of any particular institutions, except the great one of property, which is absolutely necessary in order to attain any considerable produce, no person has any claim of *right* on society for subsistence if his labour will not purchase it, *the greatest part of the mischievous declamation on the unjust institutions of society would fall* powerless to the ground . . . discontent and irritation among the lower classes of people would show themselves much less frequently than at present . . . The efforts of turbulent and discontented men in the middle classes of society, might safely be disregarded if the poor were so far enlightened.

(Malthus 1872, 191; emphases, except the first, added)

So objective facts demonstrated how government, wealth maldistribution and the existence of a rich class did not cause poverty. Similar incredible arguments made 'credible' by 'science' are popular today among conservative and many liberal politicians. They particularly use the notion of objective economic scientific laws to apportion blame for deprivation and unemployment to the deprived and unemployed themselves (by pricing their labour 'out of the market') rather than with the employers or politicians. Additionally, the British Government (advised by neo-Malthusian Crispin Tickell) predictably used neo-Malthusian arguments at the Rio environmental summit in 1992 to try to shift the blame for global environmental degradation from the West to Third World countries.

Malthus argued that human rights to the means of subsistence must inevitably be circumscribed by the 'greater' laws of nature. And, like some ecocentrics today, he insisted that if only people knew the 'facts', the scientific 'truth', then they would accept them, even if they did not particularly like them. But scientific 'truth' in Malthusianism is politically counter-revolutionary: it defends the existing order.

Legitimating political ideology: social Darwinism

If science is about nature's laws, then scientific legitimation of ideology is partly a matter of appealing to *nature* to support a political view and its implications. Herbert Spencer (1820–1903), a free-market liberal who coined the phrase 'survival of the fittest', did this. Others grafted his views with Darwinian evolutionary theory to create the ideology of *social Darwinism*. Spencer saw society, and its components, like the industrial firm, as an organism. This organic metaphor between society and nature (society *is* a natural organism) is one common aspect of all natural legitimation theories. So is the development from it, that regards nature as a template for society (the latter should model itself on the former). Also, and more subtly, such theories, which 'biologise society' usually get their view of what 'nature' is like from society in the first place. Take for example Moses Rusden's *A Further Discovery of Bees* (1679) in Figure 5.4. This shows bees in a hive organised hierarchically in the manner of human society, with a king, dukes and plebeians. But there was and is no evidence to suggest that this is how *the bees* perceive it: specialised functions they each may have, but this is not the same as a power hierarchy. It is a social view, transposed to nature. Then, the view of 'what nature is like' is often transferred *back* to justify a particular social organisation: the process is essentially circular and worthless, yet it happens repeatedly.

Darwin in fact called his thesis about the struggle for existence which followed from the plenitude of nature, 'the doctrine of Malthus applied in manifold force to the whole animal kingdom' (the social applied to nature) (Darwin 1885, 59), except that in these kingdoms the exercise of preventive checks was impossible. Because of this, his picture was an even more deterministic one than Malthus'. Malthus' 'libel on the human race' becomes even more libellous after being applied to the vegetable and animal kingdoms – and then applied back to human society. Social Darwin-

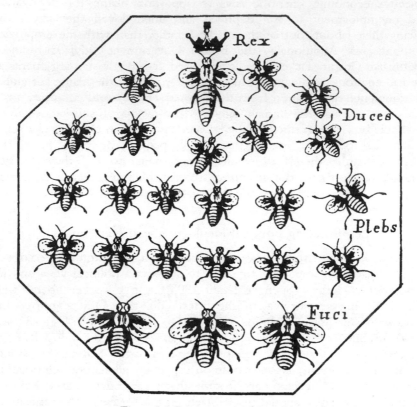

Figure 5.4 Bees organised in a hive
Source: Moses Rusden's *A Further Discovery of Bees*, 1679

ism compares human behaviour at large – not just in respect of reproduction and population size – to models of animal behaviour. The 'essentially competitive' behaviour of humans is held to derive directly from Darwin's (Malthus') model of populations growing to exceed available resources, leading to a struggle between groups and individuals in which the fittest survive, ensuring the sophistication of the species. Thomas Huxley proposed this social model in his essay 'On the struggle for existence in human society' in 1888 – an essay which provoked Kropotkin to write on mutual aid (see p. 289).

There can be little doubt that, despite the influences of free will philosophies and egalitarian movements for social reform, the ideas of social Darwinism continue to be profoundly influential today in the West (for a review of neo-Darwinism, socio-biology and socio-ecology and their uses and drawbacks, see Dickens 1992, 91–118). Schumacher (1973, 71) recognises them as part of education's 'hidden curriculum', by which we learn, without realising it, to regard humans as inherently (by *nature*) competitive, and that the fittest (should) survive the struggle while the weakest do (should) not. All this is part of a fundamentally desirable 'evolutionary' process wherein some small groups and individuals dominate and many others are dominated.

Such socialisation makes it easier for us to accept animal-like behaviour from humans, seeing it as inevitable, if not palatable, like any other phenomenon described and understood via the medium of scientific laws. As Coward (1989, 15) says: 'It is a widespread and prevalent feeling: if it's natural, then it must be good for us'. When we examine some of the 'laws' of animal behaviour which Darwin described, we can see just what it is which is allegedly 'good for us' in human societies:

> we should remember how essential it is in a flock of white sheep to destroy every lamb with the faintest trace of black.
>
> (Darwin 1885, 77)

> to my imagination it is far more satisfactory to look at such instincts as the young cuckoo ejecting its foster brothers – ants making slaves . . . not as specially endowed or created instincts but as small consequences of one general law leading to the advancement of all organic beings, namely, multiply, vary, let the strongest live and the weakest die.
>
> (p. 219)

> Natural selection in each well-stocked country, must act chiefly through the competition of the inhabitants one with another . . . Hence the inhabitants of one country, generally the smaller one, will often yield, as we see they do yield, to the inhabitants of another and generally larger country. For in the larger country there will have existed more individuals, and more diversified forms, and the competition will have been severer and thus the standard of perfection will have been rendered higher.
>
> (p. 184)

> The mental quality of our domestic animals vary, and . . . the variations are inherited.
>
> (p. 218)

Clearly the perspectives in these quotations can be used, if applied to the human world, to justify territorial aggrandisement – the state displaying 'natural' organic lusts – and slavery, via the 'scientific' theory that some groups are genetically superior to others. If one tries to argue that humans are different, and that Darwin's observations of animals are irrelevant to society, Darwin's science flatly contradicts, arguing that humans are part of nature (see Chapter 4.3), and therefore must be subject to the same laws as the rest of nature.

Social Darwinism is most familiarly applied to economic life, where the whole paraphernalia of competition, struggle and dog-eat-dog is to be discovered in the world of 'free' enterprise capitalism. The behaviour of business and industry is often morally repugnant to those who pause to reflect on it (we are familiar with examples like the arms trade or the drive to sell baby formula milk or cigarettes to the Third World). But it usually can be excused and legitimated if it is seen to result from the operation of scientific, 'economic laws' analogous with those natural laws discovered by Darwin. Thus

> Millionaires are a product of natural selection, acting on the whole body of men to pick out those who can meet the requirement of certain work to be done. They get high wages and live in luxury, but the bargain is a good one for society.
>
> (William Sumner, American political economist 1840–1910, cited in Oldroyd 1980, 214)

The bargain is 'good' because such natural selection leads to social improvement. The 'law' of competition, said business tycoon Andrew Carnegie unsurprisingly, gives us

> our wonderful material development . . . But whether the law be benign or not . . . it is here; we cannot evade it; no substitutes for it have been found and while the law may be sometimes hard for the individual it is best for the race.
>
> (cited in Oldroyd, p. 216)

Such 'substitutes' might involve governments intervening in market forces, but Sumner warned that this could do nothing but harm:

> Let it be understood that we cannot go outside this alternative: liberty, inequality, survival of the fittest; not liberty, equality, survival of the unfittest.
>
> (cited in Oldroyd, p. 215)

Like Malthus, then, social Darwinists held that meddling with 'nature's laws', that is, social 'laws', was counterproductive. 'There is no alternative' to obeying such laws – to use a favourite phrase of Margaret Thatcher and other 1980s neo-liberals who shamefully exhumed social Darwinism after its burial following the 1930s depression.

The organic metaphor is particularly strong in social Darwinism:

> The growth of a large business is merely the survival of the fittest . . . The American Beauty rose can be produced in the splendour and fragrance which bring cheer to its beholder only by sacrificing the early buds which grow up around it. *This is not an evil tendency in business. It is merely the working out of a law of nature and a law of God.*
>
> (J. D. Rockefeller, cited in Oldroyd, p. 216; emphases added)

Such ludicrous imagery abounded in the 1980s and 1990s as right-wing businesspeople and politicians sought euphemisms for the suffering caused by their economic approaches. Rationalisation and unemployment were about 'pruning out dead wood' to produce a 'leaner, fitter' more competitive industry which could create what British Chancellor Norman Lamont called in 1991 'the first green shoots of economic spring'.

While old-fashioned social Darwinism regained ground in the 1980s economic climate of possessive individualism, it had (ideologically speaking) a complementary paradigm, developed in the 'New Ecology' (Worster, p. 311–13). This was not about hand-to-hand combat for survival, but instead spoke of integrated circuitry, geochemical cycling and energy transfer (Chapter 4.7):

As a modernised economic system, nature now becomes a corporate state, a chain of factories, an assembly line. Conflict can have little place in such a well-regulated economy. Even strikes are unheard of.

Discussing one of Eugene Odum's ecosystem flow charts, Worster continues:

A traffic controller or a warehouse superintendent could not ask for a more well-programmed world . . . a kind of automated, robotised, pacified nature.

Hence, mainstream scientific ecology today offers us neither the old arcadian message nor imperialist conflict imagery (neo-social Darwinists notwithstanding). Instead it gives us the technocentric, *managerial* 'bioeconomics paradigm': nature is something to be manipulated rationally, rather than exploited, to maximise its returns to humans over time. As the leading ecologist Odum himself says (cited in Worster 1985, 311): 'The study of energy in nature does not necessarily imply an economic framework. But that is the way it has been assimilated'.

Thus the images and metaphors in nature, as represented by scientific ecology, like Darwinism before it, owe much to the larger cultural milieu, particularly though not exclusively the economic culture.

The original critics of social Darwinism often stressed this link between the 'mode of production' (capitalism) and the predominant ideas in society (i.e. Darwinist ideas). They suggested that such a link was not fortuitous but arose because Darwin's science was inherently ideological – because it was *in the first place* the transference from society to nature noted above:

The whole Darwinian theory of the struggle for life is simply the transference from society to organic nature . . . of the bourgeois economic theory of competition. Once this feat has been accomplished . . . it is very easy to transfer these theories back again from the natural world to the history of society, and altogether too naive to maintain that thereby these assertions have been proved as eternal laws of society.

(Engels 1963)

It is noteworthy that Darwin rediscovers in animals and plants his own English society with its social division of labour, competition, the opening up of fresh markets, invention and the Malthusian struggle for existence. This is the *bellum omnium contra omnes* of Hobbes.

(Letter from Marx to Engels, 18 June 1862)

Chapter 4.3 has noted, in Worster's analysis of Darwin's theory, its conflicting messages about nature. The theory has its 'arcadian' elements, for example, the 'web of life', by which species can create new niches, enabling them to avoid competition and war. However, this is at odds with the concept of competitive replacement, whereby old inhabitants of niches are removed by 'fitter' pretenders. And it is the latter which Darwin ultimately emphasised, because, Worster declares, of his own personal obsession with competition, struggle and conquest. The 'psychology of the private man' explains Darwin's ecology, says Worster (1985, 168), who, however, also goes on to emphasise the importance of social context. For Darwin was a partial result of the 'Victorian frame of mind', which in turn was linked with burgeoning capitalism: it is impossible to conceive of such a view of nature coming from a Hopi in the American Southwest, or a Hindu, even though they may live on the edge of Malthusian scarcity. Gasman (1971, xxix) confirms that ultimately it was not science which shaped Darwin's conceptions of nature and humankind, but rather national, historical and philosophical consciousness.

And there is the need to explain why the Darwinian evolutionary view became so popular, out of the many evolutionary theories on offer. This again is linked to what society was doing materially, economically. For Darwinian evolution became part of a popular view of progress as an evolution from barbarism to 'civilisation'. The latter essentially meant the norms of imperialistic, capitalist Western society. And it meant separation from the natural world:

decked out in the trappings of a positivistic science was a firm law, an inexorable movement toward civilisation that humans could not thwart, though they might affect its pace. It was an idea that gave assuring direction and meaning to a fast-changing milieu, and provided a new rationale for Anglo-American expansion over the earth's surface.

(Worster, p. 172)

Oldroyd also demonstrates how Darwinism has interacted with political culture. All political ideologies, from left to right, may show connections with it. But Oldroyd also says (p. 212), that social Darwinism is a loose amalgam of doctrines such as conservatism, militarism, racism, rejection of social welfare programmes, eugenics, laissez-faire economics and unfettered capitalism.

That is, it is mainly associated with the right.

Legitimating political ideology: environmental determinism

The metaphor of the state as an organism has been used particularly by the right to legitimate aggressive foreign policies combined with racism and territorial expansion. Peet (1985) shows how it was combined with environmental determinism in the work of geographer Ellen Semple, to justify American colonialism in the early twentieth century.

She argued (a) that the nation state is like a living organism, therefore it is natural that it should struggle against its neighbours for survival and it is natural that in this struggle the fittest should survive; (b) that geographical science, through surveys, showed that the USA is the fittest nation in North America in terms of natural resources. Hence its peoples, having inherited the superior cultural characteristics of the European nations whence they came, now had, added to this, the advantage of living in the most optimum climate for human endeavour – the USA. The Canadians also had cultural advantages, but they lived in too cold a region; the Mexicans, being partly Latins, anyway had only limited capacity for leadership, while the Indians had remained in 'savagery' because of the size and remoteness of the continent. Therefore 'Nothing could prevent the realisation of the manifest destiny of the [new] American people to occupy the American continent from ocean to ocean' said Semple in 1903, and she went on to argue that the USA would and should dominate the Caribbean. US expansionism thus was justified in terms of 'natural' law.

Hitler, of course, used similar organic theories of the state, developed partly by Haeckel, and partly by geographers, to justify his territorial aggression under the banner of the *lebensraum* thesis. Haeckel advocated an ethic of competition and struggle as the foundation for the laws of human society, he indulged in a religion of nature which meshed in with fascism (Chapter 4.6), and he conceived of 'lower' and 'higher' races of people (less and more evolutionarily advanced). The Germans belonged to higher races – a pseudo-scientific theory which Hitler also enthusiastically embraced in *Mein Kampf.*

The role of ecological science as a legitimator in another ideological battle over the use of territory, is absorbingly documented by Worster (1985, Chapter 12). He describes how botanist Frederic Clements' writings on dynamic ecology 'provided much of the scientific authority for the new ecological conservation movement' from the 1930s onwards. It argued that 'the aim of land use policy should be to leave the climax as undisturbed as possible' (p. 234). And this 'natural' climax vegetation, said Clements, was grassland.

Clements developed the concept of plant succession and climax in ecology, describing how a bare rock surface (or other medium) is colonised by successive waves of more complex plants, demanding more specific environments. The plant mix which is eventually in dynamic equilibrium with the environment, and stays there until the environment changes, is called the 'climax' vegetation. But in constructing this model he did not refer merely to nature. 'Clements was thinking about the American pioneer when he shaped his ideas of succession and climax' (Worster, p. 218). The social model which inspired him was

TRAPPER → HUNTER → PIONEER → HOMESTEADER → URBANITE ('CLIMAX')

The urbanite represented the social 'climax'. So, again, the analogy came from society originally and was imposed on to nature. After that, conservationists applied it back on to society to legitimate their views on what would constitute the best land use. Their environmentally determinist argument was, in the wake of the 1930s dustbowl, that arable homesteading, (ploughing land) was not a 'climax' activity, in equilibrium with the prairie environment, but that 'pioneer' cattle grazing was.

In the 1930s and 1950s this argument was opposed by the ecologist A. G. Tansley, and the agricultural historian Malin – the latter specifically, says Worster, because politically he wanted to justify the activities of cereal farmers – the 'sodbusters' of the dustbowl years

> underlying Malin's objections to the climax theory was a personal motive that really had little to do with issues of fact or fantasy . . . Malin refused to be hedged in by ecological laws. To obey rather than conquer nature was a surrender . . . to the chains of determinism . . . Thus Malin preached the familiar homily of Cornucopian expansion.
>
> (Worster, pp. 246–7)

Malin's view was put in the form of a scientific argument trying to disprove that the grass prairie *was* a natural climatic climax. He and Tansley argued that it was partly artificial. caused by fires ignited by humans ('red Indians'), and partly a very recent soil climax (growing on fresh glacial outwash etc.) so it was, anyway, not yet in equilibrium with the climate.

This ideological battle – against environmental (climatic) determinism and in favour of development – emerged more nakedly in another 'pioneer' context: in the newly formed USSR. 'Marxist' objections to determinism were carried to ludicrous extremes by the Soviet regime in the 1920s, 1930s and 1940s. The regime had set out to transform society radically, hence it thought it vital to reject all theories which suggested that human societies are determined by external 'non-human' forces. For such theories emphasised the limits on human action, denying society's potential to effect lasting and radical improvement. Therefore they were politically reactionary (running against the ideas and practices of the revolution).

Soviet geographers who committed this sin of 'geographical deviation' (environmental determinism) were attacked by the Party, even as late as the 1960s and 1970s. Determinism was considered to be a scientific weapon of the bourgeoisie (Matley 1982). Russian communists were so anxious in the 1930s to play up human ability to change society that this led to Stalin's disastrous dictum that nature affected only the speed of human development but not its direction, and that communist 'man' was the master of nature.

> Let the fragile green breast of Siberia be dressed in the cement armour of cities, armed with the stone muzzles of factory chimneys, and girded with iron belts of railroads. Let the taiga be burned and felled, let the steppes be trampled . . . Only in cement and iron can the fraternal union of all peoples, the iron brotherhood of all mankind, be forged.
> (V. Zazubrin, in the First Congress of Soviet Writers 1926, cited in Feshbach and Friendly 1992)

One of the practical results of this was bad husbandry in the interests of maximising production on the collectives, leading to crop failure and famine. Here, the aversion to determinism was so strong as to insist that Soviet science should be used instead of Darwinism, which was rejected for its 'bourgeois' (counter-revolutionary) implications. During 1930–60 the science of genetics in Russia was rewritten under the direction of T. D. Lysenko. He falsified data, imprisoned Soviet geneticists and set back Russian agriculture by insisting that hereditary changes in plants could be induced by exposure to controlled environments. As a result, attempts to breed plants which would have helped the effort to become self-sufficient in food failed miserably, as did so many collective farms. But this theory of evolution was ideologically acceptable, whereas Darwinist evolution was not (Medvedev 1969; 1979).

Legitimating political ideology: human nature

Perhaps the most contentious use of science and 'nature' in ideological contexts lies in the notion of 'human nature'. When a human behaviour pattern is seen as 'natural', then people are inclined to say either: 'If it is natural it is good, pure, wholesome, desirable'; or 'If it is natural, we may not like it, but there's little we can do about, therefore it is acceptable, excusable and inevitable'.

The idea of *human* nature is used, then, to legitimate behaviour – especially if it can be based on 'scientific' evidence, for example on generalised observations of how people behave, or on allegedly causal mechanisms which are regarded as part of human make-up.

The right wing ideology of human nature has been partly considered already, in social Darwinism. Left-wing critics claimed that this was just a way of justifying the bourgeois political economy and the privileges it affords. But still the modern liberal right tends to argue along Darwinist lines. It may draw on the work of Konrad Lorenz, who argued that aggression is inherent in humans. Others, like Tiger and Fox (1989), argue for the inherent differences between men and women as part of human nature. Desmond Morris (1967) considers that at heart we are still primitive and driven by self-interest. Like Dawkins (1976), such writers tend to the view that even cooperation and altruism stem from *self*-interest on the part of intrinsically selfish creatures.

The reactionary political messages of all this are as follows:

1   Inequality results due to unavoidable differences in merit between individuals. Some are biologically fitter and better than others, so equal opportunity – while desirable and worthy – will not create an equal society (liberalism).
2   Despite the veneer of civilisation we are little different from our more primitive past.
3   The fact that some fail and some succeed may be genetically determined. Hence such tendencies may well be passed on from generation to generation. A hierarchical society (in wealth, status, power) is natural and therefore justified, as it was in the past (conservatism).

According to psychologists Arthur Jensen and Hans Eysenck, people are sorted out in society by their genes. Intelligence, for instance, is inherited. It is measurable in IQ tests. If you have a high IQ you can expect, and deserve, to be prosperous.

Jensen (1969) argued that most of the differences between blacks and whites in performances on IQ tests resulted from genetic differences. The conclusion was that no programme of education could equalise the social status of blacks and whites, and blacks ought therefore to be educated for the more mechanical tasks for which their genes predisposed them. The claims of genetic inferiority for blacks were soon extended to the working classes in general (popularised by Harvard psychology professor Richard Hernstein). The Nixon administration, wanting justification for cutting welfare and education spending, used these 'scientific' arguments.

In Britain Eysenck's claim of biological differences in IQ between races became an integral part of the campaign against Asian and African immigration. The National Front has argued for genetic inferiority of Jews, Africans and Asians (see Rose, Lewontin and Kamin 1984, chapters 1 and 2).

Politically, such views are responded to in different ways. The welfare liberal position is as follows:

a) While individuals may intrinsically be limited in what they can do, the social climate can be changed to cushion those who may lose out.
b) Everyone has intrinsic merit and ability at something.

The socialist position is that:

> some may have no 'merit' or ability in any respect, but still everyone should be treated equally – 'from each according to their ability, to each according to their need' (NB equality, here, need not imply sameness).

Both welfare liberals and socialists largely oppose biological determinism, as in the genetic determinism of social Darwinism.

However, the anarchist view, compatible with that of many modern greens such as Capra, does not necessarily reject analogies comparing ourselves with nature, because anarchism was and is strongly influenced by Darwinism and modern ecology (see Pepper 1993).

Anarchists say, though, that social Darwinists have misunderstood Darwin, being heavily influenced by their conservative/*laissez-faire* liberal ideologies. They have picked out only Darwin's evidence of competition in nature, even though he wrote much about the cooperation to be found there, describing it as part of evolutionary mechanism.

Kropotkin – geographer and anarchist

Peter Kropotkin (1902) developed this alternative perspective and concluded from his observations of nature that whilst competition between and amongst species does occur, by far the most important influence on animal and human evolution is the *natural* tendency for individuals to cooperate.

Kropotkin believed mutual aid was a pre-human instinct, a law of nature (so in fact did Darwin). He was a naturalist, and *Mutual Aid* is a collection of empirical evidence for mutual aid in 'Animals: Savages/Barbarians: Medieval Society: Ourselves'.

The evidence contradicts that selected by social Darwinists, especially Thomas Huxley. Kropotkin argued that the competition within species which popularly is supposed to lead to improvement/evolution is counterbalanced by numerous examples of cooperation. He proposed that as far as human society is concerned, evolution progresses primarily through intraspecific cooperation.

Like Darwin and Marx, Kropotkin believed in evolution, as, by definition, progress. Like Darwin, Kropotkin was a natural scientist – a geographer and biologist – strongly inclined towards empiricism. He also accepted the (social Darwinian) premise that what happened in nature could be a template for human society. His book catalogues historical examples of mutual aid, for instance, animals living in *herds*, hunting in *packs* and feeding their infirm; sociability in higher vertebrates leading to enhanced intelligence (language, communication, imitation, experience) hence greater development and survivability; competition avoidance by migration; shared use of resources by early tribes; guilds in medieval cities embodying anarchist principles; trade unions, workers' cooperatives, housing cooperatives and village life in modern society.

Kropotkin concluded that mutual aid constantly reasserts itself as a factor of evolution. However, the history books do not acknowledge this because they are monopolised by those who benefit from asserting that human nature is competitive and hierarchical. For him, then, there is a human nature, and it is cooperative, not competitive.

Leading ecocentrics, including eco-anarchists, have argued much the same. Theodore Roszak (1979) wrote that Western culture relies for its morality on inculcating guilt and feelings of sin about our supposed moral nature – we believe we are naturally sinful:

> buried away in the core of Western conscience there is a festering accumulation of 'sin' that is simply unworthy of serious concern, and should never have been dinned into our children . . . this is the shallowest ballast of our moral nature and we let it hound us through a lifetime of 'good behaviour', 'high achievement' and 'respectability' as we try to prove, again and again, that we are pure, nice and loveable.

Roszak thinks that this guilt helps to bolster capitalism, through encouraging the work ethic, material acquisition and deferred gratification. Hence capitalism, which he thinks is 'toxic to the planet' will have to go before we can revert to our truly natural state – where we would be unguilty about what we are.

That state is anarchistic, Bookchin (1982) argues. He says that humans are part of nature, and nature is diverse and unhierarchical. So a human's natural state is one of freedom to be different and non-hierarchical. Bookchin seeks anthropological evidence to show that primitive 'organic' communities did and do live in freedom and non-hierarchy (Chapter 1.2).

Culture and human nature

However, Brown (1988) argues that all these anarchist arguments are flawed because they contain paradoxes. For if humans are naturally cooperative, how have they acted against their nature by setting up the state? If naturally social, how have they acted against their nature by setting up property, anti-social religion and the Church? If not guilty, how have they set up a system (capitalism) that requires them to feel guilty? If free and non-hierarchical how have they set up hierarchies? Why do they keep acting against their own nature?

A traditional Marxist view would answer this by arguing for the essentially cultural origin of supposed 'human nature'. Bertrand Russell expressed it (1914) in arguing for a society based on anarcho-syndicalism (or guild socialism) – an organisation of mutual aid based on the workplace. 'Human nature' was shaped mainly by the economic/social/political system under which people lived, he said. It was competitive at the time because the system under which people lived demanded, needed, and approved of, such behaviour. In other socio-economic systems it was not necessarily 'in people's nature' to be competitive, acquisitive and pugnacious. This sort of argument shifts the onus from natural to human sciences to explain and justify how humans do and could behave.

Can one avoid ideology altogether by following Brown's arguments further and asserting along existentialist lines that there is *no such thing as human nature*? Or, to put it another way, that we can determine our own nature for ourselves, for through our own actions we create our own, always changing, human nature. This is quite close to Marxism, but it goes further. It argues that there are no objective environmental factors with which we have to interact (except death). Rather than saying we are free to alter our culture/environment and therefore our nature, it says that we have – all of us as individuals – a much more direct control over what we are. Our ideology is our own responsibility, and cannot be explained away and justified by appeal to universal scientific laws of nature or society.

Brown cites in support Herbert Read, who argued in 1949 that anarchism and existentialism were essentially connected because they were both revolutionary militant doctrines emphasising individual freedom. Read said that existentialism allows us to create our own meaning for our lives – free from any ideas that suggest that we must be subordinate to the operation of physical or economic or social laws:

> the human person, you and I . . . and . . . everything else – freedom, love, reason, God – is a contingency depending on the will of the 'individual'.

That is, we can be something different by deciding to be so (idealism). Critics of this view point out that it is not really above ideology. Existentialism – an extreme individualist philosophy – was so popular in America precisely because any kind of socialistic theory which rests on humans having collective natures that are a product of society (and are therefore changeable by social reform) was ideologically unacceptable.

5.7 SCIENCE IN PERSPECTIVE

This chapter has suggested ways in which science does not give us an unbiased, objective view of nature (including human nature) and environmental problems. It contends that no scientist stands outside the 'facts' being observed. As Haraway (1989, 12, cited in Dickens 1992, 61) puts it:

Natural sciences, like human sciences, are inextricably *within* the processes that give them birth . . . it makes sense to ask what stakes,
methods, and kinds of authority are involved in natural scientific
accounts . . . the detached eye of objective science is ideological
fiction, and a powerful one.

Bird (1987) suggests some implications for environmentalists:

Our understanding of environmental problems is a social construction
that rests in a range of negotiated experiences. To cite the 'laws of
ecology' as a basis for understanding environmental problems is to rely
on a particular set of socially constructed experiences and interpretations that have their own political and moral grounds and implications.
There can be no recourse to 'objective' truth. The only alternative is
inter-subjective, socially negotiated moral truths achieved in the interests of (environmental) justice . . . Environmental problems are not the
result of a mistaken understanding of nature. Rather they are the results
of mis/taken (unfortunate or ill-chosen) negotiations with and constructions of nature in the shape of new socio-ecological orderings of
reality. They result from morally and politically mis/taken social
practices.

Dangers of idealism and relativism

None of this, however, should make us incline towards extremes of social
constructivism, relativism, and idealism – a possibility already identified in
Chapter 5.1. Indeed, Whitehead's trenchant criticisms of classical science
were balanced by a mistrust of those whose repudiation of its concepts
leads them to believe there is no objective world at all. The likes of Capra
and Zukav, and ecocentrics who enthuse about their subjectivist interpretations of twentieth-century science (Chapter 5.3), might reflect on Whitehead's objections to subjectivism.

First, subjectivism contradicts our direct experience. We experience a
world of colours, sounds and other sense objects related in space and time
to enduring objects such as stones, trees and human bodies. Second, if
these entities are in fact an illusion and they really depend on us, then what
of our historical knowledge, which tells us of a past time when they existed
but humans did not? And, third, subjectivism denies our human nature.
This has an instinct for action – which essentially means that we want to go
beyond ourselves and transform what is 'out there'. Believing that it is all 'in
here' logically leads us towards the unsatisfactory, idealistic, conclusion that
we must concentrate on how we (especially as individuals) interpret and feel
about nature rather than starting with political action to change what we do
to it. This is a recipe for political passivity and ineffectiveness.

The point has also been made that subjectivist interpretations of twentieth-century science did and still do divide the scientific world. Capra and company follow Niels Bohr, who wrote of an indeterminate, probabilistic universe where what is there is created by the observer, rather than Einstein, who rejected such implications about his own theory. He held that although individual observers may see things in slightly different ways, none the less, via a set of mathematical rules and transformations they can also see them from each other's point of view, so there *is* an objective universe with deterministic laws of nature. This would lead to a more balanced assessment, of the sort Martell (1994, 175) suggests: 'our knowledge of nature is . . . partly a social construction but is also dependent on objective properties of nature itself'.

As to the alleged links between mysticism and physics, sociologist Sal Restivo, according to Martin (1993), argues in *The Social Relations of Physics, Mysticism and Mathematics* that opposite conclusions can be drawn from physics to those which Capra chooses to induce. 'In fact,' says Martin (p. 354), 'by picking examples appropriately, you could find similarities between mysticism and old-style billiard-ball Newtonian physics'. Why, then, follow Capra's interpretation?, he asks. Because many people *want* to believe that nature is on their side (as in debates about human nature and social Darwinism): 'that nature – nuclear processes as well as forests and oceans – really is interactive, holistic, non-hierarchical and mysterious', and that there is a fundamental unity of humans with nature.

Separation from nature

But then we might question, as does O'Neill (1993a, 150), whether ecocentric interpretations of scientific messages about our relationship to nature are, anyway, accurate. Nothing in ecology or quantum mechanics, O'Neill believes, entails that there is 'no significant division between an individual self and its environment' – neither require that I and the world are one. Neither does classical science imply that the subject–object dualism should necessarily lead to domination of nature: there is no necessary link between measurement, prediction, technical control and domination. O'Neill argues that to reduce greenhouse gases is to try to control nature, but not to dominate it: rather, to nurture it.

He further argues (1993b), after Marx's early writings, that the 'objectification' of nature is indeed a necessary condition of appreciating its beauty and valuing it for itself. That is, we should respond to nature disinterestedly, which means not thinking constantly of what utilitarian, commercial, vested human interest it should serve. This sounds very like the deep ecology plea for recognising nature's 'intrinsic' qualities (the plea that, we have noted, sits uneasily alongside ecocentrism's subjectivist interpretations of nature). However O'Neill's (p. 141) motivation is firmly humanistic rather than biocentric:

293

A response to the objects of the non-human world for their own qualities forms part of a life in which human capacities are developed. It is a component of human well-being.

Furthermore, scientific education opens our senses to the objects around us, leading to an ethical appreciation of their value. And so (p. 145):

> The sciences and the arts form central allies in the appreciation of the value of the non-human world. Indeed it is one of the tragedies of the green movement that it has found itself caught up in an anti-scientific irrationalism fashionable in universities and 'alternative' cultures.

Where values come in

And so O'Neill (1993a, 145) rejects Yearley's view of science as an unreliable ally for radical environmentalists. Merely because it is not totally reliable does not make it completely unreliable. Indeed 'Scientific theory and evidence are a necessary condition for a rational ecological policy'. Without scientific vocabulary we could not even say what the problems are, let alone debate them. Wynne (1982, 168) reaches similar conclusions, having warned about the incomplete nature of any decision making embedded solely in the language and rituals of scientific rationality:

> 'Lapses' from the idealised standards of rationality should be regarded as normal. This is supported by a sociology and history of science which views dogma in science as an inevitable and necessary means of concentrating intellectual effort.

Bias and lack of detachment are therefore expected in science, and do not necessarily invalidate its findings. O'Neill asserts that this is possible because ethical and political values need play no part in the validation of theories. Neither need they influence the questions asked: the choice of what to investigate being that of the scientist rather than society, despite commercial pressures. Where they do, properly, come in is in debates about the aims and activities of scientists. 'That science should not be determined completely by society does not mean it should be totally determined by its inner life' (1993a, 158)

O'Neill's position may represent an acceptable compromise between realist and social constructivist polarities. First, there are values in, and social influences on, science, and the first step is to recognise them, making them explicit. Second, all of us, including scientists, should be concerned to ensure that they are the *right values*: dedicating scientific effort to the pursuit of social justice and a benign society–nature relationship. While there is room to discuss precisely what these right values are, it is unacceptable to drift so far towards relativism, subjectivism and postmodernism that we can no longer be definite about their existence and the desirability of asserting them.

Chapter 6
WAYS AHEAD

WAYS AHEAD

6.1 RADICAL IDEALISM

It is tempting to say that there are two 'extreme' perspectives among radical ecologists on how ecotopia might be attained and what it might be like. However, everyone's perspective is 'extreme' to someone else in the world: better then to talk of a spectrum of views. One end is idealist (Chapter 2.7), arguing that changes in consciousness and values will produce mass changes in how people behave. The other is materialist, concentrating on the need to change socio-economic arrangements radically as a simulta- neous or prior requisite to widespread attitudinal change. Idealism suffuses New Ageism, while eco-socialism is fundamentally materialist. Deep and social ecologists, including environmental activists, often reject both ends of the spectrum: their radical, ecocentric prescriptions are somewhere 'in the middle'. But they usually gravitate towards one or other perspective.

New Ageism's view of social change derives from a convergence of views inherent in Eastern philosophies (supposedly), postmodern science (Whitehead, Bohm, Capra, Prigogine, Sheldrake, etc. – Chapter 5.2 and 5.3) and Teilhard de Chardin's Catholic schema (Chapter 3.3). The main ideas in this convergence are summarised in Table 6.1. They are presented coherently in Russell's (1991) *The Awakening Earth*: the bible of New Ageism. Russell's idealism is unmistakable:

Consciousness precedes being, and not the other way around, as the Marxists claim. For this reason, the salvation of this human world lies nowhere else than in the human heart . . . the most important fight of all at this crucial stage in our evolution is not the fight against inflation . . . pollution . . . desertification or . . . corrupt governments. These are each very necessary and cannot be relaxed. However they will not be won until we have also won the fight within ourselves . . . [against egoistic thinking] . . . it is the enemy within that must be defeated – and urgently.

(pp. 226, 228, 229)

296

Table 6.1 The main ideas in the convergence between Eastern philosophy, postmodern science and Teilhard de Chardin's evolutionary schema

---

- Earth, humanity, universe are one integrated system
- Noosphere, geosphere and biosphere are integral parts of the total evolutionary process
- This system continuously, creatively *unfolds*
- Unfolding includes leaps to new levels of creativeness
- Love is the universal driving force in evolution
- Universal psychic energy is the all-pervading force creating spirit, matter, soul, body, energy, force
- Thus all manifestations of psychic energy are different forms of the same substance
- Hence individuals are part of, and are mutually present in, each other
- There is diversity and uniqueness, but within a unifying whole
- All manifestations have dual characteristics, but they come from and go to a unity
- Thus evil is an unavoidable aspect of goodness – they are differentiations out of the unity
- Thus sacrifices are indispensable to the creative process
- Human progress builds the noosphere, leading to progress for the whole of nature
- There is a continuous relationship between micro and macro elements in the universe
- The development of individuals and humanity is embedded in infinite time, space and universal energy – 'the way', or Tao
- As humans age they shift emphasis from earth and spirit to universe and soul
- As individuals reach harmony they contribute to harmony in all humanity, and a peaceful, ecologically sound society

---

Source: Stikker 1992, 75–7

'The real problem,' says Russell, 'lies not in the physical constraints imposed by the external world, but in the constraints of our own minds' (p. 114). Or, as Keyes (1982, 95, 98) puts it:

> The way to our survival lies in altering how we think and feel. We must use the power of collective consciousness as we learn to focus on peace . . . The bomb is not the real problem – it's only an effect of our attitudes.

Evolution

These quotations encapsulate many features of the idealistic approach to social change towards an ecological society. The starting point is with the evolution of Gaia – the earth as a living being. Notwithstanding Lovelock's own caution about the interpretation of 'living', Russell considers that Gaia has all nineteen subsystems characteristic of other living systems, and that it can also maintain internal order. Atmosphere and oceans, he says, circulate nutrients and carry away waste 'much as the blood circulates nutrients and carries away waste in our own bodies' (p. 6). There is design rather than chance in the way that Gaia evolves into new orders of complexity, diversity and connectivity: orders inherent in previous ones (cf. Bohm 1983 and Prigogine 1980).

One question is what role humanity has in Gaia's development. Are we the 'global brain' or 'cortex'? Or are we 'a planetary cancer'? With characteristic Gaian anti-urbanism and anti-humanism, Russell bemoans sprawling suburbs 'eating' across the planet: 'The analogy with cancer cannot be ignored' he asserts (p. 20).

But Russell optimistically settles on de Chardin's view of a planet evolving towards a social super-organism, consisting of interdependent individuals and societies. Although physically diverse they will grow together via their thoughts and ideas. Eventually billions of individuals form a single 'interthinking' group. This is the *noosphere*, the system comprising all conscious minds, which has evolved from material living matter (the biosphere), that in turn has evolved from 'non-living' matter (the geosphere). When the three spheres are eventually united, and there is 'planetisation' into an organic unity, this is 'point Omega', and Gaia will have emerged as a conscious thinking being or 'supermind'. After this point there is a further possibility of human global consciousness merging with the rest of the universe into the 'Gaiafield': a galactic super-organism formed by ten billion Gaias. This conceptual scheme loosely follows Hegel's (1770–1831) view of history as a process of emerging world spirit.

The process itself is not all smooth and imperceptible. There are sudden jumps across thresholds to new, unknown levels. The big leaps to date have been from orders of energy, to matter, to life, and the next one will be into the order of consciousness. Within each order there are also significant leaps – from complex molecules, to cells, to simple organisms, to organisms with nervous systems, for instance. While each new order includes the previous ones it is not just the sum of them. A synergy occurs and something new is created (Chapter 5.3).

Evolutionary leaps occurred in response to environmental *crises*. Breathing was evolved to deal with the 'pollution' of oxygen build up in the atmosphere, for instance. Russell (p. 55) thinks that 'In the present day it is readily apparent that society is also going through some major crisis': the amount of entropy or disorder it produces has shot up, hence an evolutionary response is needed to a new order of ecological harmony. Stikker (1992) thinks that we could cross a threshold within the next fifty years. He cites exponential trends in 'brainpower' and in communication speeds which all indicate a transformation threshold between 2030 and 2070. This is all rather millenarian: and familiar to science fiction buffs – one can almost hear the beginning strains of 'Also sprach Zarathustra' as one reads (although this particular vision does not evoke strange obelisks, planted by bodiless, intergalactic 'farmers' of humans, presiding over the jump from one evolutionary threshold to another, as in Arthur C. Clarke's *2001*).

In accordance with the general principles of this evolutionary schema (see Table 6.1). Russell searches for teleology, in patterns reappearing at different levels. Accelerating trends may lead to a global population stabilising at ten billion. This is 'interesting' (p. 68) because 'this figure appears to represent the approximate number of elements that need to be gathered together before a new level of evolution can emerge'. There is also this number of atoms in a cell, and this number of cells in the cortex of the human brain, hence:

> The human race may be fast approaching the stage where there are sufficient numbers of self reflective consciousnesses on the planet for the next level to emerge.

Social change

> If evolution is indeed to push on to higher levels of integration, the most crucial changes are going to take place in the realm of human consciousness and consciousness of the self in particular. In effect the evolutionary process has now become internalised within each of us.

Here again, Russell articulates the idealism behind this social change theory, and its personal-is-political message. We need to change ourselves first, to 'positive mind-sets' that do not interpret events as mostly negative. But this individualistic approach at the same time seeks to create a view of the self which embraces the collective and, even more, the universe. So when we think of ourselves we must not imagine a discrete isolated individual but a part of the whole – a mere temporary manifestation of the greater whole like the whirlpool in the stream.

In this way we can create the synergy necessary to forge a higher evolutionary social order. Hence in all working to realise ourselves as individuals we can be naturally in tune with the group, mutually supportive. This sounds like Adam Smith's 'invisible hand' theory in another context, but it is not meant as a model for selfish individualism, as in neo-liberal economics. The search for a 'pure' self here is intended to imply a revolutionary quest for ourselves as part of whole society. So:

> A person whose goal is self realisation, whether he be a yogi in a Himalayan cave or an office worker in London, is helping to change the world at the most fundamental level. Such people are perhaps the ultimate revolutionaries.
>
> (Russell, p. 141)

Perhaps. In practice, however, this approach to social change can manifest itself more in the tradition of liberal individualism than any socialistic search for collective identities and interests. For it is prone to degenerating into orgies of introspection: the search for 'self-knowledge' via therapies, consciousness-raising, 'peak experiences', meditation, ritual and all the other paraphernalia of New Ageism (hilariously lampooned by McFadden 1977 and Stott 1988). The problem about the whole approach is that in its enthusiasm for value changes through mysticism and spiritualism it can largely ignore the material dimension of environmental problems. It also places far too much onus onto individual action or thought when mass collective action would seem to be the more realistic way to break the political power of capital. And by constantly emphasising supposed fundamental unities it naively disregards the way that economic and social class divisions contribute to ecological degradation. New Ageism tends to spawn superficial and anodyne end-of-ideology banalities such as: 'We're all in this together', 'you and I are more alike than we are different', and 'Even if you drop a bomb on me your purpose is to settle arguments and create peace! These are good intentions, just like mine!' (Keyes 1982, 109, 120, 112; see Pepper 1993 for development of this critique).

Radical idealistic ecocentrism's means of translating heightened individual consciousness into heightened world consciousness have been described (Chapter 5.3). According to Sheldrake's 'morphic resonance' principal a small group of highly conscious people can have a disproportionate effect on the rest of society. Their enlightenment can be passed on by the material technologies of the 'information age' – computers, fibre-optics, satellites and the rest contribute to noosphere development. But it is also transmitted more subtly, via extra-normal sensory means. Russell offers speculative and anecdotal support for the idea of telecommunication between meditating groups a thousand miles from each other. Their synchronous higher brain 'coherence', he maintains, produces self-reinforcing and self-enhancing planet-wide resonances at 7.5 cycles/second, making the planet 'hum' with New Age energy. This is essentially a process by which the 'implicate order' becomes explicate.

Hence the more that individuals raise their own consciousness level, the easier it becomes for others to experience these states, ultimately leading to a chain reaction where everyone starts making the transition. The story of the hundredth monkey is constantly cited in evidence, despite its doubtful foundations (Chapter 5.3). We seem to be in the realm of pure idealism here: Russell (p. 178) claims that:

Preliminary statistical analysis of the crime figures in cities where one per cent of the people had learned transcendental meditation showed that crime rates in these cities fell by an average of 5.7 per cent. In other cities of the same size but with fewer people meditating, the average crime rate rose by about 1.4 per cent.

This scenario seems riddled with unsubstantiated causal assumptions. But lest we become too dismissive, we might reflect that here is but another version of a more familiar theme in Western culture: that of the power of prayer, and of faith to 'move mountains'.

6.2 RADICAL MATERIALISM: CHANGING THE ECONOMIC BASE

Aspects of eco-socialism have been described in Chapters 1.2 and 4.5. There are different varieties of eco-socialism, reflecting different emphases within socialism itself. That based on 'orthodox' Marxism could be said to have a strongly materialist perspective on social change. It holds that the way we organise ourselves materially, particularly economically, is the principal conditioning factor behind our attitudes, behaviour and social institutions. A crude model of how society works might therefore have economic arrangements and structures at the 'base' of everything else: strongly influencing everything in the social 'superstructure'. For instance, the fact that competition and consumerism are, directly and indirectly, strong motifs in education and media hinges upon the fact that they are necessary to the workings of the economic system of capitalism (Chapter 2.7 and Pepper 1993). So eradicating compulsive competition and consumerism implies eradicating capitalism itself.

Radical social change, say socialists, must come about largely through changes in people's material (economic) circumstances. Furthermore, change is unlikely to happen until it becomes clear to most people that the prevailing (capitalist) mode of production can no longer satisfy their wants, material and otherwise. The majority imagine that they lack political power, which they do unless they act as a collective. In orthodox theory the material immiseration of the proletariat (those who do not own the means of production but have only their labour to sell) is thought to be the principal factor likely to galvanise them into realising and using their latent collective power to achieve a socialist (or communist – the words are interchangeable) society. This society will also be, by definition, ecologically sound (Grundmann 1991).

The moneyless economy and the eco-socialist society

It is profit which draws men into unmanageable aggregations called towns . . . profit which crowds them up when they are there into quarters without gardens or open spaces; profit which will not take the most ordinary precautions against wrapping a whole district in a cloud of sulphurous smoke; which turns beautiful rivers into filthy sewers.

(Morris 1887b)

Orthodox Marxism sees accumulation by the bourgeoisie (owners of the means of production) of the surplus value (profit) produced by the proletariat as the key to the unacceptable relations of production in capitalism. It follows that abolishing money, thereby making its accumulation impossible, is the key to abolishing capitalist productive relations. And while 'externalisation' of social and environmental costs in production is fundamental to capitalist production, externalisation will be impossible in eco-socialism, because there will be collective, common ownership and truly democratic control of the means of production – hence no 'external' places. Ending private ownership and control of the means of production is essential to the socialist aim of creating a classless society: one where selfishly individualistic attitudes and behaviours are alien.

In a socialist society, say Buick and Crump (1986), all control the means of production on an equal basis: economics is organised so as to give social relations of equality. This in turn *means* a democratic society, which in turn *means* a classless society. Common ownership is incompatible with state ownership, since the state always represents the interests of the élite in class societies. Consequently the state has no place in socialism.

A socialist economy is not an exchange economy, neither would production be carried on for sale, since sale implies private ownership. Production would be for social 'use' (meaning both survival and enjoyment), not to create commodities (i.e. specifically intended to make money on the open market). Liberal economists always argue that a market is indispensable in a productive economy, to ensure that the supply of goods is matched to demand (via the price mechanism – e.g. Hayek 1949). They argue that this is the most efficient mechanism, yet judged by the more holistic and long-term definition of 'efficiency' that socialists, like ecocentrics, use, market economies are incredibly wasteful, illogical and *inefficient*, and by their nature cause environmental crisis.

This is underlined forcibly by Smith (1988), who argues that contrary to popular and economic wisdom people are relatively *less* well off now than they were in the Middle Ages. Nine-tenths of the working population are not producing real wealth at all, he maintains, for nine-tenths of the costs of goods stem from the fact that they are subject to the money transaction. The supposed benefits of large-scale production, division of labour and comparative advantage are largely illusory: any increased productivity they have brought has been more than swallowed up by the on-costs of administering and running capitalism.

Smith considers eleven leading branches of industry, claiming that there is an enormous amount of unproductive effort in each. Table 6.2 shows

Table 6.2 A moneyless economy

Following William Morris' Marxist-derived analysis of capitalism, and Morris' view of real wealth, which green economics owes much to, Ken Smith (1988) advocates a moneyless economy. People would take what they *need* from a central store, which would be replenished via a highly sophisticated system of planned production. (They would not take more than they need because human nature would be different under socialism to what it is under capitalism: artificial wants would not, furthermore, be encouraged by advertising.) Given all this, Smith proposes that it would be *cheaper* to distribute on the basis of to each according to need (from each according to ability) in a moneyless economy. This is because, he suggests, nine-tenths of the productive effort under capitalism and consumerism is *wasted* – it goes towards building, servicing and protecting private property and money, and towards persuading people to consume what they do not need. These are just some of the industries and services which Smith considers would be unnecessary in a moneyless economy:

- Bureaucrats, financial services, salespersons, advertising, war industries, building office blocks, wages clerks, priests and rabbis, psychiatrists and gurus
- Much transport and travel and road building
- The unemployment industry
- Packaging
- Agricultural pollution produced by agribusiness
- Fashion clothes
- Property services, estate agents
- Much energy production and consumption
- The dream factories of California
- Junk TV
- The gutter press
- Some curative rather than preventive health care
- Much of the information and IT industries
- Money industries (he lists about seventy sorts of employment, including banking and auditing, economists and debt collectors, stock exchanges, trades unions)
- The crime industry (1 million people in law enforcement and an unknown number of criminals doing mainly property-related crime)
- Tax evasion and avoidance

Presumably one could add most antipollution industries and services, because common ownership of the means of production means that there is no incentive to generate wastes which can be excluded from economic accounting by 'externalising' them.

Source: Smith 1988

just some of the occupations and activities which, he maintains, would be largely redundant in a moneyless economy, without exchange and private ownership of the means of production. Hence to concentrate on producing use values to satisfy basic needs without money or markets would be much cheaper. Already this is done on a wide scale, in vegetable and fruit gardening, home improvement, evening classes, charities, caring, conservation work and the 'black' economy generally. Voluntary organisations show that people are basically prepared to work hard for sheer love of it and to serve their fellows.

But planning is clearly essential to this eco-socialist economy, albeit achieved via 'bottom up', delegated local and regional councils rather than a state. For people would take and use goods as needed from a central store (as happens in some alternative communities today – see Pepper 1991). And work would be done by all according to their skills and abilities, as a freely chosen cooperative activity. The central problem of economics and planning, then, would be how to coordinate all this and match supply to need.

The rate and kind of technological innovation in production would be subject to its effects on the health and welfare of the producers and of the environment: this would be the measure of efficiency. Without money there would be no common unit of value but calculation in kind. Hence cost-benefit analysis would be based not on money values but on collectively decided policy.

Political-economic organisation would be at regional and world levels, not as nation states. Socialism would then have to establish a rationalised network of planned links between producers and suppliers:

> A self-regulating system of stock control will permit producers . . . (workplace councils, industry councils etc.) to ascertain more or less immediately the availability of stocks of any particular item throughout the system; the communications technology to enable this to happen is already in place . . . a buffer of surplus stock for any particular item, whether a consumer or a producer good, can be produced to allow for future fluctuations in the demand for that item, and to provide an adequate response time for any necessary adjustments . . . The relative abundance or scarcity of a good would be indicated by how easy or difficult it was to maintain such an adequate buffer stock . . . It will thus be possible to choose how to combine different factors for production, and whether to use one rather than another, on the basis of their relative abundance/scarcity. By following the rule of using the minimum necessary amounts of the least abundant factors it will be possible to ensure their efficient allocation . . . whatever shortages may persist can be tackled by some system of direct rationing.
>
> (Editorial comment in the *Socialist Standard*, no. 1075, 1994, 59)

Clearly this economy and society draws on William Morris's utopia (1890), and would also be part of ecotopia (Buick 1990).

Attitudes and values

There are many detractors of such visions and the analyses they are based on. Some come from the ranks of socialism itself. Neo- or 'humanist'

Marxists are generally uneasy about excessive economic determinism: reducing all important elements to economics while underplaying the leading importance of media, education and culture – or of attitudes and values – in social change. Some also argue that the central importance of nature and natural limits to growth are ignored in orthodox Marxism (Benton 1989). Others point to the highly developed international division of labour which capitalism has now achieved, arguing that it would be impossible to replace it by free associations of communes and producer coops of the kind that Marx originally envisaged (Sayer and Walker 1992). Or they cannot envisage that there would be no need for a state in socialism, or indeed for a market in an albeit subsidiary role (Frankel 1987).

From the ranks of liberalism come many more objections: not least the (unsubstantiated) assertion that markets are somehow equatable with general 'freedom', so the absence of markets would lead to general tyranny. There are technical objections, which cannot accept that a common unit of account is not necessary (Steele 1992). And there is the common view that the principle of 'from each according to ability: to each according to need' goes against 'human nature' – people would hoard goods if they were 'free', or they would not work without a money incentive. Socialists respond to this that 'human nature' is not universal or immutable, but results from socialisation (see Chapter 5.6). The material changes outlined above, they argue, would encourage widespread and radical changes in attitudes and values of the kind that ecocentrics want to see. The *Socialist Standard* goes on:

> the establishment of socialism presupposes the existence of a mass socialist movement and a profound change in social outlook. It is simply not reasonable to suppose that the desire for socialism and the conscious understanding of what it entails on the part of all concerned [which would come from 'doing it'], would not influence the way people behaved in socialism and towards each other.

Rationing, we might envisage, would be 'borne with forbearance – even, one might say, with a sense of altruistic restraint'. Clearly, for this to be so any revolutionary process must have involved, and been wanted by, most people, so 'communist' revolutions of the sort imposed by vanguards, as in the former Soviet Union, do not qualify as a route to eco-socialism.

6.3 PREFIGURING ECOTOPIA

Away from New Age centres and orthodox socialist groups, radical ecologism today tends towards an eclectic mix of idealist and materialist solutions. Above all, the strategy for revolutionary social change is informed by the anarchist concept of *prefiguring*. This holds that the way to create a desired society is to start living it out – thinking it and 'doing it' – here and now in the society you want to replace. By thus 'bypassing' the latter you create an example which others will follow. As they do so, the whole society changes.

The anarchy represented here is partly that of the 'new social movements' – greens, feminism, civil liberties, consumer movements – which are anti-state, though not anti-capitalist. It is liberal rather than socialist anarchism, since it focuses on the central importance of individuals, as consumers, in social change (see Scott 1990), rather than the collective, as producers (where withdrawal of labour would be a prime weapon in the struggle to overthrow capitalism – see Pepper 1993). It tends towards ideological eclecticism, and pragmatism in its actions – it will support mainstream pressure group politics as well as trying to set up radical alternatives (Wall 1990).

Conspicuous elements of its practical programme are compatible with liberal, small-business capitalism. But when these are presented within the theoretical framework of *social ecology*, greens (paradoxically?) regard them as part of an 'alternative' society.

Those elements include small-scale organisation, with communes or city neighbourhoods as basic socio-political units; strong localism – expressed, for instance, in LETS schemes (local employment and trading systems); and a determination to organise and live with the grain of ecological principles, for instance in bioregions, practising permaculture.

Small scale and bioregionalism

Central to ecocentrism is a belief that revising the *scale* of living will solve, *at root*, many theoretical and practical problems. This is because in small communities people should more easily see the effects of their own actions on the community and the environment. The *areal division of power* is what will best promote equality, efficiency, welfare and security in all society. This will produce more cohesion, less crime, more citizen participation in government and sensitivity to the needs of others. Papworth (1990) suggests it will militate against destruction and for the development of human creativeness: 'The only form [of power] which holds aggression in check and liberates the creative genius of the human spirit is not united power but equally divided power'.

Leopold Kohr's writings have inspired this important 'small is beautiful' theme (see especially Schumacher 1973). Breaking nations into smaller units is

not only a matter of expediency but of divine plan, and . . . it is on this account that it makes everything soluble. It constitutes, in fact, nothing but the political application of the most basic and organising balance in nature. The deeper we penetrate into its mystery the more are we able to understand why the primary cause of historic change . . . lies not in the mode of production, the will of leaders, or human disposition, but the size of the society within which we live.

(Kohr 1957, 97–8)

In thus believing in scale as the key factor in explaining society and social change, Kohr, in turn, revives earlier writers such as Lancelot Hogben. Writing in the 1930s, 'Hogben thought that scale was a decisive consideration, over-riding the difference between socialism and capitalism' (Martinez-Alier 1990, 150).

Some ecocentrics propose *bioregions* as the basic geographical unit of small-scale ecological society. A bioregion is 'any part of the earth's surface whose rough boundaries are determined by natural characteristics rather than human dictates' (Sale 1985, 55). Bioregions are found within size hierarchies ranging down from an 'ecoregion', such as the Ozark Plateau, which is several hundred thousand square miles and is defined by 'native' vegetation and soils. A 'georegion', such as California's Central Valley, is defined especially by its physiography, for instance river basins, mountain ranges and watersheds. Georegions have 'distinctive boundaries'.

The principles of bioregionalism are liberating the self, reducing the importance of impersonal market forces and bureaucracies, opening up local political and economic opportunities, enjoying communitarian values of cooperation, participation, reciprocity and confraternity, and having roots. This is all in order to develop the region's potential towards self-reliance and to develop a sense of place in people who live there – the latter involves learning about folklore and history and the technologies that 'traditional' people had.

Bioregionalism revives old 'regions' such as the 'Shasta nation' (south Oregon and north California – Callenbach's Ecotopia), or 'Middle England', a pre-Norman-English concept. The small communities in self-sufficient bioregions would be insulated from boom and bust economic cycles initiated from far away, being free from distant economic control. They would, Sale believes, be richer, not having to pay for imports or high levels of transport. They would control their own currencies and economic policies. Their people would be healthier, more 'cohesive' and self-regarding. Their locally made decisions would be based on a cooperation born of sharing the same problems.

In Sale's essentially anarchist-communist bioregional society (surely modelled on Kropotkin) there would be production for need, value according to social usefulness, labour without wages, common ownership of the means of production and a planned economy, to produce enough for everyone. The wealth of nature would be the wealth of all, through common ownership of the commons. Though each bioregion would trade as little as possible, relying on natural assets and finding substitutes for

307

absent materials, there would be federation into 'morphoregions' to sustain hospitals, universities and symphony orchestras.

The 'bioregional mosaic', 'would seem to be made up basically of communities, as textured, developed and complex as we would imagine' (Sale, p. 66). 'Earth is best described as a mosaic of coevolving, self-governing communities' (Engel 1990, 15) – neighbourhood unit, block or street, or urban or rural commune.

There are several problems about this vision of society. First, it is difficult to conceive of Ecotopia without more, not less, planning and a lesser or non-existent role for markets. There must therefore be supra-regional and, indeed, global bodies. This, amid the breakdown of nation states, risks excessive economic and political *centralisation*, as may be happening in Europe today: the resistance of regions to a continent-scale centre may be less effective than that of the original nation states. Second, so much impetus for bioregionalism has come from areas like the 'Shasta bioregion' where the Planet Drum Foundation started in 1973. This is, in fact, California, which is unusually well-endowed with natural advantages. But what of less advantaged regions? If, for example, the Lombard League were to break the north away from Italy today, this would leave the poor south depending on the centralised EC institutions to effect considerable resource redistribution.

Third, regionalism and decentralism can so easily be a force for the extreme right, encouraging regional chauvinism and racism (Chapter 5.6). Historically, demands for self-determination on the basis of race or territory have led to more conflict than harmony. In Europe they helped to set the working class against itself in World War I, and to legitimate the Third Reich. Greens worry about this: for instance flurries of anxious correspondence followed an article supporting regionalism in *Green Line* (Kinzley 1993).

Fourth, the vision of a mosaic of self-determining communes, neighbourhoods and bioregions only produces global ecotopia if *all* regions share the same values (at least about environment). This contradicts self-determination. What happens about the community or region that wants to pollute or have nuclear power or a fascist dictatorship (see Chapter 6.4)? It was thanks to the much-vaunted principle of 'subsidiarity' in the European Community – allowing as many decisions as possible to be devolved downwards – that the British Government was able in 1993 to continue to destroy the ecologically valuable site of Twyford Down to build a motorway despite earlier EC attempts to curb this vandalism.

Lastly, there is the problem that even when it is possible to define regions based on allegedly 'natural' features those regions have limited social or economic meaning in a modern world. The quasi-government conservation organisation, English Nature (1993), recently divided England up into 76 'natural areas' as a framework for setting conservation objectives rather than administrative counties. These were based on geology, soils, topography and climate (little natural vegetation or wildlife being left). English Nature claimed (p. 3) that

> These 'natural areas' not only coincide with tracts of the countryside traditionally recognised by local communities . . . they also correspond to agricultural land use and settlement patterns.

This is quite erroneous. Land use is now substantially determined by the pattern of EC subsidies and grants in agriculture, while the settlement of Greater London covers at least five 'natural areas'. English Nature and bioregionalists here might recognise what the old regional geographers had to come to terms with in the first half of the twentieth century:

> The regionalists' error . . . was the notion of the uniqueness of location. But there was another dimension to the critique . . . the presumed intimacy of these ecological bonds was admirably suited to the historical geography of Europe before the industrial revolution . . . with the final disappearance of the old, local, rural, largely self sufficient way of life the centrality of regional work to geography has been permanently affected.
>
> (Gregory, in Johnston 1981, 287, citing other geographers)

Where there still is regional differentiation in industrial society it is based far less on watersheds than on economic function, and other criteria such as language and religion (see Alexander 1990). To suggest reidentifying and reinstating old cultural, let alone physical, regions appears grossly unrealistic. Indeed, the British Bioregional Development Group, in realising several permaculture-based sustainable development projects in the 'loose Surrey bioregion' has to admit of the paradox:

> Though bioregionalism is not based on explicit political or administrative boundaries, the county councils (with their existing countryside and woodland fora) may represent the most suitable vehicles for carrying forward its ideas.
>
> (Desai 1993, 8)

LETS and localism

LETS schemes constitute an expression of radical ecologism's localism and 'lifestyle politics par excellence' (Wilding 1991). Their basic features are shown in Tables 6.3 and 6.4 (see also Dauncey 1988). They are multilateral trading rather than straight bartering schemes, and although there are no visible notes or coins they constitute local currency in the sense of an information system recording transactions, debts ('commitments') and credits ('acknowledgements') (see Figures 6.1, 6.2, 6.3). By 1994 there

Table 6.3 Characteristics of LETS schemes

LETS is a local, multi-choice bartering system. It enables local people to give and receive all kinds of services from one to another, without the need to spend money.

The principles of LETS
1 People form a club to barter amongst themselves, using their own local system of accounts.
2 A members' directory of skills, goods and resources is drawn up and circulated.
3 Members trade with each other whenever they wish, writing out credit notes to acknowledge a service.
4 The notes are sent to a central office which records all transactions and sends out regular statements, directory updates and news.
5 The scheme is non-profit making with no interest charged to those 'in debit', or paid to those 'in credit'.
6 All members are entitled to know the balance and turnover of everyone else.

LETS offers
- Ways to save cash and survive on a tight budget
- Goods and services without the need to spend money
- Opportunities to borrow equipment in exchange for credits
- No worry about bad debts
- Interest free credit
- A local network of social contacts
- Recognition that your skills are valuable, whatever they are
- The chance to make creative use of all your talents
- Tuition and training opportunities in new skills

Features
Direct exchanges are not necessary: you can knit a sweater for one person, in order to have your shed repaired by another. There is no obligation to trade, but without trading the scheme becomes moribund. You can spend in deficit, the system depends on some members being in deficit. There are no physical tokens of currency, but recorded acknowledgements of transactions. Values of currency units depend on what is available in the LETS system. Most services are not taxable, but tax arrangements are possible with the Inland Revenue. Services and goods can be paid for with a mix of LETS units and conventional money.

Source: Letslink information package on LETS schemes 1993

were 200 UK systems with 10,000 members and as many in Australia, but only 10 each in the USA and Canada, where the contemporary idea started (local currencies actually have a long history, see Seyfang 1994).

Because they are not valid outside their area, and because there are no notes and coins, LETS schemes overcome many disadvantages associated with universal currencies. There can be no commoditisation of money itself, currency speculation or fluctuations in value engendered by specula-tion. And removing wealth from a locality by 'repatriating' profits, so characteristic of capitalism, becomes impossible. LETS therefore shields communities against external economic processes. And because LETS are

Table 6.4 Examples of goods and services traded in a LETS scheme

---

Hedge trimming, 4B/hour
Basic gardening 4B/hr
White water lilies 4B
Taxi service 4B/hr plus petrol
Car washing 4B/car
Light van with driver 4B/hr plus running costs
Huge adjustable spanner 2B/day
Car and van clutch replacement and servicing 8B/hr
Arc welding and fabrication 8B/hr
Creative dance tuition 10B/session
Photography family portraits 5B/hr plus cost of film
Indian devotional singing tuition 4B/hr
Childminding 1B/hr
Babysitting 4B/hr
Babysitting 5B/hr
Dry-stone walling 8B/hr
General household repairs 6B/hr
Architectural plans 6B/hr
Painting and decorating 8B/hr
Pair of wrought iron gates 15B
Computer tuition 8B/hr
Typing 4B plus £1
Ghost writing CVs 8B/hr
Potatoes 4B for 25 kilos
Windchimes 10B plus £10
Horse riding and lessons 6B/hr
Slave for half a day 4B/hr
Storytelling 8B/hr
Cafe space for meetings 5B/hr
Amstrad PC 20B plus £150

Source: *Lets Barter Directory*, February/March 1993, the LETS scheme for Malvern and district, UK

Note: Units are called 'Beacons'

closed, local systems, any environmental or social 'externalities' that come from producing goods and services cannot go unnoticed (Robertson 1989).

It is the social, community aspects of LETS schemes that make them particularly attractive to green economists. They are a way of networking in a local community, of developing community awareness, of finding out about local needs, of recompensing otherwise unpaid volunteer labour, of helping low-income people to gain self-esteem, self-reliance and access to goods and services otherwise beyond them, and of nurturing small businesses – all aims of green economics (Chapter 2.3). LETS is 'not an alternative money system but an alternative *to* the money system', says Stott (1993). If orthodox socialists should ask, why then bother with a LETS at all rather than a moneyless economy?

L E T S stands for Local Exchange Trading System. Each LETS is a nonprofit-making local scheme in which people can trade goods, services or other neighbourhood resources with one another without the need for money, interest payments, or credit. In addition, LETS allows you to do all kinds of things that would not otherwise be possible:

My first love is gardening but I'd like to learn about pottery as well.

I wish someone could sort out my garden then I'd have time to do that typing for Dave.

I can't type but I'm into cars.

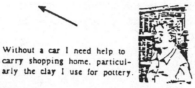

Without a car I need help to carry shopping home, particularly the clay I use for pottery.

I want someone to keep my car serviced then Sally could rely on me to take her with me when I go shopping.

LETS is more flexible than barter. You can receive any goods/ services offered on the LETS. You can repay whenever it suits you to do so, by providing goods/services to any member (or members) of the system.

Each member:
* makes a list of goods, services or skills they can offer, and
* a list of goods, services or skills they wish to request,
* puts a price, in local units, on their offers
* pays a small joining fee (to cover costs) of £............

Each member receives:
  * A Directory of all the members' offers and requests.
  * An account, starting at zero.
  * A LETS chequebook.

Starting with your account at zero, you can either:
* use a service, paying by cheque in local units from your account. Your account is debited, but no interest is ever charged.
* carry out a service, receive a cheque and send it to the LETS accountant, who credits your account for that amount.

A local unit is not a physical token but simply a measure of value, such as £1, 50p, or an hour's work. Each scheme has its own unit which is valued and named by the local group (eg Green Pounds, Links, Locals). The unit adopted by this scheme is the....................which is worth......................

To join the local scheme or find out more, contact:

Figure 6.1 LETS
Source: Letslink information pack 1993

312

Green Pounds reach the Parts Other Currencies Can't

Figure 6.2 A LETS advertisement
Source: Letslink information pack 1993

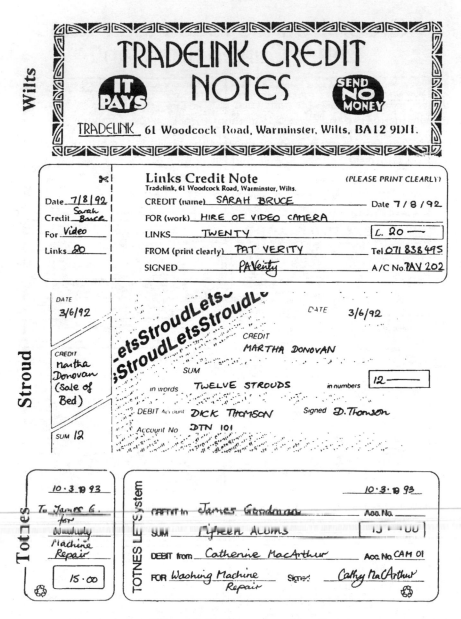

Figure 6.3 Credit note samples
Source: Letslink information pack 1993

Stott reminds us that

> Although it is not a capitalist system in that it does not involve the accumulation of wealth, there is no particular desire to undermine capitalist society.

At the same time people do not seem to behave in LETS as they do in capitalism, for example running up bad debts (by leaving the system in deficit, whereupon all remaining members would share the loss). Members usually justify the trust invested in them to behave considerately:

> The trust is greatly helped by the fact that a large proportion of the group, including myself, don't mind if somewhere along the line we give more than we get.
> (an Australian LETS coordinator cited in *The Independent*, 1.5.92, 16)

In fact LETS have many different virtues, appealing to a variety of ideologies. This is perhaps symptomatic of the ideological diversity (or incoherence) displayed by ecologism in general, thinks Seyfang. She has assessed LETS alongside the criteria for different theories of value (Table 2.1, p. 49), and finds elements of *all* mainstream theories present.

There is apparent the view of currency as the medium through which consumers' express their subjective preferences: a form of information enabling production to match demand. Goods and services are therefore created only when money (tokens) is to be exchanged, and although some members are in debt and others in credit, total 'money' supply matches demand. This is entirely consistent with the neo-liberal monetarist perspective. Hayek himself, the inspiration behind Milton Friedman, Reaganomics and Thatcherism in the 1980s, argued in his original writings for concurrent privatised currencies. Monetarism's overwhelming concern with the system's stability, says Seyfang, is in turn compatible with bioregionalism in ecologism.

Another common view of money, as a *store* of value, is absent in LETS, so that interest payments and charges are also absent. This, says Seyfang, is compatible with Keynesianism. So, too, is the wish to encourage as much economic activity as possible and the willingness to 'print' more 'money' (recording more transactions) in accordance with the increased activity. Furthermore LETS, like welfare liberal economics, wants to manage capitalism for the greater community good, including internalising environmental externalities.

And LETS also militates against the idea of currency having intrinsic value as a commodity, to be accumulated or exchanged for gain. LETS recognises only use value, and is halfway towards the moneyless economy. For these reasons, says Seyfang, there are compatibilities with Marxist theories of value, although the labour theory of value itself is not immediately recognisable in LETS. (Note how in Table 6.4 babysitting is offered for four *or* five Beacons per hour. Presumably the higher offer will be bought only if babysitters are in short supply: there is no suggestion that the labour of one babysitter is greater than that of the other.)

Despite the key position of LETS in any green strategy for social change, Seyfang concludes that the only theory of value which is not inherent in LETS is the green theory! However, it is fair to point out that LETS is highly compatible with the 'green corollaries' – those social features which stem logically, according to Goodin (1992), from the green value theory.

Permaculture, eco-cities and communes

Ecological ethics have been translated into design principles governing all production and living, through the science of *permaculture* (see Table 6.5), which is another manifestation of alternative technology. The term derives from 'permanent', meaning indefinitely sustainable, and 'culture', which was originally derived from 'agriculture' but now refers to all cultural activity. Permaculture is conscious design and maintenance of productive ecosystems to give them the diversity, stability and resilience of natural ecosystems. It is the harmonious integration of the landscape and people

Table 6.5 Principles of permaculture design

Based on the ethics of caring for the Earth (reducing resource consumption and pollution), caring for people (equal wealth distribution in small democratic communities) and giving away surpluses.

1 **Work with nature**: it should not be necessary to use fertilisers, pesticides or excessive energy, e.g. site buildings to minimise heat loss.
2 **Everything gardens**: all nature plays a part in working land, and this action should be used to a maximum, e.g. worms aerating soil and composting organic matter.
3 **Use minimum effort for maximum effect**: produce high yields with minimum maintenance, e.g. multiple cropping of perennial plants without ploughing.
4 **Unlimited yields**: it should be possible continuously to increase the yield of a permaculture system, e.g. by finding new cultivation methods.
5 **Outputs become inputs**: the products of one part of the system should provide materials for another part, e.g. human manure for methane generation and fertiliser.
6 **Each function should be supported by many elements**: there are back ups for all parts of the system, e.g. diversity of (local) sources for energy generation.
7 **Each element performs several functions**: e.g. a garden pond supplies irrigation, fish, wildlife habitat, reflects light into buildings, etc.
8 **Relative location**: each element in the system should be located in the most beneficial place for the whole system, e.g. a wildlife zone shaped as a corridor through cultivated areas benefits the whole system aesthetically, ecologically and in terms of pest reduction.

Source: Mollison 1988

providing their food, energy, shelter and other material and non-material needs in a sustainable way (Mollison 1988).

Permaculture involves arranging plants and animals in mutually beneficial arrays, to create systems that are self-fertilising, self-perpetuating, self-watering, self-mulching, self-pollinating, disease and pest resistant and require low energy inputs.

> The most fundamental of the permaculture design principles is the Principle of Self-Regulation . . . [this places] elements or components in such a way that each serves the needs (forms the inputs) and accepts the products (uses the outputs) of other elements.
>
> (Desai 1993, 7)

Thus charcoal manufacturing could be regenerated in Britain (cutting out imports) by heating the waste from local woods in retorts powered by methane gas from landfill sites.

The aim here is not to provide self-sufficiency (therefore insularity), but self-sustainability. Nor should permaculture require massive labour inputs. But, permaculturalists claim, there could be up to 25 times conventional agricultural yields through the system, so the planet could feed itself from 6 per cent of current agricultural land (Dixon 1991).

Permacultural design is practised by some ecocentric gardeners and farmers, although it is more theorised about than done, and some of its theories either are not properly tested or are unreliable. Harper (1992), for instance, has seen 'no system that shows unequivocally that permanent plantings outperform row crops', while his experience shows that small, rather than being beautiful, 'is often inefficient, resource-intensive and extremely expensive per unit of output'.

Permacultural principles also apply to city design, and permaculturalists have 'targeted' the local government Agenda 21 initiatives with this in mind to press for urban design working with nature. This includes, says Davey (1992), such practices as using sewage to produce methane and compost, and making parks, gardens and lawns productive as well as ornamental (grazing sheep on grass, growing trees with harvestable fruit and nuts). Domestic and commercial buildings could be re-plumbed to collect and use rainwater and to recycle 'grey' (bath) water for flushing lavatories. That way reservoir space could be reduced and prioritised for drinking water. And there should be insulation and energy conservation in building design: renewable energy sources being built in (photoelectric cell arrays on external surfaces for instance).

Green theorists often imagine that rural and urban alternative communities, or 'communes', constitute the best way of prefiguring ecological society. In them one could start almost with a blank sheet of paper, designing appropriate physical, social and economic (through producer cooperatives) structures in accordance with permacultural principles and practice. Leading theorists like Bookchin, Bahro, Sale, and Goldsmith have all made communes a cornerstone of their revolutionary strategies.

There are at least a hundred alternative communities in England and Wales (Coates *et al.* 1993). Many were founded during the first wave of popular ecological concern in the 1970s. There is evidence that they do foster ecological lifestyles. There is resource sharing and recycling; home (organic) food production and high consciousness of the politics of food (see Goodman and Redclift 1991). Often of necessity, communes de-emphasise consumerism, with consequent lowering of environmental impact. What is consumed is thought about: communards attempt ethical and environmentally sound lifestyles. Deep ecology conscious-ness often manifests itself in holistic medicine, self-awareness and self-realisation therapies, nature awareness and even anti-urbanism and anti-industrialism. Many communards attempt also to share work and incomes, to achieve participatory (perhaps consensus) democracy, to hear and tolerate most viewpoints and to be non-hierarchical, non-sexist and non-racist (Pepper 1991).

But this is far from alternative communities constituting a vanguard for ecotopia, a position, indeed, which most communards would not claim. For one thing, their motives for communal living are not, today, primarily ecological: failed marriages, loneliness and inability to afford rising house prices are often the reasons why people seek a logical way out and live together. For another, the 'counterculture' of which communes are a part is not uninfluenced by the mainstream culture it opposes. It is not an independent beacon shining forth with a steady light of unchanging revolutionary values. Its values change as mainstream values change. In the 1980s and 1990s privatisation, individualism, consumerism, mana-gerialism and the values of the market place, of commercial viability and of the nuclear family have all made inroads in alternative communities. Core members, some of whom 'dropped out' in the 1960s, are dropping back in to mainstream values. They are respectable, like their own space, may not share meals, dislike free riders, drugs or 'anything goes' anarchism, see the need to balance budgets by earning income from the outside world and like cars, televisions and compact discs (and who can blame them). 'Rather than being an attack on our culture of capitalism', says Cock (1985, 13):

> alternative lifestylers [in Australia] are generally affirming the rightist predominant concern with freedom and individualism, although not defining it so exclusively in materialistic terms . . . most of the secular experiments of the last thirteen years have failed to sustain their impetus towards commitment to community and the development of a deep ecologically based lifestyle . . . All revealed a gradual decline of community cohesion and purpose together with an erosion of environ-mental sensitivity.

Cock identifies mainstream Australian society as the problem:

> We have been socialised to live essentially private lives within imper-
> sonal worlds. We are devoid of the experience of intimate sharing
> beyond family . . . Attempts to overcome the power of this social-
> isation have been severely limited by an understandable ambivalence to
> hang on to what we know and the security it brings.

This suggests severe limitations with the prefiguring strategy, even as
practised by the more educated, motivated and aware groups in society
(who form the majority of communards). Revolutionary social change,
whether it be gradual or sudden, peaceable or violent, needs to be accomplished in the mainstream if it is to be sustained: anarchistic-socialist
experiments from the Diggers onwards have been stifled because of the
hostile broader context in which they occurred.

Social ecology and confederal municipalism

Green radicalism needs to become a mass movement for social change of
itself, or an integral part of an associated mass movement. Since the former
is unlikely, this means looking either to the new social movements or to the
established labour movement as agents for creating ecotopia. Deep and
social ecologists alike continue to eschew labour movements: their anarcho-
commun(al)ist ecotopia has more in common with *populism*, a form of
politics which appeals to people to exercise direct pressure on governments and

> emphasises the virtues of the uncorrupt and unsophisticated common
> people against the double-dealing and selfishness to be expected of
> professional politicians . . . It can therefore manifest itself in left, right
> or centrist forms.
>
> (Bullock and Stallybrass 1988, 668)

Populism originated in Russia, among intellectuals who thought that
socialism could be established directly on the basis of existing peasant
communes. It also has a strong tradition in America, where it can be rural
or urban, peasant or middle class. It is well-matched, then, to the ideological ambiguity (in traditional terms) of ecologism. Murray Bookchin and
others have attempted to fuse it coherently with Kropotkin's (1892, 1899)
anarcho-communism, into what they call 'confederal municipalism'.

This tries to chart a course between various 'evils'. It rejects the idealism,
irrationalism, nature mystification and future primitivism of deep ecology
and bioregionalism (Bookchin 1987). It rejects, too, the monopolisation of
power by labour, either through state 'socialism' or anarcho-syndicalism. At
the same time it abhors capitalism. Its kind of decentralisation is not to be
conflated with the decentralisation of industry and commerce in

contemporary capitalism. That is really *dependent decentralisation*, since economic power stays firmly concentrated in core areas. Confederal municipalism also rejects green economics' appeals to a localised state, or, alternatively, total autarky (economic self-sufficiency). 'Decentralisation, localism, self-sufficiency and even confederation,' says Bookchin (1992a), 'each taken singly do not constitute a guarantee that we will achieve a rational ecological society'. It does not follow that small is *necessarily* beautiful or democratic, he reminds us: small communalist villages and towns formed the basis for despotic regimes in India and China (one might also think of the potential repressiveness of Goldsmith's (1977) 'de-industrialised' small-scale ecological society).

Instead, confederal municipalism represents a synthesis of the autonomous, democratic, libertarian socialist and radical green strategies. It tries to reconcile the need for planning with that for true democracy, also accepting that there cannot be limitless abundance of material goods.

Fotopoulos (1992) pictures confederal municipalism as a stateless, moneyless market economy, ultimately ruled by Athenian (face-to-face democracy in town meetings) community assemblies. Each village or town's assembly confederates with others in administrative regional and supra-regional councils, attended by rotating delegates *mandated* to vote in particular ways by their members (it is not a 'democracy' as found in a parliamentary or congressional systems, where elected 'representatives' vote as they think fit). The councils would lease productive resources from the

Table 6.6 Key elements of confederal municipalism

- Not equatable with a state, or possible in presence of a state
- Instead, a network of administrative councils
- Their delegates/members elected from popular face-to-face democratic assemblies in villages, towns and city neighbourhoods
- They are strictly mandated by assemblies that chose them
- Are recallable
- They are purely administrative, not policy-making 'representatives'
- Policy making is the right of the popular assemblies
- Confederal councils link up towns villages and cities
- Power flows from the bottom up instead of top down
- Communities are interdependent, not self sufficient
- They share resources and products and policy making
- Local community manages its own economic resources within the interlinked network of communities
- Produce is distributed among communities according to need
- Special interests of work, workplace, status and property relations are transcended
- Moral education and character building for active citizenship
- Hence *democratic interdependence without surrendering local control*: a 'dialectical development of independence and dependence'

Source: Bookchin 1992a

320

whole community to producers functioning, for instance, as producer coops. Production would be for basic needs. Bookchin proposes similar features (Table 6.6). Fotopoulos further suggests a system of work vouchers (really a form of money based on the labour theory of value). These would be earned in different amounts by different forms of work (more per hour for the less fulfilling jobs), and would be tradeable for the fruits of production. This is not a new idea, having been postulated by Skinner (1948) and tried in the American 'Walden II' community in the 1970s.

The basic principles of confederal municipalism are community self-reliance and ownership of productive resources, and confederal allocation of what is produced. *Democracy* is the key, then, to this version of an ecological society. Citizens' assemblies ensure direct participation and equal distribution of political power. There must, too, be equality of *economic* power, through every community member's full involvement in economic decisions. Some of the components of this economic democracy are shown in Table 6.7, where Fotopoulos clearly advances a money economy: indeed all these components also feature in 'mainstream' green, capitalistic, economics (see Ekins 1992a).

It is significant that Fotopoulos describes the *components* of confederal municipalism as a *strategy* for accomplishing it. Here, again, is the anarchist pre-figuring idea: to achieve the given society you start building its components here and now. Bookchin (1992b, 97) is aware of an obvious problem about this. 'I don't expect any national government willingly to "grant power" to a confederal municipalist movement without resistance of one sort or another', he says, however declining to suggest what to do about it: 'the relation between [the nation state and this movement] is a matter for the future and for another generation to decide'. But he does make it clear that like some eco-socialists (and unlike some others, see Pepper 1993) he

Table 6.7 Economic features of confederal municipalism

1 Decentralised economic power, via local currencies, community banks, coops and credit unions.
2 These to be networked where appropriate, for instance federations of the banks belonging to each municipality for the purpose of absorbing savings, financing investment in modern production and offering other financial services.
3 Trade for mutual self-reliance, instead of 'free' trade.
4 Community participation in production decisions via collective self-managed enterprises producing according to patterns decided by *all* members of the community.
5 All decisions about production and development to be based on alternative economic indicators to those used at present.
6 Community ownership of productive resources, via such means as community land-buying trusts.
7 Worker self-management to be genuine, not by employees owning stocks.

Source: Fotopoulos 1992

does not consider it possible to democratise the state by transferring power to a local state that operates by consulting citizen and neighbourhood groups (O'Connor 1992). By definition, the state is considered hierarchical and oppressive.

Neither do confederal municipalists want to centre on labour as the locus of grassroots power, as eco-socialists might. The working class cannot be given this position, says Fotopoulos, because in the 'post-industrial' era that class is fast disappearing (this is Gorz's (1982) thesis, which has been influential amongst greens). And in any case producers' vested interests might not coincide with those of the wider community. This debate between worker-oriented communism, or anarchist-syndicalism (see Devine 1988), and community-oriented anarchist-communism mirrors old splits in left anarchist movements, as Hawkins (1992) reminds us. It is reflected in some vituperative exchanges between the social ecologists Bookchin (1993) and Purchase (1993) today.

It is the new social movements rather than the labour movement that the confederal municipalist social ecologists regard as the principal agents for social change to an ecological society. But the evidence of these movements' potential effectiveness in this role is at least ambiguous, even contradictory. Fuentes and Frank (1992) characterise them as middle class in the West, working class in the South and containing both in the East. They have in common a deep sense of social justice, opposition to the state and disillusion with conventional economics. While they can be agents for social transformation, we should not overestimate their power to resist the effects, adverse and otherwise, of global modernisation. Indeed, Fuentes and Frank suggest that while they take up a challenging pose to the system, 'very few social movements are antisystemic in their attempt, and still less their success, to destroy the system': they can end up being coopted by it. So, too, can the labour movement. But an eco-socialist would point out that whatever accommodation it might reach with labour, inherently capitalism works against that movement, which is not necessarily true of new social movements.

6.4 ECOTOPIA OR DYSTOPIA? A PERSONAL VIEW

The strategies and social experiments described in Chapter 6.3 span many differences between factions within ecologism. Confederal municipalism may even be the basis for a rapprochement between deep and social ecology (Bookchin and Foreman 1991), as well as between the latter and eco-socialism. LETS schemes also build bridges – between ecologism and reformist technocentrism. Purists from deep or social/socialist ecology may complain about ideological incoherence in these strategies but others make a virtue of it.

Ferris (1993), for instance, argues that pragmatic social experiments which are part of a fragmented ideology might represent the most realistic way forward: far more than espousing dogma. Green prescriptions may feature no distinctive social policy, being a scattered set of issues and demands, but they do at least unambiguously demand 'ecological rationality'. This, after Drysek (1987), would order all society by the priorities of low entropy and the nature-knows-best principle. Ferris (p. 149) recognises two strands in ecologism: this ecological rationalism and also the demand for convivial, human-scale communities. He desires a fruitful 'postmodernist' synthesis of these green and social democratic ideals in a way that avoids the old 'Achilles heel of modernism'.

That Achilles' heel was the tendency, visible in all fundamentalist doctrines – Marxism and neo-liberalism alike, Ferris argues – to promise *liberation* from social problems but paradoxically to deliver *authoritarianism* and further social problems. Ferris here repeats the complaint which liberals commonly make about all fundamentalism and utopianism (Goodwin and Taylor 1982): dogmatic belief that you have found the one, true, way to the good society inevitably leads to intolerance and suppression of those who espouse other views.

Liberalism has a point here, and one which all fundamentalist green-socialists, -anarchists and -feminists must grapple with. Liberalism believes that there is a variety of possible 'goods', which are not all reducible into one 'good society'. This, it maintains, is because individual citizens all differ in their conception of the good life and of what constitutes value – therefore no one set of values is objectively better than another. If all individuals are to be equal then this pluralism must be maintained: the state particularly should not try to foist on everyone some common view. And ways must be found to avoid a majority 'tyrannising', as Orwell put it, the minority.

Many greens argue this liberal view in the social sphere. Ecotopia, they say, should be a culturally diverse place where each small community lives out its own values. But herein lies their dilemma, shared by all fundamentalists who also want to be democratic. What if a community felt that it did not want to be green? What if people decided, in their face-to-face democratic town meetings, that they wanted to release plumes of sulphurous smoke into the atmosphere? Or that they wanted to live in an inegalitarian patriarchy? Ecocentrics not only want true democracy and cultural diversity; they also insist on the primacy of ecological laws and the imperative of 'ecological rationalism' as defined above. 'Externalising' wastes by definition affects not just the waste producers, so those producers cannot be allowed to decide completely for themselves what to do: they cannot be allowed to reach, democratically in their small communities, the 'wrong' decision.

The classic liberal 'get out', espoused by socialists too, lies in the very term *ecological rationalism*. The argument runs that if only society were truly democratic, and if only its members were all educated to be rational, then ecological ills and social injustice simply would not happen, because these evils are fundamentally irrational: as 'common sense' would show – if only people were commonly sensible.

The irrationality of social injustice and environmental degradation stems from the fact that they do not produce the most good or satisfaction for most people. As an observational and inductive fact, ecological ills and social injustice produce less-than-optimal aggregate good, so it is irrational to support them. (Porritt's (1984) utilitarian arguments for enlightened self-interest as the solution to ecological ills constitute a form of this position.)

It follows that if rational social, political and natural science were allowed full rein, then eventually there would be one, universal, rational view of what is best. People could be free to believe what they liked, but, being rational, all would be likely to believe in the same, green, principles. Many left-liberal new social movements take this view. Their efforts often become populist campaigns for more open and democratic government and society – on the assumption that a 'true' educated democracy where the majority's 'common sense' prevailed would be green/feminist/peaceful, etc. But this analysis is often limited because it sidesteps the realities of political and economic power, refusing to acknowledge the possibility that an ecological and socially just society is also about *curtailing* some liberal 'freedoms' (free markets and trade, freedom to own privately the means of production, freedom to externalise costs) as much as ensuring other freedoms.

If it is conceded that the 'required' ecological consensus among ecotopia's inhabitants might not be ensured merely by establishing an Athenian democracy where all are educated and rational, then the argument may take a different turn. It is asserted that a basic unselfishness and communalism which is in human nature would come to the fore in ecotopia's unalienated society, as it has never been allowed to do in industrial capitalism – and this is why people would in practice all think and behave in ecologically appropriate ways (Chapter 5.6).

But these arguments, too, may seem simply hard to believe. (Perhaps this is because we look at the arguments about human nature being communal from the perspective of a very non-communal society.)

O'Neill (1993a) attempts to resolve this dilemma of wanting freedom at the same time as ecologically and socially 'correct' thinking and behaviour. He believes that we can have a common vision of what a 'good life' should be, reached after debate and consensus, and that this good life can allow for diverse ways of living. This is because anyone's happiness can be complete only in relation to the happiness of others: we cannot be self-sufficient and always have exactly our own way and at the same time be fulfilled individuals. Hence even though we think our values are 'correct', we will always want to debate and accommodate and live with other, less meritorious values.

And if this eventually leads towards everyone believing in the same things, this is not necessarily bad, in the sense of being inherently totalitarian. After all, a plurality of beliefs or no one believing anything: neither particularly have intrinsic value. Indeed the very object of political debate, which is healthy, is to persuade people towards a shared set of values. O'Neill reminds us that the arch-liberal J. S. Mill himself had a particular conception of the good society: it was that which realised most fully our distinctive human capacities. This, rather than difficult arguments about 'intrinsic value' and bioethics, may represent the best reason for seeking diverse and robust ecosystems. At the same time it may be a good argument for not condoning cultural diversity where that means condoning oppressive cultures (and this includes the culture of capitalism).

O'Neill also thinks that cultures where all believe in similar things in an unthinking way – through blind religious faith, or mere tradition, or through imposition and repression – are not desirable, because they are undeveloped. Developed cultures do not necessarily contain many different perspectives at odds with each other, but if they have converged on a common 'truth' it must have been as a result of much argument and democratic discourse.

This is an attractive argument for liberal/socialistic intellectuals who want both freedom and democracy and specific values and behaviour to characterise their ideal society. It contains all sorts of value positions, of course, for instance that 'development' is good, and that thinking and rational debate, in order to reach a consensus, are good too, and essential to development. Indeed it could be seen as a form of special pleading by intellectuals, comparable to that of other groups, such as industrial and commercial entrepreneurs, who always want us to believe that *their* particular world view and behaviour patterns constitute 'progress' for all. You read many academic books which suggest that more education – thinking, debate, rationality, logic, knowledge, erudition, information and research – are needed to solve environmental or other 'crises'. They inescapably conclude that more academics with more resources must be a good thing: a key to ecotopia.

Indeed, I tend towards this position myself, but I consider it a far from sufficient condition for ecological enlightenment. There must also be ways of ensuring clear ecological and socialist values, action and behaviour which most people share and subscribe to. We must accept that there *are* dangers here of incipient authoritarianism, and try to be better able than in the past to combat these dangers. My hope is that by making humanist, egalitarian and socialist aspirations a prerequisite for an ecological society, rather than something that is supposed automatically to follow, we can avoid making ecological society a repressive dystopia.

That ecotopia does have a dystopian potential is underlined by Michael Carson's amusing short story read on BBC Radio, entitled 'The Punishment of Luxury'. Is Carson's tale a serious warning about the kind of society we may soon have to create? Perhaps not: maybe it is just a reminder that ecocentrics, like all political radicals, should never lose their sense of their own potential ridiculousness. If readers at least heed this reminder, then this book will not have been written totally in vain.

The Dark Green government demanded that extensive coverage be given to the first execution of a citizen convicted under the new transportation act. Broadcasting equipment was brought to Trafalgar Square by bicycle, tricycle and solar-powered scooter, and set up around Vehicle Compacter Number One, to the south of Sperm Whale column.

A crowd gathered from early morning. Around eight, six Dark Green Ecological Enforcers pushed a black Jaguar into the square. Its owner, the condemned man, Dr Robert Stone of Cattawade, Essex, had been arrested, convicted and was now about to be executed for being found in possession of an automobile. This crime would have been sufficient to assure that the doctor spent the rest of his life in uncomfortably natural surroundings. But what had brought down the full rigours of the new law upon his head was that he'd actually been caught *driving* his car . . . Dr Stone had been arrested, chained and brought to the Old Bailey in a police rickshaw.

The Jaguar was placed next to the compactor. The doctor sat in it stoically as the ecological enforcers pushed the vehicle into the machine. The public executioner pushed the switch. The compactor rumbled into life. The peace of rush-hour London was shattered. Goat-hands had trouble controlling their herds on the Whitehall allotments. Shire ponies shied. Thousands of pigeons, unused to a mechanical sound, took to the air in panic, wheeling and flapping. Even the ripening wheat in Kensington Gardens seemed to tremble. The doomed doctor looked up and caught sight of the steeple of Saint Martin's with birds soaring all around it. Then he saw the passenger door to his left coming towards him. The roof of the car crumpled, approaching and retreating.

Five minutes later, the compacter opened its jaws to reveal a solid black and grey cube. This was man-handled to a corner of Trafalgar Square and placed on a plinth directly opposite the National Gallery of Batik. A sign was placed below the compacted Jaguar-and-Doctor Stone, which read:

THE PUNISHMENT OF LUXURY

. . . And the Good Earth heaved a sigh of satisfaction on seeing Humankind put firmly – at long last – in its place.

GLOSSARY OF SOME TERMS
USED IN THIS BOOK

(NB The definitions are not necessarily complete, but are intended to be relevant to how the terms are used in this book.)

Most commonly used sources include:

B = Button, J. (1988)
B and B = Button, J. and Bloom, W. (1992)
B and S = Bullock, A. and Stallybrass, O. (1988)
L = Lacey, A. (1986)
OED = Oxford English Dictionary
R = Russell, B. (1946)

Alienation: a complex concept, used in many different contexts to mean separation and estrangement. In ecocentrism it implies separation of society from nature so that the former no longer lives according to 'natural' principles. By extension alienation includes the supposed loss of 'original' human nature and characteristics in over-sophisticated, artificial Western societies. In Marxism, particularly, alienation denotes separation from aspects of *oneself*. In capitalist industrial organisation, for instance, automation, the production line and division of labour separate workers from their own creativity. In capitalism the process of producing for an anonymous market separates producers from consumers, so one part of society is unable to identify with another part. In capitalist exchange the reduction of everything to money values hides the social relations behind the production of commodities (we might see an electronic watch as an object worth so much money, rather than as a produce of sweatshop assembly line labour), so again society is separated from knowing and affecting aspects of itself. And alienation from nature, in Marxism, implies separation from the true understanding that nature is a product of interaction with human society. The two are not and cannot be separate.

Analysis: opposite of **synthesis**. Resolving something into its simpler elements. Finding out how something is constituted by breaking it down into its parts.

Anarchism: a political movement, largely on the left, which opposes all government except self-government, and the state. Instead it advocates non-hierarchical self-organisation, often in small collectives, communes or neighbourhoods, where all can participate in consensus decisions. The forms of democracy particularly favoured are bottom-up, and direct rather than representative. Any 'rules' and agreements represent the genuine collective wish of the people, and last only as long as people want them to. Anarchy would replace the state by the free association and voluntary cooperation of individuals and groups. Anarchism is for freedom of the individual to live 'naturally', and natural societies are regarded as those where individuals are most fulfilled by association with others. The different forms of anarchism partly revolve around how this association is to be achieved. In anarcho-syndicalism, for instance, the focus of association is the workplace and the trade union; in anarcho-communism it is the commune where people live. Anarchism's small-scale rural communes, small towns and regions can federate for organisational purposes above and beyond the local.

Animism: attribution of living souls to plants, inanimate objects and natural phenomena (OED). A pre-modern belief, resurrected in New Ageism and some deep ecology.

Anthropocentrism: a world view placing humans at the centre of all creation – one which is 'taken for granted by most Westerners' (B). It sees humans as the source of all value (i.e. it is they who bestow value on other parts of nature), since the concept of value itself is a human creation. Hence anthropocentrism opposes **ecocentrism** and **bioethics**.

Arcadia: a rural-agricultural mountain district of the Peloponnese 'which townsmen imagined to be idyllic but which really was full of ancient barbaric horrors' (R). In proverb and myth it was a place of pastoral simplicity and harmony between people and nature – an ideal middle landscape between town and wilderness which held none of the fears or disadvantages of either.

Bioethic: an ethical principle which holds that the biosphere has **intrinsic value**. It therefore has a right to existence for itself, regardless of its usefulness or otherwise to humans.

Bourgeoisie: in the Marxist sense, those who own and control the means of production (including land and resources), distribution and exchange, in opposition to the **proletariat**.

Cartesianism/Cartesian dualism: see **Objectivity**

Catastrophism: the eighteenth/nineteenth century theory (later replaced by uniformitarianism) that sudden isolated upheavals and events brought to an end each geological/stratigraphical era, killing off all creatures where the catastrophe occurred, after which new and different living forms arose in the stricken area. Noah's flood was thought to be one such catastrophe.

Cosmography: the mapping and description of the features of the cosmos.

Cosmology: the system of ideas, or world view, by which the universe is ordered and understood.

Creationism: accepts literally the account of the creation of the Earth given in Genesis. One seventeenth-century version was based on an estimate by Bishop

Ussher that the Earth and all in it were created in 4004 BC. Creationism also holds that all living species have separate origins and were created simultaneously, rather than having evolved from common roots according to Darwinism.

Deduction: a form of reasoning in which conclusions about what will be found in the real world are drawn from premises about how the world works. It is inferring something particular from general principles. If the premises are correct then the conclusion must also be; if incorrect then the conclusion may be incorrect. If observations of the real world contradict the premises, then the latter need to be revised. See **Induction**.

Determinism: the view that every event has a cause. Some forms of determinism imply that the final link in the causal chain determining human actions and behaviour, and the fate of humankind, is beyond human control. So human action and human destiny are not the outcome of human free will. There are outside forces, such as the laws of God or nature, or economics or history, which limit what we can and cannot do and achieve. As well as economic, historical, environmental, etc. determinism, there is biological determinism which holds that the final cause of human actions lies in the genes. An extreme determinism would be fatalistic, arguing that all events can be predicted if only we know fully the laws and processes that cause them. Nothing is left to chance or free will, so we might as well accept whatever happens knowing we could not have changed anything.

Dialectic/dialectical relationships: these terms have many shades of meaning. These include the idea of dialectical 'opposites' such as society and nature, which cannot be defined except in terms of each other. Hence a dialectical relationship between society and nature is very intimate and organic – neither makes sense as a separate entity. And changes in one constantly cause changes in the other, which react back to cause further change continuously. This circular, interpenetrating, mutually-affecting relationship is, then, much more complex than the deterministic relationships of classical science. Dialectics also constitute a theory about the nature of logic and reasoning, and about the nature of the world, history and social change; different versions were developed by, among others, Hegel and Marx.

Diluvial theory: held that the superficial deposits on top of solid rocks (drift), except for the alluvium deposited by rivers, were 'diluvium' – sediments deposited in the waters of Noah's flood. Later, many of them were recognised as glacial in origin.

Dualism: see **Objectivity**

Ecocentrism: a 'mode of thought' (O'Riordan 1981) which regards humans as subject to ecological and systems laws. Essentially it is not human-centred (anthropocentric), but centred on the natural ecosystems, of which humans are reckoned to be just another component. There is a strong sense of respect for nature in its own right (bioethic) as well as for pragmatic reasons. Ecocentrics lack faith in modern large-scale technology and society, and the technical, bureaucratic, economic and political élites.

Ecologism: the political philosophy of radical environmentalism, including ecocentrism (Dobson 1990).

Elitism: a model of decision making in society which is distorted away from **pluralism**, because the interests of a particular élite are disproportionately represented and influential. The resulting decisions are therefore undemocratic, and may leave some groups completely unsatisfied.

Empiricism: the theory that all valid knowledge stems from experience. Empirical knowledge comes solely from observation rather than theory or hypothesis. It derives from experience and experiment; from, that is, the evidence of the senses. Or it is knowledge which can be derived from such evidence by methods of inductive logic, including mathematics.

Enlightenment: a 'European and North American movement flourishing in the eighteenth century which stressed tolerance, reasonableness, common sense and the encouragement of science and technology' (B and S). Technological optimism underlay the central belief that by understanding (through observation and experiment and rationality) and applying nature's laws the material position of all humankind could be indefinitely improved, as part of progress and evolution.

Entropy: a measure of the degree of randomness in a closed system; the degradation or disorder of the universe. 'The second law of thermodynamics states that the entropy of a closed system must always increase with time: thus hot drinks cool to room temperature' (B).

Essentialism: a number of meanings, but in the context of this book, the doctrine that at least some objects have essential characteristics, which serve to identify them and to explain their properties (B and S). These characteristics are universal, manifesting themselves throughout space and time. They form the inherent nature of things, which reappears through history (e.g. what is in human nature, men's nature, women's nature. Some say that relationships of hierarchy, domination or patriarchy are an essential aspect of human society).

Existentialism: a philosophy which holds that reality is created by the free acts of humans: there are no 'external', uncontrollable determining factors, such as laws of nature or society which constrain us if we set out to achieve what we want and behave as we want – except the unalterable fact that we are born and one day will die. If we imagine there are such constraints, we are living an inauthentic existence. Really, we are free to create our own world as we want, which means that we are also responsible for the world we have created, whether it is good or bad. Existentialism holds that 'individual existence must be the starting point of any belief system: the universe itself has no fixed or pre-ordained meaning . . . the emphasis which existentialism puts on individual responsibility and experience fits well with green thinking, but the [associated] idea that only human intervention can give true meaning to nature is anathema to green thinkers, especially to deep ecologists' (B).

Exponential growth: growth by a constant fraction of the growing quantity during a constant time period, e.g. x per cent per year. The series 2, 4, 8, 16, 32, 64, 128 is exponential (constant increase by 100 per cent). It illustrates that an exponential growth curve is gradual over much of its length, but then gets very steep very quickly.

Gaia: Greek name for the Earth Goddess. James Lovelock's 'Gaia' theory proposes the earth as a complex system which is 'alive', in the sense of being

330

'autopoietic', that is, continually able to reconstitute and repair itself, via a series of complicated feedback mechanisms which respond to environmental changes.

GATT: the General Agreement on Tariffs and Trade, which results from international negotiations to reduce tariff barriers and liberalise world trade. These negotiations have taken place in a series of 'rounds' since World War II. The eighth, 'Uruguay' round took place in the early 1990s, and it brought Third World countries much more fully into the global market. Many environmentalists think that this is harmful to the interests of the majorities in these countries and to their environments.

Gemeinschaft: term used by sociologist Ferdinand Tönnies in 1887 to denote an ideal type of community, based on social relationships of solidarity, where the community means more than just the sum of the individuals in it, and there are unalienated organic face-to-face relationships. Religion, hierarchy, status inequality may be binding forces in the organic society. Medieval society was this kind of organic totality. See ***Gesellschaft***.

Gesellschaft: term used by sociologist Ferdinand Tönnies in 1887 to denote an ideal type of society based on contractual relationships between people as isolated individuals and operating according to self-interest. The society is merely the sum of the individuals in it, and the relationships are atomistic. All individuals have equal rights. See ***Gemeinschaft***.

Holism: the view that wholes are more than just the sums of their parts, and that it is not possible to define wholes merely as a collection of their basic constituents. Holistic medicine, for example, would look at the whole body and mind to explain an illness or malfunction of any part. Reductionist medicine might just look at the part and seek to repair that.

Idealism: philosophical theory that there are no material things that exist independently of minds (e.g. the minds of God or humans). Consequently idealist interpretations of history see events and social change as resulting especially from the introduction of new ideas and/or the development of old ideas and values. Change in the modern period might be seen as a result of the growth of reason. Idealist greens might attribute environmental 'crisis' to sets of wrong attitudes and values (in society and in individuals), thus emphasising the importance of education and value changes in creating an ecological society. (Contrasted with **materialism**.)

Ideology: broadly speaking, a world view, or set of ideas which renders the world more comprehensible. There are within the set some assumptions which are taken for granted ('common sense'), and never questioned. Many of the ideas of **social Darwinism**, for instance, form part of capitalist ideology. More narrowly defined, the term denotes a set of ideas, beliefs and ideals which are at the basis of an economic or political theory or system. If the assumptions are brought to the surface they are generally seen to reflect the material vested interests of those who share the ideology (for instance the assumption that competition is natural often underpins the ideology of those who do well in economic competition). Ideologies are presented as statements of universal truth, but reflect these narrower interests. As such, an ideology supports an already-reached position, which will contain bias and prejudice.

Induction: drawing out from a mass of empirical evidence general principles which explain the evidence. Inferring a law governing how the world works from observations of particular instances.

Intrinsic value: value which is inherent within something, and does not depend on the ideas, preferences or prejudices of an external valuer. Hence it is objectively present – it is essential. The bioethical idea that nature has intrinsic value over and above any value bestowed by humans means that if humans were wiped from the face of the earth there would still be value and purpose in the continuation of the rest of life and of the earth.

Invisible hand: the notion, usually attributed to Adam Smith, that if each individual in society seeks to maximise their own economic gain, this will be to the maximum benefit of society as a whole. It will be as if the activities of individuals in aggregate were guided by an invisible hand towards this wider social benefit, which is unintended by the individuals.

Keynesianism: describing the theories of the economist John Maynard Keynes. In particular, Keynesianism has been seen as an alternative strategy to monetarism (where government regulates the supply of money but otherwise leaves the economy largely free) in the latter part of the twentieth century. Keynesianism recognises that unemployment stifles some of the latent demand in the economy, repressing that very growth in production which would mop up unemployment. In order to break out of the vicious circle of unemployment–lower demand–more unemployment, government should intervene to increase aggregate demand, by creating jobs through 'public works' (e.g. road building). This in turn will eventually increase demand for the products of private industry.

Market-based incentive: a way of modifying free markets by centrally deciding (by government) what is to be the value of environmental services. These values are then incorporated into the prices of goods and services traded on the open market. A pollution charge or tax is an MBI if built into prices (Pearce *et al.* 1989).

Materialism: philosophy which holds that everything which exists is material, occupying some space at some time. Thus it denies substantial existence to minds as abstract entities (rather than as an aspect of the material organism of the brain). In explaining history and social change, materialist analyses (for instance Marxism) regard development as a function principally of material factors – such as the way we organise our productive activity to gain material subsistence (economic mode of production). Materialism holds that the predominant ideas and values in society do not derive from abstract thinking and reasoning, but that they are related to material events and ways of organising society. Hence any change to an ecological society which requires radically different values must be accompanied by radical social-economic-political change. (Contrasted with **idealism**.)

Millenarianism: believing in a future period of ideal happiness on earth, especially relating to the prophesy of Christ's thousand-year reign in person on earth.

Modernism: 'an international tendency', arising in the arts and architecture of the West in the late nineteenth and twentieth centuries (B and S). The term has been extended to mean, in opposition to **postmodernism** the approaches to knowledge, the scientific, technological and industrial developments and the hopes and aspirations of the period from the eighteenth to the twentieth century. Underlying all these was the 'Enlightenment project' – the central beliefs of the **Enlightenment**, which held that it was possible to discern general underlying principles governing nature and society, it was possible to use and manipulate these principles, and that by so doing we could improve the material lot of all humankind. Modernism also holds that there are absolute values which *should* underpin all modern societies. What these are may be debated (e.g. social justice, sanctity of human life, respect for the individual), but there is a danger that anyone believing they have discovered the 'right' values and principles may use authoritarian means to ensure they are applied universally. The modern period has been characterised by constant change, innovation and crises, and whether it has resulted in the universal progress which had been thought possible is doubted by 'postmodernists'.

Monism: 'any view which claims that where there appear to be many things or kinds of things there is really only one or only one kind' (L). Different monistic theories hold all the things in the universe as different aspects of the same material 'stuff', or alternatively of the same mental idea.

Monoglacial theory: held that all the glacial drift deposits on solid rocks originated from only one ice sheet belonging to one glacial episode in the Quaternary, rather than, as now accepted, from several episodes separated by warmer periods.

New Age: a wide range of concepts, beliefs and groups, united by the common conviction that the world is about to enter a new era involving a radical shift (a 'turning point' as Capra calls it) in human consciousness and relationships. The notion of a New Age can be traced at least as far back as the eighteenth century (B), and much further back in its relationship to **millenarianism**. New Age 'culture' was resurrected in the 1960s and again in the environmental movement, where it has many affinities with deep ecology. It emphasises spiritualism, mysticism and new technology, and has a highly **idealistic** view of social change.

Objectivity: the property of being separate from whatever is being studied, so that the subject who is studying the object does not influence in any way the properties of the object. Also the property of being separate from the interests, perspectives and points of view of any *particular social group* or from any *preconceived notion*, as distinct from being unbiased and value free – identifying with no interests and perspectives or with those which can be agreed between everyone. Descartes proposed that subject and object, mind and matter, humans and the rest of nature can indeed be separated. 'Objective' or 'primary' qualities, such as position, size, shape and motion, will theoretically appear the same to everyone, so they can universally be agreed on, whereas 'subjective', 'secondary' qualities, like colour, smell, goodness, will vary according to the disposition of the observer (the subject) because they are projections of the human mind, hence they will be slightly different for each individual. Not everyone is persuaded that this **Cartesian dualism** represents reality. Some dispute the very idea of conceiving things in terms of separate, polar opposites – of **dualisms** - such as hot and cold, good and bad, society and nature. **Monists** would argue that such 'opposites' are really but different manifestations of the same thing.

Paganism: a religion which revives the pre-modern, pre-Christian worship of nature, holding the earth (the Great Mother), its cycles and seasons as sacred, regarding plants and trees as kinsfolk and living in peace and harmony with the earth (B). (Incorrectly defined by OED as 'unenlightened or irreligious person'). Some pagans also describe themselves as witches or Wiccan: they practise magic for healing (B and B).

Pantheism: the belief that God is everything and everything is part of God – a view to which 'almost all mystics are attracted' (R).

Paradigm: when related to a scientific or academic discipline, the word means a set of rules, assumptions or procedures under which people study and investigate the subject matter of the discipline. The paradigm for the discipline is what most of its practitioners will agree are the proper subjects for the discipline to study and questions for it to ask.

Phenomenology: a method of enquiry developed by Edmund Husserl, which focuses on the nature of phenomena as experienced through human consciousness and perception. It begins with 'exact, attentive inspection of one's mental, especially intellectual processes in which all assumptions about the causation, consequences and wider significance of the mental processes under inspection are eliminated' (B and S). Because we can never know the world around us objectively, only through the medium of our own consciousnesses (individual and group) it focuses attention on *how we perceive and structure* that world. Since each of us experiences the world differently, it is impossible to make law-like generalisations about it, as classical science attempts to do. Rather than so-called 'objective facts', then, we attempt to grasp intuitively the inner experiences of people and what is significant to them about the world immediately around them: their 'lifeworld'. Having done so, we can then describe that lifeworld and our own.

Plenitude: fullness, completeness, plenty. The 'principle of plenitude' originates with Plato's *Timaeus*. It observes that there is an abundance and rich variety of creatures and species on earth, and recognises a natural tendency for this to be so. Individuals and species tend to multiply freely so as to fill every niche and habitat which the earth provides.

Pluralism: in politics, a term to describe a society where there are no politically or economically or socially dominant groups. Hence decisions are (or should be) reached via an open, equal and democratic process of competition between many groups, each advocating and debating their particular position and viewpoint. The competition is (or should be) adjudicated by relatively impartial people (the judiciary, public inquiry inspectors, government officers). The outcome is a decision which may offer something to each group – a compromise for each party. In a pluralist society, change thus comes about in response to 'stresses' articulated by particular, unsatisfied, groups, and these stresses are met to a degree by the rest of society accommodating to the views of each group. There is an analogy between this social model and the systems model of the natural world. See also **Elitism**.

Populism: a form of politics which appeals to people to exercise direct pressure on governments and 'emphasises the virtues of the uncorrupt and unsophisticated common people against the double-dealing and selfishness to be expected of professional politicians . . . It can therefore manifest itself in left, right or centrist forms' (B and S). Third candidates in US Presidential elections often run populist campaigns (e.g. Ross Perot 1992).

Positivism: the view that the only valid form of knowledge about nature and human affairs is scientific knowledge, in the sense of being based on observation, experiment and rational thinking. Used as abbreviation for the Vienna Circle of philosophers in the 1920s known as 'logical positivists'.

Postmodernism: 'the overall character or direction of experimental tendencies in Western arts, architecture since the 1940s or 1950s and particularly more recent developments associated with post-industrial society' (B and S). More generally, and by extension, postmodernism denies the validity or feasibility of the 'Enlightenment project', and that there are universal principles which are worth striving for. There are no valid totalising truths or universal political ambitions, so ideologies like liberalism or socialism will lead only to the negative results for humankind that they ostensibly set out to avoid. Postmodernism celebrates the equal validity of all points of view and all movements and periods. It also denies that there are deep underlying economic, social or any other structures and universal principles at work which explain what we see in the world around us. What we see on the surface of society is all there is: 'superficial' images and experiences are the reality of life. Postmodernism is an 'amorphous body of developments marked by eclecticism, pluriculturalism and often a post-industrial high-tech frame of reference coupled with a sceptical view of technical progress' (B and S).

Proletariat: in Marxism, that class of people who do not own and control the means of production, distribution and exchange, but have only their labour to sell. Within this strict definition, then, are included most 'middle' as well as 'working' classes, who, even if they own shares, seldom have real control over what is produced, when and how.

Reductionism: the view that a whole can be understood by breaking it down (reducing it) to its more elementary and basic constituents. Thus an organism might be defined as a collection of cells, a cell as a collection of molecules, a molecule as a collection of atoms, so it is strictly possible to describe humans as a collection of atoms. Again, life may be reduced to matters of molecular biology, itself reducible to chemistry, itself reducible to physics, reducible to maths. Or social structures and processes might be reduced to the interactions of individuals. As B and S put it, 'reductionism is seldom an uncontentious activity', and ecocentrism rejects it in favour of **holism**.

Relations of production: a term used particularly in Marxism to describe the relationships between people, and between people and nature, which arise from the particular arrangements that are made by a society to gain its means of material existence. The relations of production are different in different modes of production. Thus capitalism is a mode of production where goods and services are exchanged predominantly via a market place, using money as a common unit of value. This makes for relationships in which the values of people and of nature are expressed substantially in money terms – they are commoditised. People's labour, and consequently their life, is bought and sold in markets, hence they compete to sell themselves. Producers compete to supply cheaper products and services than other producers. All this makes for competitive relations of production. In

non-money communism, by contrast, production is via relationships of cooperation and community, and goods and services, and nature are valued in terms of
their 'use' alone. ('Use' is here a broad term incorporating notions of aesthetic and
existence value.)

Senescence/natural senescence: the theory that the earth has been ageing and
deteriorating since it was created: possibly as a result of human sin. This decline is
manifest in declining population, soil erosion, moral decline, etc. The analogy is
between the earth and a living organism, which also decays as it ages. Natural
senescence was supplanted in the eighteenth century by those who argued for a
constancy in nature, and that population was not declining.

Social Darwinism: application of Darwin's evolutionary scheme of nature to the
historical development of human societies, especially in the spheres of economics
and geopolitics. The ideas of competition for (allegedly) scarce resources, struggle
for existence and survival of the fittest are emphasised with approval, as the
processes whereby species are improved. Social Darwinism draws heavily on
concepts of society developed by Malthus and Herbert Spencer.

Structuralism: the concept that what we see in the way of social events and
individual and group behaviour is connected to deeper and less apparent underlying structures in the human mind and/or in society. Any theory is 'structuralist' if
it holds that deep, unobservable and only subconsciously apprehended realities
give rise to observed realities.

Subsidiarity: the principle that decisions should be taken at the most local level
which is appropriate, furthest removed from the centre of power.

Synthesis: opposite of **analysis**. Building something up into an interconnected
whole from simple constituents – bringing the parts together.

Technocentrism: a 'mode of thought' (O'Riordan 1981) which recognises environmental problems but believes either unrestrainedly that society will always solve
them through technology and achieve unlimited material growth ('cornucopian'),
or, more cautiously, that by careful economic and technical management the
problems can be negotiated (the 'accommodators'). In either case considerable
faith is placed in the ability and usefulness of classical science, technology and
conventional economic reasoning. There is little desire for genuine public participation in decision making, in favour of leaving decisions to politicians advised by
technical élites ('experts').

Teleology: having to do with design or purpose, being directed to an end. An
explanation of something, for example an aspect of nature, in terms of its purpose.

Thermodynamics, laws of: the first says that in an energy system with a constant
mass energy cannot be created or destroyed, though it can change form – heat, for
instance, is a form of energy (law of the conservation of energy). The second law
says that energy always moves from hotter to colder parts of the system (the law
of increasing entropy, or disorder in a closed system). The third law says that it is
impossible to cool an energy system right down to the absolute zero of
temperature.

Uniformitarianism: in opposition to **catastrophism,** the view that the earth as it now is was shaped by processes essentially the same as processes that can be seen to operate today, working gradually and over a very long time. Hence 'the present is key to the past'.

Vitalism: the idea that there is a 'vital principle' of life which distinguishes living from dead things, and which cannot be reduced to the same basic constituents of dead things (e.g. their chemical, physical, etc. properties). Neither can living processes be 'explained in terms of the material composition and physico-chemical performances of living bodies' (B and S).

RECOMMENDED FURTHER READING

In making these recommendations, the criteria of readability and enjoyability have been uppermost in my mind. Bibliographical details are given only for books not in the list of references.

THE GREEN PERSPECTIVE

There are two approaches to reading and learning about the green perspective. One is to get it 'from the horse's mouth' – from activists involved in the green movement; the other is to read more academic overviews, like this book.

The horse's mouth route could involve some books from the early 1970s, such as Goldsmith *et al. Blueprint for Survival*, which, with Meadows *et al*.'s *Limits to Growth* and Schumacher's *Small is Beautiful* form a classic trio.

Limits caused scares at the time (1972), with its easy-to-read and ostensibly scientific approach and its graphs forecasting Malthusian gloom and doom. It was slated by many scientific and humanistic critics, however, but, nothing daunted, its basic message reappeared as Meadows *et al. Beyond the Limits* in 1992. A more dramatic, emotional doomster was Paul Ehrlich, whose equally readable late 1960s *Population Bomb* became a *Population Explosion* in 1990. Edward Goldsmith never gives up either, and his unchanging message comes through in 1988, in a collection of essays spanning the 1970s and 1980s, *The Great U-Turn*. The mantle of pessimism was taken up in 1980 by the *Global 2000* report (Council on Environmental Quality, *The Global 2000 Report to the President*, 1982 edition, Harmondsworth: Penguin): a very indigestible mountain of figures and commentary which is not recommended. However, Simon and Kahn's (1984) repudiation of *Global 2000* is worth reading: in a series of clearly presented and sometimes well-argued essays (and sometimes not well argued) you learn the technocentric, cornucopian perspective. Also likely to excite controversy is Richard North's upbeat perspective on how we can cope with environmental problems and maintain Enlightenment ideals of progress, in

Life on a Modern Planet: a manifesto for progress, 1995, Manchester University Press. Today, blockbuster doomsaying is generally out, but every year Lester Brown's Worldwatch Institute launches a *State of the World* (New York: W.W. Norton) report which makes for sobering reading: a good means of keeping up to date.

Blueprint wanted to do more than scare people. It presented a detailed picture of what an ecological society would be like, in a series of numbered paragraphs, like a planning report almost. It also laid out a programme and timetable for radical social change: interesting to look at now to see how much has actually come to pass. A more palatable version of Blueprint's utopia is Callenbach's novel *Ecotopia*: not very well constructed or written, but just about entertaining enough to sweeten the pill of dryish detailed description. Indeed, it is slightly surprising that it was never filmed.

Schumacher's *Small is Beautiful* was much better written, and more profound. It influenced a generation, including, some say, Mrs Thatcher. Its liberal message is that we must look to our values and our education to find the way out of the wrong economic thinking that has caused ecological 'crisis'. Philosophical on the one hand and practical on the other, this book is the most worth revisiting of all. Though students today often don't find it that easy to read, they should at least grapple with Chapter 6 on education, which is an eye opener for those who have never come across Illich's notions of deschooling society, the hidden curriculum, etc.

Perhaps the ultimate among earlier books which placed values at the centre of environmental questions was Capra's *The Turning Point*. Its greatest worth is in drawing connections between different facets of life: economics, medicine, war and peace and ecology. His systems view purports to draw them together and examine the 'wrong' values now underlying them and what would be the 'right' values for the new ecological age which is dawning. Surprisingly, perhaps, the chapter with least clarity is that which summarises his own earlier writings in *The Tao of Physics*. But this is an important book even though you may eventually reject much of what Capra says. Capra's New Ageism is moderate by comparison with Peter Russell's *The Awakening Earth*. But this, too, is a must, since it is almost the New Age bible. It synthesises the ideas of twentieth-century physics with those of Prigogine, Bohm, Sheldrake and the rest into a coherent and followable, if not necessarily believable, theory of social change and evolution.

The most quoted British green activist of the 1980s was Jonathan Porritt, and his very approachable *Seeing Green* was a basic green manifesto of the quite-radical-but-respectable type that middle-class liberal audiences love. It all sounds such 'common sense': in fact it is rather glib about problems like where political power actually lies, and how to get it. Kemp and Wall's *A Green Manifesto for the 1990s* is more inclined to deal with such structural issues as political power and social and redistributive justice. It, too, is very clear and readable, though it is a bit of a shopping list.

For deep ecology, Devall and Sessions' *Deep Ecology* is still a basic text: a more populist deep ecological moan about the way the world is was Bill McKibben's *The End of Nature* (1990, Harmondsworth: Penguin), which got much publicity, though it does not seem very fully thought out. Social ecology still awaits a really accessible text. Murray Bookchin is often difficult and confusing to read, though he can be abusive and withering with great economy and skill, if you like that sort of thing. Best try Bookchin's *Remaking Society: paths to a green future* (1990, Boston: South End Press).

One of the most approachable of the academic texts is Dobson's *Green Political Thought*, a good review, updated and revised in 1995, of both 'light' and (mainly) 'dark' green positions. The basis of green political theory (the green theory of value), and green strategies, is also clearly and interestingly laid out and discussed in Goodin's *Green Political Theory*. John O'Neill's *Ecology, Policy and Politics* is worth a try: he discusses issues relevant to green political philosophy, such as intrinsic value, cost-benefit analysis, pluralism and the status of science. His style is very clear and his perspectives interesting and thought-provoking, although if you do not come from a background in politics or philosophy you could find aspects of it hard going.

Sociologists have increasingly become interested in environmental matters, and the best review of environmentalism from a sociological perspective is Martell's *Ecology and Society*. It is clear, comprehensive and very reader friendly. For sociology students, Dickens' *Society and Nature* goes deeper, and is interesting for its attempts to create a theory of society that acknowledges that humans are creatures who are biologically and ecologically constrained, but does not throw out the more anthropocentric perspectives of contemporary sociology.

There has been a plethora of green economics books in the past five to ten years also. By far and away the easiest to understand is Ekins' *Wealth Beyond Measure*, written for the non-specialist, comprehensive and lavishly and entertainingly illustrated. Jacobs' *The Green Economy* goes deeper, but maintains readability, and will be on many student reading lists. He approaches the subject from a little left of centre, while Pearce *et al.'s Blueprint for a Green Economy*, and *Blueprints Two* (Pearce, D. (1991) *Blueprint 2: greening the global economy*, London: Earthscan) and *Three* (Pearce, D. *et al.* (1993) *Blueprint 3: measuring sustainable development*, London: Earthscan) are more to the right. Indeed Pearce's reports take a market-based incentive approach, and have formed the basis of British Conservative Government pronouncements on what measures it might consider implementing. For a radically different economic analysis of what is going wrong (but not of what to do), then Johnston's *Environmental Problems: nature, economy and state* is a must. Based on Marxism's analysis of capitalism, it focuses on the central role of modes of production in understanding environmental degradation and rejects individual-centred responses (lifestylism, green consumerism) so beloved by some greens. Its admirable clarity, eloquence and conciseness renders complex political-economic matters accessible to undergraduates and others with no prior knowledge in these areas.

One other academic survey, which has no particular disciplinary bias and which is more North American based, is Merchant's *Radical Ecology*: a clear and easy to understand text.

HISTORIES OF ENVIRONMENTAL IDEAS

No one interested in the history of ecological ideas, especially natural and environmental scientists, can afford to miss Worster's *Nature's Economy: a history of ecological ideas*. It discusses leading figures from the eighteenth century onwards, the evolution of 'arcadian' and 'imperial' strands in ecology, and the influence of both classical scientific thinking and organicism on ecology. It is extremely well written and quite fascinating. Absorbing, too, is Oldroyd's *Darwinian Impacts*: a history of ideas about evolution and of how Darwin's theory has been used in and has permeated so many areas of thought and endeavour such as politics, music, psychology, literature and anthropology. Another book that is a pleasure to read is Nash's *Wilderness and the American Mind*, an account of changes in attitudes towards the American wilderness – from a place to be feared and conquered to a national asset to be conserved and cherished. Here, you can read about the growth of the American Romantic movement and early environmentalism. All of these books could be holiday reading because of their accessibility and interest, but they are also very substantial works of scholarship.

More difficult to approach is Bramwell's *Ecology in the Twentieth Century: a history*. This is an exploration of what Bramwell claims to be the forerunners of the ecology movement in the 1920s and 1930s particularly. Critics have disputed whether her central characters were really 'ecocentric', perhaps because many of them were far to the right, and/or very mixed up. She does not make her material or her arguments that easy to digest, but this is none the less a good book to dip into, and full of interest. Merchant's *The Death of Nature* has also had some adverse comment, perhaps because she may overstate or oversimplify the links between the development of the classical scientific world view, the subordination of women, and the growth of capitalism. However, the book shows much scholarship, and is sharp and clearly written. It is always stimulating and provocative. If you are interested in animal welfare then you must get Keith Thomas' *Man and the Natural World*, a pleasant book to read, although so crammed with references, notes, quotations and anecdotes that it is a bit bitty. Thomas has also been criticised by fellow historians, but the value of his study of changing attitudes to nature over the period 1500–1800 is considerable, and it is certainly thought-provoking.

There is, too, a huge literature of cultural studies which examines how ideas about nature and society–environment relations are translated into literature, painting and music, and also into the landscape. John Short's *Imagined Country* is a very absorbing, useful, brief and easy to understand introduction to such studies for those who have not been near them before. Daniels' *Fields of Vision* is more specifically about how images of landscape and ideas about nature, and of national identity, are represented in painting in England and the USA. It is a very good read.

Anarchist-socialist ideas have had a considerable impact on ecocentrism, and two accessible accounts linking socialist history with ecologism are well worth the time. Coleman and O'Sullivan's *William Morris and News from Nowhere: a vision for our time* is an edited collection of essays about the 'first British Marxist' and his visions of a socialist society. What the essays do, clearly and provocatively, is to illustrate how very relevant Morris' thought is for today's problems, contrary to those who think that socialism is irrelevant. Gould's title *Early Green Politics: back to nature, back to the land and socialism in Britain* is self-explanatory: he traces the back-to-the-land movement of the late nineteenth and early twentieth centuries and discusses how socialists were involved in this and other back-to-nature ideas. He covers some of the ground taken in by Bramwell, but sticks to socialist thought.

REFERENCES

Abbey, E. (1975) *The Monkey Wrench Gang*, New York: Avon Books.

Achterberg, W. (1993) 'Can liberal democracy survive the environmental crisis?', in Dobson, A. and Lucardie, P. (eds) *The Politics of Nature: explanations in green political theory*, London: Routledge, 81–101.

Adam, B. (1993) 'Time and environmental crisis: an exploration with special reference to pollution', *Innovation in Social Sciences Research*, 6(4), 399–413.

Adams, W. (1990) *Green Development: environment and sustainability in the Third World*, London: Routledge.

Agarwal, A. and Narain, S. (1990) *Towards Green Villages: a strategy for environmentally sound and participatory rural development*, New Delhi: Centre for Science and the Environment.

Albury, D. and Schwartz, J. (1982) *Partial Progress: the politics of science and technology*, London: Pluto Press.

Alexander, D. (1990) 'Bioregionalism: science or sensibility?', *Environmental Ethics*, 12, 161–73.

Allaby, M. (1989) *Guide to Gaia*, London: Optima.

Allison, L. (1991) *Ecology and Utility: the philosophical dilemmas of planetary management*, Leicester: Leicester University Press.

Anderson, T. and Leal, D. (1991) *Free Market Environmentalism*, San Francisco: Pacific Research Institute for Public Policy.

Anderson, V. (1991) *Alternative Economic Indicators*, London: Routledge.

Anton, A. (1992) in critical discussion of paper by Johnson and Johnson (see below), *Capitalism, Nature, Socialism*, 10, 111–14.

Ash, M. (1980) *Green Politics: the new paradigm*, London: Green Alliance.

Ash, M. (1987) *New Renaissance: essays in search of wholeness*, Bideford: Green Books.

Atkinson, A. (1991) *Principles of Political Ecology*, London: Belhaven.

Attfield, R. (1983) 'Christian attitudes to nature', *Journal of the History of Ideas*, 44(3), 369–86.

Barbour, I. (1980) 'Technology, environment and human values', in Barbour, I. (ed.) *Western Man and Environmental Ethics*, Reading, Mass.: Addison Wesley.

Barnes, B. (1985) *About Science*, Oxford: Blackwell.

Barrell, J. (1980) *The Dark Side of the Landscape: the rural poor in English painting 1730–1840*, Cambridge: Cambridge University Press.

Barrows, H. (1923) 'Geography as human ecology', *Annals of the Association of American Geographers*, 13, 1–14.

Barry, J. (1994) 'The limits of the shallow and the deep: green politics, philosophy, and praxis', *Environmental Politics*, 3(3), 369–94.

REFERENCES

Bate, J. (1991) *Romantic Ecology: Wordsworth and the environmental tradition*, London: Routledge.

BBC (1993) 'Putting market values on the environment', *File on Four*, transmitted on Radio 4.

BBC (1994) 'Fear of frying', *Analysis*, transmitted on Radio 4.

Bennett, J. (1987) *The Hunger Machine: the politics of food*, Cambridge: Polity Press.

Benson, J. (1978) 'Duty and the beast', *Philosophy*, 53, 541.

Benton, T. (1989) 'Marxism and natural limits: an ecological critique and reconstruction', *New Left Review*, 178, 51–87.

Benton, T. (1993) *Natural Relations: ecology, animal rights and social justice*, London: Verso.

Beresford, P. and Croft, S, (1992) 'Beyond welfare', in Ekins, P. and Max-Neef, M. (eds) *Real Life Economics: understanding wealth creation*, London: Routledge, 283–9.

Bergson, H. (1911) *Creative Evolution*, translated from the 1907 edition by A. Mitchell, London: Macmillan.

Biehl J. (1991) *Rethinking Feminist Politics*, Boston, Mass.: Southend Press.

Biehl, J. (1993) '"Ecology" and the modernisation of fascism in the German ultra-right', *Society and Nature*, 2(2), 130–70.

Bird, E. A. (1987) 'The social construction of nature: theoretical approaches to the history of environmental problems', *Environmental Review*, 11 (4), 255–64.

Blowers, A. (1984) *Something in the Air: corporate power and the environment*, London: Harper and Row.

Blowers, A. (1987) 'Transition or transformation? Environmental policy under Thatcher', *Public Administration*, 65, 277–94.

Blowers, A. and Lowry, D. (1987) 'Out of sight: out of mind: the politics of nuclear waste in the UK' in Blowers, A. and Pepper, D. (eds) *Nuclear Power in Crisis*, London: Croom Helm, 129–63.

Blowers, A. and Pepper, D. (eds) (1987) *Nuclear Power in Crisis*, London: Croom Helm.

Bohm, D. (1983) *Wholeness and the Implicate Order*, London: Ark.

Bookchin, M. (1979) 'Ecology and revolutionary thought', *Antipode*, 10(3)/11(1), 21–32.

Bookchin, M. (1980) *Towards an Ecological Society*, Montreal: Black Rose Books.

Bookchin, M. (1982) *The Ecology of Freedom: the emergence and dissolution of hierarchy*, Palo Alto, Calif.: Cheshire Books.

Bookchin, M. (1987) 'Social ecology versus "deep ecology" a challenge for the ecology movement', *The Raven*, 1(3), 219–50.

Bookchin, M. (1990) *The Philosophy of Social Ecology*, Montreal: Black Rose Books.

Bookchin, M. (1992a) 'The meaning of confederalism', *Society and Nature*, 1(3), 41–54.

Bookchin, M. (1992b) 'The transition to the ecological society: an interview by Takis Fotopoulos', *Society and Nature*, 1(3), 92–105.

Bookchin, M. (1993) 'Deep ecology, anarchosyndicalism and the future of anarchist thought', in *Deep Ecology and Anarchism: a polemic*, London: Freedom Press (no editor given), 47–58.

Bookchin, M. and Foreman, D. (1991) *Defending the Earth: a dialogue between Murray Bookchin and Dave Foreman*, Montreal: Black Rose Books.

Bottomore, T. (ed.) (1985) *A Dictionary of Marxist Thought*, Oxford: Blackwell.

Boyle, G. and Harper, P. (eds) (1976) *Radical Technology*, London: Wildwood House.

Bradford, G. (1989) *How Deep is Deep Ecology?*, Haley, Mass.: Times Change Press.

Bradley, I. (1990) *God is Green*, London: Dorton, Londman and Todd.

REFERENCES

Bramwell, A. (1985) *Blood and Soil: Walther Darre and Hitler's 'Green Party'*, Bourne End, Bucks: Kensal.

Bramwell, A. (1989) *Ecology in the Twentieth Century: a history*, London: Yale University Press.

Bramwell, A. (1994) *The Fading of the Greens: the decline of environmental politics in the West*, New Haven, Conn.: Yale University Press.

Brennan, A. (1988) *Thinking About Nature: an investigation of nature, value and ecology*, London: Routledge.

Briggs, J. P. and Peat, F. D. (1985) *Looking Glass Universe: the emerging science of wholeness*, London: Fontana.

Brown, L. S. (1988) 'Anarchism, existentialism and human nature' *The Raven*, 2(1), 49–60.

Buchanan, K. (1973) 'The white north and the population explosion', *Antipode*, 5(3), 7–15.

Buchanan, A. (1982) *Food, Poverty and Power*, Nottingham: Spokesman Books.

Buick, A. (1990) 'A market by the way: the economics of nowhere', in Coleman, S. and O'Sullivan, P. (eds) *William Morris and News From Nowhere: a vision for our time*, Bideford: Green Books, 151–68.

Buick, A. and Crump, J. (1986) *State Capitalism: the wages system under new management*, London: Macmillan.

Bullock, A. and Stallybrass, O. (eds) (1988) *The Fontana Dictionary of Modern Thought*, London: Fontana.

Bunyard, P. (1988) 'Gaia: its implications for industrialised society' in Bunyard, P. and Goldsmith, E. (eds) *Gaia, the Thesis, the Mechanisms and the Implications*, Wadebridge, Cornwall: Wadebridge Ecological Centre, 201–21.

Button, J. (1988) *A Dictionary of Green Ideas*, London: Routledge.

Button, J. and Bloom, W. (1992) *The Seeker's Guide: a New Age resource book*, London: Aquarian/Thorsons.

Callenbach, E. (1978) *Ecotopia*, London: Pluto Press.

Callenbach, E. (1981) *Ecotopia Emerging*, Berkeley, Calif.: Banyan Tree Books.

Callicott, J. B. (1982) 'Traditional American Indian and Western European attitudes toward nature: an overview', *Environmental Ethics* 4, 293–318.

Callicott, J. B. (1985) 'Intrinsic value, quantum theory and environmental ethics', *Environmental Ethics*, 7, 293–318.

Callicott, J. B. (1989) *In Defense of the Land Ethic: essays in environmental philosophy*, Albany, NY: State University of New York Press.

Capra, F. (1975) *The Tao of Physics*, London: Fontana.

Capra, F. (1982) *The Turning Point*, London: Wildwood House.

Carruthers, R. G. (1939) 'On northern glacial drifts: some peculiarities and their significance', *Quarterly Journal of the Geological Society of London*, 95(3), 299–333.

Carruthers, R. G. (1953) *Glacial Drifts and the Undermelt Theory*, Newcastle on Tyne: Harold Hill and Son, 38 pp..

Carson, R. (1962) *Silent Spring*, Boston, Mass.: Houghton Miflin.

Central TV (1994) *Death of a Nation*, documentary with J. Pilger.

Chapman, P. (1975) *Fuel's Paradise: energy options for Britain*, Harmondsworth: Penguin.

Chase, A. (1980) *The Legacy of Malthus: the social costs of the new scientific racism*, Urbana, Ill.: University of Illinois Press.

Church, C. (1988) 'Great chief sends modified word', *ECOS*, 4, 40–1.

Cini, M. (1992) 'Science and sustainable society', *Society and Nature*, 1(2), 32–48.

Clark, J. (1990) 'What is social ecology', in Clark, J. (ed.) *Renewing the Earth: the promise of social ecology*, London: Green Print, 5–11.

REFERENCES

Clarke, J. J. (1993) *Nature in Question: an anthology of ideas and arguments*, London: Earthscan.

Coase, R. (1960) 'The problem of social cost', *Journal of Law and Economics*, 3(1).

Coates, C., How, J., Jones, L., Morris, W. and Wood, A. (1993) *Diggers and Dreamers: the 1994/5 guide to communal living*, Winslow, Bucks: Communes Network.

Cock, P. H. (1985) 'Sustaining the alternative culture: the drift towards rural suburbia!', *Social Alternatives*, 4(4), 12–16.

Cole, K., Cameron, J. and Edwards, C. (1983) *Why Economists Disagree*, London: Longman.

Coleman, D. A. (1994) *Ecopolitics: building a green society*, New Brunswick, NJ: Rutgers University Press.

Coleman, S. and O'Sullivan, P. (1990) *William Morris and News from Nowhere: a vision for our time*, Bideford: Green Books.

Coleman, W. (1971) *Biology in the Nineteenth Century: problems of form, function and transformation*, London: Wiley.

Collard, A. (1988) *Rape of the Wild*, London: The Women's Press.

Commoner, B. (1972) *The Closing Circle*, New York: Bantam.

Commoner, B. (1990) *Making Peace With the Planet*, New York: Pantheon.

Cook, I. (1990) 'Anarchistic alternatives: an introduction', *Contemporary Issues in Geography and Education*, 3(2), 9–21.

Cooper, T. (1990) *Green Christianity*, London: Spire/Hodder and Stoughton.

Cooter, W. S. (1978) 'Ecological dimensions of medieval agrarian systems', *Agricultural History*, 52, 438–77.

Corbridge, S. (1993) *Debt and Development*, Oxford: Blackwell.

Cosgrove, D. (1984) *Social Formation and Symbolic Landscape*, London: Croom Helm.

Cosgrove, D. (1990) 'Environmental thought and action: pre-modern and post-modern', *Transactions of the Institute of British Geographers*, 15(3), 344–58.

Cosgrove, D. (1994) 'Contested global visions: one-world, whole-earth, and the Apollo space photographs', *Annals of the Association of American Geographers*, 84(2), 270–94.

Costanza, R., Daly, H. E. and Bartholomew, J. (1991) 'Goals, agenda and policy recommendations for ecological economics', in Costanza, R. (ed.) *Ecological Economics: the science and management of sustainability*, New York: Columbia University Press.

Coward, R. (1989) *The Whole Truth: the myth of alternative medicine*, London: Faber and Faber.

Cox, G. (1988) '"Reading" nature: reflections on ideological persistence and the politics of the countryside', *Landscape Research*, 13(3), 24–34.

Cox, S. J. (1985) 'No tragedy on the commons', *Environmental Ethics*, 7, 49–61.

Daly, H. (1991) *Steady State Economics*, Washington DC: Island Press.

Daly, H. and Cobb, J. (1990) *For the Common Good: redirecting the economy toward community, the environment and a sustainable future*, London: Green Print.

Daly, M. (1987) *Gyn/Ecology*, London: The Women's Press.

Daniels, S. (1993) *Fields of Vision: landscape imagery and national identity in England and the US*, Oxford: Polity Press.

Darby, H. C. (1956) *The Draining of the Fens*, Cambridge: Cambridge University Press.

Darby, H. C. (ed.) (1973) *The New Historical Geography of England*, Cambridge: Cambridge University Press.

Darwin, C. (1885) *The Origin of Species*, sixth edition, London: Murray.

Dauncey, G. (1988) *After the Crash*, London: Green Print.

Davey, B. (1992) 'Eco cities', *Permaculture*, 1(1), 13–14.

Dawkins, R. (1976) *The Selfish Gene*, Oxford: Oxford University Press.

REFERENCES

de Chardin, T. (1965a) *The Phenomenon of Man*, London: Fontana.

de Chardin, T. (1965b) *The Hymn of the Universe*, London: Collins.

Del Sesto, S. (1980) 'Conflicting ideologies of nuclear power: Congressional testimony on nuclear reactor safety', *Public Policy*, 28(1), 39–70.

Desai, P. (1993) *Bioregional Surrey*, Carshalton, Surrey: Bioregional Development Group.

Devall, W. and Sessions, G. (1985) *Deep Ecology: living as if nature mattered*, Salt Lake City, Utah: Gibbs M. Smith.

Devine, P. (1988) *Democracy and Economic Planning: the political economy of a self-governing society*, Boulder, Col.: Westview Press.

Devlin, J. and Yap, N. (1993) 'Structural adjustment programmes and the UNCED agenda: explaining the contradictions', *Environmental Politics*, 2(4), 65–79.

Dickens, P. (1992) *Society and Nature: towards a green social theory*, Hemel Hempstead: Harvester Wheatsheaf.

Dickson, D. (1974) *Alternative Technology and the Politics of Technical Change*, London: Fontana.

Dietz, F. .J and Straaten, J. van der (1993) 'Economic theories and the necessary integration of ecological insights', in Dobson, A. and Lucardie, P. (eds) *The Politics of Nature: explanations in green political theory*, London: Routledge, 118–44.

Dixon, C (1991) 'Calling a spade a spade', *Green Line*, 87, 8–9.

Dizard, J. E. (1993) 'Going wild: the contested terrain of nature', in Bennett, J. and Chaloupka, W. (eds) *In the Nature of Things: language, politics and the environment*, Minneapolis, Minn.: University of Minnesota Press, 111–35.

Dobb, M. (1946) *Studies in the Development of Capitalism*, London: Routledge.

Dobson, A. (1990) *Green Political Thought*, London: Unwin Hyman, second edition 1995.

Doughty, R. (1981) 'Environmental theology: trends and prospects in Christian thought', *Progress in Human Geography*, 5(2), 234–48.

Drysek, J. (1987) *Rational Ecology*, Oxford: Blackwell.

Eckberg, D. L. and Blocker, T. J. (1989) 'Varieties of religious involvement and environmental concern', *Journal for the Scientific Study of Religion*, 28(4), 509–17.

Eckersley, R. (1992) *Environmentalism and Political Theory: towards an ecocentric approach*, London: University College London Press.

Eckersley, R. (1993) 'Free market environmentalism: friend or foe?', *Environmental Politics*, 2(1), 1–19.

Economist, The (1981) 'The nature of knowledge', 26 December.

Edwards, P. (ed.) (1972) *The Encyclopaedia of Philosophy*, vol. 7, London: Collier-Macmillan. pp. 206–9.

Ehrlich, P. (1969) 'Eco-catastrophe', *Ramparts*, 8(3), 24–8.

Ehrlich, P. and Ehrlich, P. (1990) *The Population Explosion*, New York: Simon and Schuster.

Ehrlich, P. and Hoage, R. (eds) (1985) *Animal Extinction: what everyone should know*, Washington DC: Smithsonian Institute Press.

Ekins, P. (1992a) *Wealth Beyond Measure: atlas of new economics*, London: Gaia Books.

Ekins, P. (1992b) 'Towards a progressive market', in Ekins, P. and Max-Neef, M. (eds) *Real Life Economics: understanding wealth creation*, London: Routledge, 322–7.

Ekins, P. (1992c) *A New World Order: grassroots movements for global change*, London: Routledge.

Ekins, P. (1993) '"Limits to growth" and "sustainable development": grappling with ecological realities', *Ecological Economics*, 8(3), 269–88.

Ekins, P. (1994) 'Environmental sustainability of economic processes: a framework for analysis', in Bergh, J. van den and Straaten, J. van der (eds) *Concepts, Methods and Policy for Sustainable Development*, Washington DC: Island Press, 25–55.

Ekins, P. (forthcoming) 'Green economics' in Paehlke, R. (ed.) *The Encyclopaedia of Conservation and Environmentalism*, New York: Garland Publishing.

Elkington, J. and Burke, T. (1987) *The Green Capitalists*, London: Gollancz.

Elliot, J. A. (1994) *An Introduction to Sustainable Development: the developing world*, London: Routledge.

Elsom, D. (1992) *Atmospheric Pollution: a global problem*, second edition, Oxford: Blackwell.

Engel, J. R. (1990) 'The ethics of sustainable development', in Engel, J. R. and Engel, J. G. (eds) *Ethics of Environment and Development; global challenge and the international response*, London: Belhaven, 1–23.

Engel, J. R. and Engel, J. G. (eds) (1990) *Ethics of Environment and Development; global challenge and the international response*, London: Belhaven.

Engel, M. (1994) 'Gone with the wind', *Guardian 2*, 11 March, pp. 2–3.

Engels, F. (1963) *The Dialectics of Nature*, Moscow: Foreign Languages Publishing.

English Nature (1993) *Natural Areas: English Nature's approach to setting nature conservation objectives: a consultation paper*, Peterborough: English Nature.

Erisman, F. (1973) 'The environmental crisis and present day romanticism: the persistence of an idea', *Rocky Mountain Social Science Journal*, 10, 7–14.

Ette, A. and Waller, R. (1978) 'The anomaly of a Christian ecology', *Ecologist Quarterly*, Summer, 144–8.

Etzioni, A. (1992) 'The I and we paradigm', in Ekins, P. and Max-Neef, M. (eds) *Real Life Economics: understanding wealth creation*, London: Routledge, 48–53.

Evans, D. (1992) *A History of Nature Conservation in Britain*, London: Routledge.

Evans, J. (1993) 'Ecofeminism and the politics of the gendered self', in Dobson, A. and Lucardie, P. (eds) *The Politics of Nature: explanations in green political theory*, London: Routledge, 177–89.

Faber, D. and O'Connor, J. (1989) 'The struggle for nature: environmental crisis and the crisis of environmentalism in the US', *Capitalism, Nature, Socialism*, 2, 12–39.

Ferguson,. M (1981) *The Aquarian Conspiracy: personal and social transformation in the 1980s*, London: Granada.

Ferkiss, V. (1993) *Nature, Technology and Society: cultural roots of the current environmental crisis*, London: Adamantine Press.

Ferris, J. (1993) 'Ecological versus social rationality: Can there be green social policies?', in Dobson, A. and Lucardie, P. (eds) *The Politics of Nature: explanations in green political theory*, London: Routledge, 145–58.

Ferris, T. (1990) *Coming of Age in the Milky Way*, London: Bodley Head/Vintage.

Fieshbach, M. and Friendly, M. (1992) *Ecocide in the USSR*, London: Aurum Press.

FoE (Friends of the Earth) (1993) 'Sapping the forests', *Earth Matters*, 20, 8–9.

FoE (1994) *Planning for Wind Power: guidelines for project developers and local planners*, London: Friends of the Earth.

Fotopoulos, T. (1992) 'The economic foundations of an ecological society', *Society and Nature*, 1(3), 1–40.

Fox, W. (1984) 'Deep ecology: a new philosophy of our time?', *The Ecologist*, 14, 194–200.

Fox, W. (1989) 'The deep ecology-ecofeminism debate and its parallels', *Environmental Ethics*, 11, 5–25.

Fox, W. (1990) *Towards a Transpersonal Ecology*, London: Shambhala.

Francis, L. P. and Norman, R. (1978) 'Some animals are more equal than others', *Philosophy*, 53, 507–27.

Frank, A. G. (1989) 'The development of underdevelopment', *Monthly Review*, 41(2), 44.

Frankel, B. (1987) *The Post Industrial Utopians*, Cambridge: Polity Press.

Frey, R. G. (1980) *Interests and Rights: the case against animals*, Oxford: Clarendon Press.

Fuentes, M. and Frank, A. G. (1992) 'Ten theses on social movements', *Society and Nature*, 1(3), 131–57.

Galbraith, J. K. (1958) *The Affluent Society*, Harmondsworth: Penguin.

Gandy, M. (1992) *The Environmental Debate: a critical overview*, Sussex: University of Sussex Geography Research Paper No. 5.

Gasman, D. (1971) *The Scientific Origins of National Socialism: social Darwinism in Ernst Haeckel and the German Monist League*, London: MacDonald.

George, S. (1976) *How the Other Half Dies: the real reasons for world hunger*, Harmondsworth: Penguin.

George, S. (1989) *A Fate Worse Than Debt*, Harmondsworth: Penguin.

George, S. (1992) *The Debt Boomerang*, London: Pluto Press.

Georgescu-Roegen, N. (1971) *The Entrophy Law and Economic Process*, Cambridge, Mass.: Harvard University Press..

Ghai, D. and Vivian, J. (eds) (1992) *Grassroots Environmental Action: people's participation in sustainable development*, London: Routledge.

Glacken, C. (1967) *Traces on the Rhodian Shore*, Berkeley, Calif.: University of California Press.

Gold, M. (1984) 'A history of nature', in Massey, D. and Allen, J. (eds), *Geography Matters!*, Cambridge: Cambridge University Press, 12–33.

Goldman, M, and O'Connor, J. (1988) 'Ideologies of enviromental crisis: technology and its discontents', *Capitalism, Nature, Socialism*, 1(1), 91–106.

Goldsmith, E. (1977) 'Deindustrialising society', *Ecologist*, 7, May, 128–43.

Goldsmith, E. (1978) 'The religion of a stable society', *Man–Environment Systems*, 8, 13–24.

Goldsmith, E. (1987) in 'Choices', a discussion on the environment, televised by the BBC.

Goldsmith, E. (1988) *The Great U-Turn*, Bideford: Green Books.

Goldsmith, E., Allan, R., Allaby, M., Davoll, J. and Lawrence, S. (eds) (1972) 'Blueprint for survival', *The Ecologist*, 2(1), 1–43.

Goldsmith, E., Hildyard, N., Bunyard, P. and McCully, P. (eds) (1992) 'Whose common future?', *The Ecologist*, 22(4), July–August.

Goldsmith, E., Hildyard, N., Bunyard, P. and McCully, P. (eds) (1993) 'Cakes and caviar? The Dunkel draft and third world agriculture', *The Ecologist*, 23(6), 219–22.

Gombrich, E. H. (1989) *The Story of Art*, fifteenth edition, London: Phaidon.

Goodin, R. E. (1992) *Green Political Theory*, Cambridge: Polity Press.

Goodman, D. and Redclift, M. (1991) *Refashioning Nature: food, ecology and culture*, London: Routledge.

Goodwin, B. and Taylor, K. (1982) *The Politics of Utopia*, London: Hutchinson.

Gordon, J. (1993) 'Letting the genie out: local government and UNCED', *Environmental Politics*, 2(4), 137–55.

Gorz, A. (1982) *Farewell to the Working Classes: an essay on post-industrial socialism*, London: Pluto.

Gosse, E. (1907) *Father and Son*, Harmondsworth: Penguin Classic edition, 1986.

Gosse, P. H. (1857) *Omphalos: an attempt to untie the geological knot*, London.

Gould, P. (1988) *Early Green Politics: Back to Nature, back to the land and socialism in Britain*, Brighton: Harvester Press.

REFERENCES

Gould, S. J. (1982) 'Punctuated equilibrium: a new way of seeing', *New Scientist*, 15 April, 137–41.

Gray, J. (1993) *Beyond the New Right: markets, government and the common environment*, London: Routledge.

Greeley, A. (1993) 'Religion and attitudes toward the environment', *Journal for the Scientific Study of Religion*, 32(1), 19–28.

Green, K. and Yoxen, K. (1993) 'The greening of European industry: what role for biotechnology?', in Smith, D. (ed.) *Business and the Environment: implications of the New Environmentalism*, London: Paul Chapman Publishing, 150–71.

Green Party (1994) *European Election Manifesto 1994*, London: Green Party.

Greene, O. (1993) 'International environmental regimes: verification and implementation review', *Environmental Politics*, 2(4), 156–73.

Gregory, D. (1981) 'Regional geography', entry in Johnston, R. J. (ed.) *The Dictionary of Human Geography*, Oxford: Blackwell, 286–8.

Griffin, D. R. (1993) 'Whitehead's deeply ecological worldview', in Tucker, M. E. and Grim, J. A. (eds) *Worldviews and Ecology*, Lewisburg, Pa.: Bucknell University Press, 190–206.

Griffin, R. (1991) *The Nature of Fascism*, London: Pinter Publishers.

Griffiths, J. (1990) 'The collective unfairness of laissez faire', *The Guardian*, 14 June.

Grimston, M. (1990) 'A critique of green "science"', *Atom*, 408, 17–20.

Grinevald, J. (1988) 'A history of the idea of the biosphere', in Bunyard, P. and Goldsmith, E. (eds) *Gaia, the Thesis, the Mechanisms and the Implications*, Wadebridge, Cornwall: Wadebridge Ecological Centre,.

Grove, R. (1990) 'The origins of environmentalism', *Nature*, 345, 3 May, 11–14.

Grove-White, R. and Michael, M. (1993) 'Nature conservation: culture, ethics and science', in Burgess, J. (ed.) *People, Economies and Nature Conservation*, London: University College Ecology and Conservation Unit Discussion Paper No. 60, 139–52.

Gruffudd, P. (1991) 'Reach for the sky: the air and English cultural nationalism', *Landscape Research*, 16(2), 19–24.

Grundmann, R. (1991) *Marxism and Ecology*, Oxford: Clarendon Press.

Guha, R. (1991) 'Lewis Mumford: the forgotten American environmentalist: an essay in rehabilitation', *Capitalism, Nature, Socialism*, 2(3), 67–91.

Hales, M. (1982) *Science or Society?*, London: Pan Books/Channel 4.

Hamer, M. (1987) *Wheels Within Wheels*, Andover: Routledge and Kegan Paul.

Haraway, D. (1989) *Primate Visions: gender, race and nature in the world of modern science*, London: Routledge.

Hardin, G. (1968) 'The tragedy of the commons', *Science*, 162, 1243–8.

Hardin, G. (1974) 'Living on a lifeboat', *BioScience*, 24, 10.

Hardin, G. (1993) 'Carrying capacity', *Real World*, Spring, 12–13.

Hardy, D. (1979) *Alternative Communities in Nineteenth-Century England*, London: Longman.

Hargrove, E. C. (1979) 'The historical foundations of American environmental attitudes', *Environmental Ethics*, 1, Fall, 209–39.

Harper, P. (1990) 'I told you so', *Clean Slate*, 2, 4–5.

Harper, P. (1992) 'p-p-Permaculture?', *Clean Slate*, 7, 9.

Harrison, C. M. and Burgess, J. (1994) 'Social constructions of nature: a case study of conflicts over the development of Rainham Marshes', *Transactions of the Institute of British Geographers*, New Series, 19, 291–310.

Hartstock, N. (1987) 'The feminist standpoint: developing the ground for a specifically feminist historical materialism', in Harding, S. (ed.), *Feminism and Methodology*, Bloomington, Ind.: Indiana University Press.

REFERENCES

Harvey, D. (1990) *The Condition of Postmodernity*, Cambridge: Polity.

Harvey, D. (1993) 'The nature of environment: the dialectics of social and environmental change', in Miliband, R. and Panitch, L. (eds) *Socialist Register*, London: Merlin, 1–51.

Hawken, P. (1975) *The Magic of Findhorn*, London: Fontana Books.

Hawkins, H. (1992) 'Community control, workers' control and the cooperative commonwealth', *Society and Nature*, 1(3), 55–85.

Hayek, F. A. von (1949) *Individualism and Economic Order*, London: Routledge and Kegan Paul.

Hecht, S. and Cockburn, A. (1990) *The Fate of the Forest: developers, destroyers and defenders of the Amazon*, Harmondsworth: Penguin.

Heilbroner, R. (1980) *Marxism, For and Against*, London: Norton.

Heizer, R. and Elsasser, A. (1980) *The Natural World of the California Indians*, Berkeley, Calif.: University of California Press.

Henderson, H. (1981) *The Politics of the Solar Age*, New York: Doubleday.

Hirsch, F. (1976) *The Social Limits to Growth*, Cambridge, Mass.: Harvard University Press.

HMSO (1979) *The Control of Radioactive Wastes: a review of Command 884*, London: HMSO.

Hollingsworth, T. (1973) Introduction to Malthus' *Essay on the Principle of Population*, seventh edition 1872, London: Dent.

Holton, G. (1956) 'Johannes Kepler's universe: its physics and metaphysics', *American Journal of Physics*, 24, 340–51.

Hornsby-Smith, M. P. and Proctor, M. (1993) 'Environmental concerns in Britain in the 1990s: evidence from the European Values Survey', *Proceedings of the Conference on Values and the Environment*, Guildford: University of Surrey, 36–41.

Hoskins, W. G. (1955) *The Making of the English Landscape*, Harmondsworth: Penguin.

Hueting, R. (1992) 'The economic functions of the environment', in Ekins, P. and Max-Neef, M. (eds) *Real Life Economics: understanding wealth creation*, London: Routledge, 61–9.

Hughes, D. (1992) *Environmental Law*, second edition, London: Butterworths.

Illich, I. (1975) *Tools for Conviviality*, London: Fontana.

Ingham, A. (1993) 'The market for sulphur dioxide permits in the USA and UK', *Environmental Politics*, 4(2), 98–122.

Irvine, S. (1989) 'Consuming fashions: the limits of green consumerism', *The Ecologist*, 19(3), 88–93.

Jackson, B. (1990) *Poverty and the Planet: a question of survival*, Harmondsworth: Penguin.

Jacobs, M. (1991) *The Green Economy: environment, sustainable development and the politics of the future*, London: Pluto Press.

Jacobs, M. (1993) "Free market environmentalism": a response to Eckersley', *Environmental Politics* 2(4), 238–41.

Jacobs, M. Levett, R. and Stott, M. (1993) 'Sustainable development and the local economy' in *Local Economy*, Luton: The Local Government Management Board.

James, P. (1979) *Population Malthus: his life and times*, London: Routledge and Kegan Paul.

Jantsch, E. (1980) *The Self-Organising Universe: scientific and human implications of the emerging paradigm of evolution*, Oxford: Pergamon Press.

Jeans, D. (1974) 'Changing formulations of the man–environment relationship in Anglo-American geography', *Journal of Geography*, 73(3), 36–40.

Jensen, A. R. (1969) 'How much can we boost IQ and scholastic achievements?', *Harvard Educational Review*, 39, 1–123.

351

REFERENCES

Joad, C. M. (1933) *The Book of Joad*, London: Faber.

Johnson, D. K. and Johnson, K. R. (1992) 'Humans must be so lucky: moral prejudice, species and animal liberation', *Capitalism, Nature, Socialism*, 10, 83–109.

Johnston, R. J. (1981) *The Dictionary of Human Geography*, Oxford: Blackwell.

Johnston, R. J. (1989) *Environmental Problems: nature, economy and state*, London: Belhaven.

Joll, J. (1979) *The Anarchists*, London: Methuen.

Jones, A. K. (1990) 'Social symbiosis: a Gaian critique of contemporary social theory', *The Ecologist*, 20(3), 108–13.

Jukes-Browne, A. J. (1895) 'Origins of the valleys of the chalk Downs of North Dorset', *Proceedings of the Dorset Natural History and Antiqities Field Club*, 16.

Jungk, R. (1982) *Brighter Than a Thousand Suns*, Harmondsworth: Penguin.

Kamenka, E. (1982) 'Community and the socialist ideal', in Kamenka, E. (ed.) *Community as a Social Idea*, London: Arnold.

Kay, J. (1988) 'Concepts of nature in the Hebrew Bible', *Environmental Ethics*, 10(4), 309–27.

Keating, M. (1993) *The Earth Summit's Agenda for Change: a plain language version of Agenda 21 and the other Rio agreements*, Geneva: Centre for Our Common Future.

Kemball-Cook, D. Baker, M. and Mattingly, C. (1991) *The Green Budget*, London: Green Print.

Kemp, P. and Wall, D. (1990) *A Green Manifesto for the 1990s*, Harmondsworth: Penguin.

Keyes, K. (1982) *The Hundredth Monkey*, Coos Bay, Oreg.: Vision Books.

Kimber, R. and Richardson, J. (eds) (1974) *Campaigning for the Environment*, London: Routledge and Kegan Paul.

King, Y. (1989) 'The ecology of feminism and the feminism of ecology', in Plant, J. (ed.) *Healing the Wounds*, London: Green Print.

Kinzley, M. (1993) 'Here come the regionalists', *Green Line*, 111, 10–11.

Kitses, J. (1969) *Horizons West*, London: Thames and Hudson.

Koestler, A. (1964) *The Sleepwalkers*, Harmondsworth: Penguin.

Kohr, L. (1957) *The Breakdown of Nations*, London: Routledge and Kegan Paul.

Kropotkin, P. (1892) *The Conquest of Bread*, London: Freedom Press.

Kropotkin, P. (1899) *Fields, Factories and Workshops Tomorrow*, London: Freedom Press.

Kropotkin, P. (1902) *Mutual Aid*, London: Freedom Press.

Kuhn, T. (1962) *The Structure of Scientific Revolutions*, Chicago, Ill.: Chicago University Press.

Kulctz, V. (1992) 'Eco-feminist philosophy: interview with Barbara Holland-Cunz', *Capitalism, Nature, Socialism*, 3(2), 63–78.

Kumar, K. (1978) *Prophecy and Progress*, Harmondsworth: Penguin.

Lacey, A. R. (1986) *A Dictionary of Philosophy*, London, Routledge.

Landes, D. S. (1969) *The Unbound Prometheus: technological change and industrial development in Western Europe from 1750 to the present*, Cambridge: Cambridge University Press.

Lang, T. and Hines, C. (1993) *The New Protectionism*, London: Earthscan.

Laszlo, E. (1983) *Systems Science and World Order*, Oxford: Pergamon Press.

LeGuin, U. (1975) *The Dispossessed*, London: Grafton Books.

Leopold, A. (1949) *A Sand County Almanack*, New York: Oxford University Press.

Levin, M. G. (1994) 'A critique of ecofeminism', in Pojman, L. P. (ed.), *Environmental Ethics: readings in theory and application*, London: Jons and Bartlett Publishers, 134–40.

352

REFERENCES

Lindeman, R. (1942) 'The trophic–dynamic aspect of ecology', *Ecology*, 23, 399–417.

Lone, O. (1992) 'Environmental and resource accounting', in Ekins, P. and Max-Neef, M. (eds) *Real Life Economics: understanding wealth creation*, London: Routledge, 239–54.

Loske, R. (1991) 'Ecological taxes, energy policy and greenhouse gas reductions: a German perspective', *The Ecologist*, 21(4), 173–6.

Lovejoy, A. (1974) *The Great Chain of Being*, Cambridge, Mass.: Harvard University Press.

Lovelock, J. (1989) *The Ages of Gaia: a biography of our living earth*, Oxford: Oxford University Press.

Lovelock, J. (1990) in conversation with Roger Dounda, Findhorn, Forres, Scotland: Whole World Productions video.

Lowe, P. (1983) 'Values and institutions in the history of British nature conservation' in Warren, A. and Goldsmith, F. (eds) *Conservation in Perspective*, New York: Wiley.

Lowe, P. and Goyder, J. (1983) *Environmental Groups in Politics*, London. George Allen and Unwin.

Lucardie, P. (1993) 'Why would egocentrists become ecocentrics? On individualism and holism in green political theory', in Dobson, A. and Lucardie, P. (eds) *The Politics of Nature: explanations in green political theory*, London: Routledge, 21–35.

Luke, T. W. (1993) 'Green consumerism: ecology and the ruse of recycling', in Bennett, J. and Chaloupka, W. (eds) *In the Nature of Things: language, politics and the environment*, Minneapolis, Minn.: University of Minnesota Press, 154–72.

Lutz, M. (1992) 'Humanistic economics: history and basic principles', in Ekins, P. and Max-Neef, M. (eds) *Real Life Economics: understanding wealth creation*, London: Routledge, 90–120.

Lyotard, J.-F. (1984) *The Postmodern Condition: a report on knowledge*, Manchester: Manchester University Press.

McCloskey, J., Smith, D. and Graves, R. (1993) 'Exploring the green sell: marketing implications of the environmental movement', in Smith, D. (ed.) *Business and the Environment: implications of the new environmentalism*, London: Paul Chapman Publishing, 84–97.

McCully, P. (1991) 'Discord in the greenhouse: how WRI is attempting to shift the blame for global warming', *The Ecologist*, 21(4), July/August, 157–65.

McDonagh, S. (1988) *To Care for the Earth: a call to a new theology*, New York: Cornell.

McEvoy, A. F. (1987) 'Towards an interactive theory of nature and culture: ecology, production and cognition in the California fishing industry', *Environmental Review*, 11(4), 289–305.

McFadden, C. (1977) *The Serial: a year in the life of Marin County*, New York: Knopf.

Mackintosh, M. and Wainwright, H. (1992) 'Popular planning in practice', in Ekins, P. and Max-Neef, M. (eds) *Real Life Economics: understanding wealth creation*, London: Routledge, 358–68.

McRobie, G. (1982) *Small is Possible*, London: Abacus.

Magner, L. N. (1979) *A History of the Life Sciences*, New York: Marcel Dekker.

Mahlberg, A. (1987) 'Evidence of collective memory: a test of Sheldrake's theory', *Journal of Analytical Psychology*, 32, 23–34.

Malthus, T. (1872) *An Essay on the Principle of Population*, seventh edition, London: Dent.

Manuel, F. E. and Manuel, F. P. (1979) *Utopian Thought in the Western World*, Oxford: Blackwell.

Martell, L. (1994) *Ecology and Society: an introduction*, Cambridge: Polity.

353

REFERENCES

Martin, B. (1993) 'Is the "new paradigm" of physics inherently ecological?', *The Raven*, 24, 353–6.

Martin, C. (1978) *Keepers of the Game: Indian animal relationships and the fur trade*, Berkeley, Calif.: University of California Press.

Martin, J. (1978) *The Wired Society*, New York: Prentice Hall.

Martin, P. (1973) 'The discovery of America', *Science*, 969–74.

Martinez-Alier, J. (1989) 'Ecological economics and eco-socialism', *Capitalism, Nature, Socialism*, 2, 109–22.

Martinez-Alier, J. (1990) *Ecological Economics: energy, environment and society*, Oxford: Blackwell.

Marx, L. (1973) 'Pastoral ideals and city troubles', in Barbour, I. G. (ed.) *Western Man and Environmental Ethics*, London: Addison-Wesley, 93–115.

Matley, I. (1982) 'Nature and society: the continuing Soviet debate', *Progress in Human Geography*, 6(3), 367–96.

Max-Neef, M. (1992) 'Development and human needs', in Ekins, P. and Max-Neef, M. (eds) *Real Life Economics: understanding wealth creation*, London: Routledge, 197–213.

Meadows, D. H., Meadows, D. L. and Randers, J. (1992) *Beyond the Limits: global collapse or a sustainable future*, London: Earthscan.

Meadows, D. H., Meadows, D. L., Randers, J. and Behrens, W. (1972) *Limits to Growth*, London: Earth Island.

Medvedev, Z. (1969) *The Rise and Fall of T. D. Lysenko*, New York: Columbia University Press.

Medvedev, Z. (1979) *Soviet Science*, Oxford: Oxford University Press.

Mellor, M. (1992) 'Dilemmas of essentialism and materialism', *Capitalism, Nature, Socialism*, 3(2), 43–62.

Mercer, J. (1984) *Communes: a social history and guide*, Dorset: Prism Press.

Merchant, C. (1982) *The Death of Nature: women, ecology and the scientific revolution*, London: Wildwood House.

Merchant, C. (1987) 'The theoretical structure of ecological revolutions', *Environmental Review*, 11(4), 265–74.

Merchant, C. (1992) *Radical Ecology*, New York: Routledge.

Midgley, M. (1979) *Beast and Man*, Brighton: Wheatsheaf.

Midgley, M. (1983) *Animals and Why They Matter*, Athens, Ga.: Georgia University Press.

Miles, I. (1992) 'Social indicators for real-life economics', in Ekins, P. and Max-Neef, M. (eds) *Real Life Economics: understanding wealth creation*, London: Routledge, 283–99.

Mills, W. (1982) 'Metaphorical vision: changes in Western attitudes to the environment', *Annals of the Association of American Geographers*, 72(2), 237–53.

Mishan, E. J. (1967) *The Costs of Economic Growth*, London: Staples Press

Mishan, E. J. (1993) 'Economists versus the greens: an exposition and critique', *Political Quarterly*, 64(2), 222–42.

Mollison, B. (1988) *Permaculture: a designer's manual*, Tyalgum, NSW: Tagari Publications.

Montague, R. (1992) 'Shelley – poet and socialist', *Socialist Standard*, 88, 1058, 156–7.

Moore, R. (1990) 'A new Christian reformation', in Engel, J. R. and Engel, J. G. (eds) *Ethics of Environment and Development; global challenge and the international response*, London: Belhaven 104–16.

Morris, B. (1993) 'Reflections on deep ecology', in *Deep Ecology and Anarchism: a polemic* (no editor given), London: Freedom Press.

Morris, D. (1967) *The Naked Ape*, New York: McGraw Hill.

Morris, D. (1990) 'Free trade: the great destroyer', *The Ecologist*, 20(5), 190–5.

Morris, Wm (1885) 'Useful work versus useless toil', Socialist League pamphlet, in *Collected Works*, Penguin edition, 117–36.

Morris, Wm (1887a) 'The society of the future' in Morton, A. L. (ed.) (1979) *The Political Writings of William Morris*, London: Lawrence and Wishart, 188–203.

Morris, Wm (1887b) 'How we live and how we might live', *Collected Works*, XXIII, London: Socialist Party of Great Britain edition, 3–26.

Morris, Wm (1889) 'Under an elm tree; or thoughts on the countryside', in Morton, A. L. (ed.) (1979) *The Political Writings of William Morris*, London: Lawrence and Wishart, 214–18.

Morris, Wm (1890) *News from Nowhere*, London: Routledge and Kegan Paul, 1970 edition.

Morton, A. L. (ed.) (1979) Introduction to *The Political Writings of William Morris*, London: Lawrence and Wishart, 214–18.

Muir, J. (1898) 'The wild parks and forest reservations of the West', *Atlantic Monthly*, LXXXI, 483.

Mulberg, J. (1993) 'Economics and the impossibility of environmental evaluation', *Proceedings of the Conference on Values and the Environment*, Guildford: University of Surrey, 107–12.

Mulgan, G. and Wilkinson, H. (1992) 'The enabling (and disabling) state', in Ekins, P. and Max-Neef, M. (eds) *Real Life Economics: understanding wealth creation*, London: Routledge, 340–52.

Mumford, L. (1934) *The Future of Technics and Civilisation* (second half of *Technics and Civilisation*), London: Freedom Press edition 1986.

Mumford, L. (1938) *The Culture of Cities*, London: Secker and Warberg.

Naess, A. (1973) 'The shallow and the deep, long-range ecology movement: a summary', *Inquiry*, 16, 95–100.

Naess, A. (1988) 'The basics of deep ecology', *Resurgence*, 126, 4–7.

Naess, A. (1989) *Ecology, Community and Lifestyle: outline of an ecosophy*, Cambridge: Cambridge University Press.

Naess, A. (1990) 'Sustainable development and deep ecology', in Engel, J. R. and Engel, J. G. (eds) *Ethics of Environment and Development: global challenge and the international response*, London: Belhaven, pp. 86–96.

Nash, R. (1974) *Wilderness and the American Mind*, New Haven, Conn.: Yale University Press.

Nash, R. (1977) 'Do rocks have rights?', *The Center Magazine*, November/December, 1–12.

Nelkin, D. (1975) 'The political impact of technical expertise', *Social Studies of Science*, 5, 35–54.

Newby, H. (1985) *Green and Pleasant Land? social change in rural England*, second edition, London: Wildwood House.

Newby, H. (1987) *Country Life*, London: Weidenfield and Nicholson.

New Consumer (1993) study of Trans National Corporations, summarised in *New Consumer Briefing*, Autumn, 1–3.

Norgaard, R. (1992) 'Co-evolution of economics, society and environment', in Ekins P. and Max-Neef, M. (eds) *Real Life Economics: understanding wealth creation*, London: Routledge, 76–88.

O'Connor, J. (1992) 'A political strategy for ecology movements', *Capitalism, Nature, Socialism*, 3(1), 1–6.

Oldroyd, D. R. (1980) *Darwinian Impacts: an introduction to the Darwinian revolution*, Milton Keynes: Open University Press.

Omo-Fadaka, J. (1976) 'Escape route for the poor', in Boyle, G. and Harper, P. (eds) *Radical Technology*, London: Wildwood House, 249–53.

Omo-Fadaka, J. (1990) 'Communalism: the moral factor in African development', in Engel, J. R. and Engel, J. G. (eds) *Ethics of Environment and Development: global challenge and the international response*, London: Belhaven, 176–82.

O'Neill, J. (1993a) *Ecology, Policy and Politics: human wellbeing and the natural world*, London: Routledge.

O'Neill, J. (1993b) 'Science, wonder and the lust of the eyes', *Journal of Applied Philosophy*, 10(2), 139–46.

O'Neill, J. (1994) 'Humanism and nature', *Radical Philosophy*, 66, 21–9.

Opie, J. (ed.) (1971) *Americans and Environment: the controversy over ecology*, Lexington, Mass: D. C. Heath.

O'Riordan, T. (1981) *Environmentalism*, second edition, London: Pion.

O'Riordan, T. (1989) 'The challenge for environmentalism', in Peet, R. and Thrift, N. (eds), *New Models in Geography*, London: Unwin Hyman, 77–102.

O'Riordan, T. and Turner, K. (1983) introductory essay on the commons theme in O'Riordan, T. and Turner, K. (eds) *An Annotated Reader in Environmental Planning and Management*, Oxford: Pergamon, 265–88.

Ormerod, P. (1994) 'I see, said the blind man', *Independent on Sunday*, 13 March, 21, extract from *The Death of Economics*, London: Faber and Faber.

Osbourne White, H. J. (1909) 'Geology of the countryside round Basingstoke', *Memoirs of the UK Geological Survey*, London: HMSO.

O'Sullivan, P. (1990) 'The ending of the journey: Wm Morris, News from Nowhere and ecology', in Coleman, S. and O'Sullivan, P. (eds), *William Morris and News from Nowhere: a vision for our time*, Bideford: Green Books, 169–81.

Owen, D. (1993) 'The emerging green agenda: a role for accounting?', in Smith, D. (ed.) *Business and the Environment: implications of the new environmentalism*, London: Paul Chapman Publishing, 55–74.

Pacey, A. (1983) *The Culture of Technology*, Oxford: Blackwell.

Palmer, M. (1990) 'The encounter of religion and conservation', in Engel, J. R. and Engel, J. G. (eds) *Ethics of Environment and Development: global challenge and the international response*, London: Belhaven, 50–62.

Papworth, J. (1990) 'The fourth world: the world of small units', *Noah's Ark*, 2, Leopold Kohr Extra, unpaginated.

Parsons, H. L. (1977) *Marx and Engels on Ecology*, London: Greenwood.

Passmore, J. (1980) *Man's Responsibility to Nature*, London: Duckworth.

Pearce, D. (1993) *Economic Values and the Natural World*, London: Earthscan.

Pearce, D. and Turner, R. K. (1990) *The Economics of Natural Resources and the Environment*, Hemel Hempstead: Harvester Wheatsheaf.

Pearce, D. Markandya, A. and Barbier, E. (1909) *Blueprint for a Green Economy*, London: Earthscan.

Pearce, D. Markandya, A. and Barbier, E. (1991) *Sustainable Development: economics and environment in the Third World*, London: Earthscan.

Pearce, F. (1992) 'Corporate shades of green', *New Scientist*, 136, 21–2.

Peet, R. (1985) 'The social origins of environmental determinism', *Annals of the Association of American Geographers*, 75(3), 309–33.

Peet, R. (1991) *Global Capitalism: theories of societal development*, London: Routledge.

Pepper, D. M. (1980) 'Environmentalism, the "lifeboat ethic" and anti-airport protest', *Area*, 12(3), 177–82.

Pepper, D. M. (1988) 'The geography and landscapes of an anarchist Britain', *The Raven*, 1(4), 339–50.

REFERENCES

Pepper, D. M. (1991) *Communes and the Green Vision: counterculture, lifestyle and the New Age*, London: Green Print.

Pepper, D. M. (1993) *Eco-socialism: from deep ecology to social justice*, London: Routledge.

Petersen, W. (1979) *Malthus*, London: Heinemann.

Petulla, J. M. (1988) *American Environmental History*, Columbus, OH: Merril Publishing Co..

Piercy, M. (1979) *Woman on the Edge of Time*, London: The Women's Press.

Pietila, H. (1990) 'The daughters of Earth: women's culture as a basis for sustainable development', in Engel, J. R. and Engel, J. G. (eds) *Ethics of Environment and Development; global challenge and the international response*, London: Belhaven, 235–44.

Pirsig, R. (1974) *Zen and the Art of Motor Cycle Maintenance*, London: Transworld Publishers (Corgi).

Plumwood, V. (1990) 'Women, humanity and nature', in Sayers, S. and Osborne, P. (eds), *Socialism, Feminism and Philosophy: a radical philosophy reader*, London: Routledge.

Plumwood, V. (1992) 'Beyond the dualistic assumptions of women, men and nature', *The Ecologist*, 22(1), 8–13.

Popper, K. (1965) *The Logic of Scientific Discovery*, New York: Harper and Row.

Porritt, J. (1984) *Seeing Green*, Oxford: Blackwell.

Porritt, J. (1992) 'Facts of life', *BBC Wildlife*, March, 55.

Powers, J. (1985) *Philosophy and the New Physics*, London: Methuen.

Prigogine, I. (1980) *From Being to Becoming: time and complexity in the physical sciences*, San Francisco, Calif.: W. H. Freeman.

Prior, M. (1954) 'Bacon's man of science', *Journal of the History of Ideas*, XV, 41–54.

Purchase, G. (1993) 'Social ecology, anarchism and trades unionism', in *Deep Ecology and Anarchism: a polemic* (no editor given), London: Freedom Press, 23–35.

Ravetz, J. (1988) 'Gaia and the philosophy of science', in Bunyard, P. and Goldsmith, E. (eds) *Gaia, the Thesis, the Mechanisms and the Implications*, Wadebridge, Cornwall: Wadebridge Ecological Centre, 133–44.

Rawls, J. (1971) *A Theory of Justice*, Cambridge, Mass.: Harvard University Press.

Redclift, M. (1986) 'Redefining the environmental "crisis" in the South' in Weston, J. (ed.) *Red and Green: a new politics of the environment*, London: Pluto Press, 80–101.

Rees, R. (1982) 'Constable, Turner and views of nature in the nineteenth century', *Geographical Review*, 72(3), 251–69.

Regan, T. (1982) *All That Dwell Therein: animal rights and environmental ethics*, Berkeley, Calif.: University of California Press.

Regan, T. (1988) *The Case for Animal Rights*, Berkeley, Calif: University of California Press.

Richards, S. (1983) *Philosophy and Sociology of Science: an introduction*, Oxford: Blackwell.

Rifkin, J. (1980) *Entropy: a new world view*, London: Bantam.

Ritchie, M. (1992) 'Free trade versus sustainable agriculture: the implications of NAFTA', *The Ecologist*, 22(5), 221–7.

Robertson, J. (1989) *Future Wealth: a new economics for the twenty first century*, London: Cassell.

Robson, B. T. (1981) 'Geography and social science: the role of Patrick Geddes', in Stoddart, D. (ed.) *Geography, Ideology and Social Concern*, Oxford: Blackwell.

Roddick, A. (1988) preface to Elkington, J. and Hailes, J. *The Green Consumer Guide*, London: Gollancz.

Roddick, J. and Dodds, F. (1993) 'Agenda 21's political strategy', *Environmental Politics*, 2(4), 242–8.

Rodman, J. (1977) 'The liberation of nature', *Inquiry*, 20, 83–131.

357

Roelofs, J. (1993) 'Charles Fourier: proto red green', *Capitalism, Nature, Socialism*, 4(3), 69–88.

Rolston, H. III (1989) *Philosophy Gone Wild: essays in environmental ethics*, Buffalo, NY: Prometheus Books.

Rose, S. (1992) in critical discussion of paper by Johnson and Johnson (see above), *Capitalism, Nature, Socialism*, 10, 117–20.

Rose, S. Lewontin, L. and Kamin, R. (1984) *Not in Our Genes*, Harmondsworth: Penguin.

Rostow, W. W. (1960) *The Stages of Economic Growth: a non-communist manifesto*, Cambridge: Cambridge University Press.

Roszak, T. (1979) *Person/Planet*, London: Gollancz.

Ruether, R. (1975) *New Woman, New Earth: sexist ideologies and human liberation*, New York: Seabury Press.

Russell, B. (1914) *Roads to Freedom*, third edition 1948, London: Unwin 1977.

Russell, B. (1946) *History of Western Philosophy*, London: Unwin, 1980 edition.

Russell, P. (1991) *The Awakening Earth: the global brain*, London: Arkana.

Ryle, M. (1988) *Ecology and Socialism*, London: Radius.

Sacher, E. (1881) *Foundations of a Mechanics of Nature*, Jena: Gustav Fischer.

Sagoff, M. (1988) *The Economy of the Earth: philosophy, law and the environment*, Cambridge: Cambridge University Press.

Sahlins, M. (1972) *Stone Age Economics*, Chicago, Ill.: Aldine Atherton.

Sahtouris, E. (1989) 'The Gaia controversy: a case for the earth as a living planet', in Bunyard, P. and Goldsmith, E. (eds) *Gaia and Evolution: the second Wadebridge Ecological Centre symposium*, Camelford, Cornwall: Wadebridge Ecological Centre, 55–65.

Sale, K. (1985) *Dwellers in the Land: the Bioregional Vision*, San Francisco, Calif.: Sierra Club.

Sandbach, F. (1980) *Environment, Ideology and Policy*, Oxford: Blackwell.

Saurin, J. (1993) 'Global environmental degradation, modernity and environmental knowledge', *Environmental Politics*, 2(4), 46–64.

Sayer, A. and Walker, R. (1992) *The New Social Economy: reworking the division of labour*, Oxford: Blackwell.

Schnaiberg, A. (1980) *The Environment: from surplus to scarcity*, New York: Oxford University Press.

Schumacher, E. F. (1973) *Small is Beautiful: economics as if people really mattered*, London: Abacus.

Schwarz, M. and Thompson, M. (1990) *Divided we Stand: redefining politics, technology and social choice*, London: Harvester Wheatsheaf.

Scott, A. (1990) *Ideology and the New Social Movements*, London: Unwin Hyman.

Sen, A. (1981) *Poverty and Famine: an essay on entitlement and deprivation*, Oxford: Clarendon Press.

Seyfang, G. (1994) 'The Local Exchange Trading System: the political economy of local currencies; a social audit', University of East Anglia: unpublished MSc thesis.

Shaiko, R. G. (1987) 'Religion, politics and environmental concern: a powerful mix of passions', *Social Science Quarterly*, 68, 244–62.

Sheldrake, R. (1982) *A New Science of Life: the theory of formative causation*, Los Angeles, Calif.: J. P. Tarcher.

Shiva, V. (1988) *Staying Alive: women, ecology and survival in India*, London: Zed Books.

Shiva, V. (1992) 'The seed and the earth: women, ecology and biotechnology', *The Ecologist*, 22(1), 4–7.

Shoard, M. (1980) *The Theft of the Countryside*, London: Temple Smith.

REFERENCES

Shoard, M. (1982) 'The lure of the moors', in Gold, J. and Burgess, J. (eds), *Valued Environments*, London: George Allen and Unwin.

Short, J. (1991) *Imagined Country: society, culture and environment*, London: Routledge.

Sikorski, W. (1993) 'Building wilderness', in Bennett, J. and Chaloupka, W. (eds) *In the Nature of Things: language, politics and the environment*, Minneapolis, Minn.: University of Minnesota Press, 24–43.

Simmons, P. (1992) '"Women in Development": a threat to liberation', *The Ecologist*, 22(1), 16–21.

Simon, J. (1981) *The Ultimate Resource*, Oxford: Martin Robertson.

Simon, J. and Kahn, H. (1984) *The Resourceful Earth*, Oxford: Blackwell.

Singer, C. (1962) *A History of Biology: to about the year 1900*, New York: Abelard-Schuman.

Singer, P. (1983) *Animal Liberation: towards an end to man's inhumanity to animals*, Wellingborough: Thorsons.

Singer, P. (ed.) (1985) *In Defence of Animals*, Oxford: Basil Blackwell.

Singh, N. (1989) *Economics and the Crisis of Ecology*, third edition, London: Bellew.

Skinner, B. F. (1948) *Walden II*, New York: Macmillan, 1976 edition.

Skolimowski, H. (1990) 'Reverence for life', in Engel, J. R. and Engel, J. G. (eds) *Ethics of Environment and Development: global challenge and the international response*, London: Belhaven, 97–103.

Skolimowski, H. (1992) *Living Philosophy*, London: Arkana.

Smith, D. (1993) 'The Frankenstein syndrome: corporate responsibility and the environment', in Smith, D. (ed.) *Business and the Environment: implications of the new environmentalism*, London: Paul Chapman Publishing, 172–89.

Smith, D. M. (1981) 'Neoclassical economics', in Johnston, R. *et al.* (eds) *The Dictionary of Human Geography*, Oxford: Blackwell, 233–8.

Smith K. (1988) *Free is Cheaper*, Gloucester: The John Ball Press.

Smuts, J. (1926) *Wholism and Evolution*, 1973 reprint, Westport, Conn.: Greenwood.

Snyder, G. (1969) *Turtle Island*, New York: New Directions.

Snyder, G. (1977) *The Old Ways*, San Francisco, Calif.: City Lights.

Steele, D. (1992) *From Marx to Misers: post capitalist society and the challenge of economic calculation*, Chicago, Ill.: Open Court.

Stein, D. (1993) 'Be fruitful and multiply', *Real World*, Spring, 13.

Stikker, A. (1992) *The Transformation Factor: towards an ecological consciousness*, Shaftsbury, Dorset: Element.

Stillman, P. G. (1983) 'The tragedy of the commons: a reanalysis', in O'Riordan, T. and Turner, K. (eds) *An Annotated Reader in Environmental Planning and Management*, Oxford: Pergamon, 299–303.

Stirling, A. (1993) 'Environmental valuation: how much is the emperor wearing?', *The Ecologist*, 23(3), 97–103.

Stoddart, D. R. (1965) 'Geography and the ecological approach: the ecosystem as a geographic principle and method', *Geography*, 50, 242–51.

Stoddart, D. R. (1966) 'Darwin's impact on geography', *Annals of the Association of American Geographers*, 56, 683–98.

Stone, C. (1974) *Should Trees Have Standing? Toward Legal Rights for Natural Objects*, Los Angeles, Calif.: William Kaufman.

Storm, R. (1991) *In Search of Heaven on Earth: a history of the New Age*, London: Bloomsbury Press.

Stott, K. (1993) 'Discovering an alternative economy', *Oxford Times*, 28 May, 12.

Stott, M. (1988) *Spilling the Beans: a style guide to the New Age*, London: Fontana.

Sylvan, R. (1985a) 'A critique of deep ecology, part 1', *Radical Philosophy*, 40, 2–12.

Sylvan, R. (1985b) 'A critique of deep ecology, part 2', *Radical Philosophy*, 41, 10–22.

359

Taylor, B. (1991) 'The religion and politics of Earth First!', *The Ecologist*, 21(6), 258–66.

Thackray, A. (1974) 'Natural knowledge in a cultural context: the Manchester model', *American Historical Review*, 79, 672–709.

Thomas, C. (1993) 'Beyond UNCED: an introduction', *Environmental Politics*, 2(4), 1–27.

Thomas, K. (1983) *Man and the Natural World: changing attitudes in England, 1500 to 1800*, London: Allen Lane.

Thompkins, P. and Bird, C. (1972) *The Secret Life of Plants,* London: Harper and Row.

Thompson, E. P. (1968) *The Making of the English Working Classes*, London: Gollancz, 1980 edition.

Thoreau, H. D. (1974) *Walden*, New York: Collier Books, 8th printing.

Tiger, L. and Fox, R. (1989) *The Imperial Animal*, second edition, New York: McGraw Hill.

Tofler, A. (1980) *The Third Wave*, London: Collins.

Tokar, B. (1987) *The Green Alternative: creating an ecological future*, San Pedro, Calif.: R. and E. Miles.

Tönnies, F. (1887) *Community and Association*, London: Harper and Row, 1963 edition.

Tuan, Yi Fu (1968) 'Discrepancies between environmental attitudes and behaviour', *Canadian Geographer,* 12(3), 176–91.

Tuan, Yi Fu (1970) 'Our treatment of the environment in ideal and actuality', *American Scientist,* 58(3), 244, 247–9.

Tuan, Yi Fu (1971) *Man and Nature*, Resource Paper No. 10, Washington DC: Association of American Geographers.

Tuan, Yi Fu (1972) 'Structuralism, existentialism and environmental perception', *Environment and Behaviour,* 4(3), 319–31.

Tuan, Yi Fu (1974) *Topophilia: a study of environmental perception, attitudes and values,* Englewood Cliffs, NJ: Prentice-Hall.

Tucker, M. E. and Grim, J. A. (eds) (1993) *Worldviews and Ecology*, Lewisburg, Pa.: Bucknell University Press.

UN (1987) World Commission on Environment and Development, *Our Common Future* (The Brundtland Report), Oxford: Oxford University Press.

Vincent, A. (1993) 'The character of ecology', *Environmental Politics*, 2(2), 248–76.

Vogel, S. (1988) 'Marx and alienation from nature', *Social Theory and Practice* 14(3), 367–88.

von Bertalanffy, L. (1968) *General System Theory: foundations, development, applications*, revised edition, New York: George Braziller.

von Hildebrand, M. (1900) 'An Amazonian tribe's view of cosmology', in Bunyard, P. and Goldsmith, E. (eds) *Gaia, the Thesis, the Mechanisms and the Implications*, Wadebridge, Cornwall: Wadebridge Ecological Centre, 186–200.

Wall, D. (1989) 'The Green Shirt effect', *Searchlight*, 168.

Wall, D. (1990) *Getting There: steps to a Green society*, London: Green Print.

Wall, D. (1994) *Green History: a reader in environmental literature, philosophy and politics*, London: Routledge.

Wallerstein, I. (1974) *The Modern World-System: capitalist agriculture and the origins of the European world-economy in the sixteenth century*, New York: Academic Press.

Warnock, M. (1970) *Existentialism*, Oxford: Oxford University Press.

Warren, K. (1990) 'The power and the promise of ecological feminism', *Environmental Ethics*, 12, 125–46.

Warren, M. A. (1983) 'The rights of the nonhuman world', in Elliot, R. and

Gare, A. (eds), *Environmental Philosophy*, Milton Keynes: Open University Press, 109–34.

Watson, L. (1980) *Lifetide: the biology of consciousness*, New York: Simon and Schuster.

Watson, R. (1983) 'A critique of non-anthropocentric biocentrism', *Environmental Ethics*, 3, 245–56.

Watts, A. (1968) *The Wisdom of Insecurity*, New York: Random.

Watts, A. and Huang C A (1975) *Tao: the watercourse way*, Harmondsworth: Penguin.

Weber, M. (1976) *The Protestant Ethic and the Spirit of Capitalism*, London: George Allen and Unwin.

Webster, F. and Robbins, K. (1986) *Information Technology: a Luddite analysis*, New Jersey: Ablex.

Weston, A. (1985) 'Beyond intrinsic value: pragmatism in environmental ethics', *Environmental Ethics*, 7, 321–39.

Weston, J. (1989) *The FoE Experience: the development of an environmental pressure group*, Oxford: Oxford Polytechnic School of Planning Working Paper No. 116.

White, L. (1967) 'The historical roots of our ecologic crisis', *Science*, 155, 1203–7.

Whitehead, A N (1926) *Science and the Modern World*, Cambridge: Cambridge University Press. London: Free Association Books edition 1985 edited by Robert Young.

Wiener, M. J. (1981) *English Culture and the Decline of the Industrial Spirit 1850–1980*, Cambridge: Cambridge University Press.

Wignaraja, P. (1992) 'People's participation: reconciling growth with equity', in Ekins, P. and Max-Neef, M. (eds) *Real Life Economics: understanding wealth creation*, London: Routledge, 392–9.

Wilding, N. (1991) 'Green money that makes the world go round', *Green Line*, 90, 15–16.

Williams, M. (1993) 'International trade and the environment: issues, perspectives and challenges', *Environmental Politics* 2(4), 80–97.

Williams, R. (1975) *The Country and the City*, St Albans: Paladin.

Williams, R. (1983) *Keywords: a vocabulary of culture and society*, London: Flamingo.

Winner, L. (1986) *The Whale and the Reactor*, Chicago, Ill.: University of Chicago Press.

Woodcock, G. (1975) *Anarchism*, Harmondsworth: Penguin.

Wooldridge, S. W. and Ewing, C. (1935) 'The Eocene and Pliocene deposits of Lane End, Bucks', *Quarterly Journal of the Geological Society*, London, 91, 293–317.

Worster, D. (1985) *Nature's Economy: a history of ecological ideas*, Cambridge: Cambridge University Press.

Worster, D. (ed.) (1988) *The Ends of The Earth: perspectives on modern environmental history*, Cambridge: Cambridge University Press, Introduction: 'The Vulnerable Earth', 1–20.

Wright, P. (1985) *On Living in an Old Country*, London: Verso.

Wynne, B. (1982) *Rationality and Ritual: the Windscale Inquiry and nuclear decisions in Britain*, Chalfont St Giles, Bucks: British Society for the History of Science.

Wynne, B. (1989) 'Sheepfarming after Chernobyl: a case study in communicating scientific information', *Environment*, 31, 11–39.

Yearley, S. (1989) 'Bog standards: science and conservation at a public enquiry', *Social Studies of Science*, 19, 421–38.

Yearley, S. (1991) *The Green Case: a sociology of environmental issues, arguments and politics*, London: Harper Collins.

Young, J. (1990) *Post-Environmentalism*, London: Belhaven.

REFERENCES

Zaring, J. (1977) 'The romantic face of Wales', *Annals of the Association of American Geographers,* 67(3), 397–418.

Zimmerman, M. E. (1983) 'Toward a Heideggerean *ethos* for radical environment-alism', *Environmental Ethics,* 5, 99–131.

Zohar, D. and Marshall, I. (1993) *The Quantum Society: mind, physics and a new social vision,* London: Bloomsbury.

Zukav, G. (1980) *The Dancing Wu Li Masters: an overview of the new physics,* London: Fontana.

INDEX